Covers all Dual
Award and

Longman GCSE Chemistry

Longman

JIM CLARK

Contents

Section D: Air, Water and Earth

Section E: Organic Chemistry

Section F: Sums

Appendices

Contents

Introduction

This book covers everything you would need to know for any GCSE Chemistry exam set by AQA, Edexcel or OCR. This means that you will not need to know everything in this book – only the parts which are relevant to the GCSE specification that you are following. It is important to find out exactly what your particular examiners are likely to ask you. To do this you will need a copy of your specification and as much other useful information as you can get.

If your teacher hasn't given you a copy of your specification, you can download it from your Awarding Body's website. Find out from your teacher which Awarding Body your school is using. If you are doing Chemistry as a single subject and are using OCR, find out whether you are doing Extension Block A or B.

The web addresses are:

www.aqa.org.uk

www.edexcel.org.uk

www.ocr.org.uk

Find your way to the GCSE Chemistry specification you want and download it. Downloading won't take more than a few minutes. The OCR Chemistry specification contains both Extension Blocks.

While you are on your Awarding Body's site, see what other useful things you can find. You should be able to find examples of exam papers and mark schemes. These are important so that you can see exactly what sort of questions your examiners ask, and how they expect you to answer them. You will also be able to find material written to help teachers. There's no reason why students can't make good use of this as well. Much of it will be free.

To do well in Chemistry at GCSE you don't need to know everything in this book, but you do need to understand exactly what your examiners want.

About this book

This book has several features to help you with GCSE Chemistry.

Introduction
Each chapter has a short introduction to help you start thinking about the topic and let you know what is in the chapter.

End of chapter checklists
These lists summarise the material in the chapter. They could also help you to make revision notes because they form a list of things that you need to revise. (You need to check your specification to find out exactly what you need to know.)

Section A: Particles

Chapter 2: Bonding

This chapter looks at what happens when atoms combine together – whether into small groups or very large ones.

The elements sodium and chlorine are dangerous. The compound sodium chloride (salt) isn't.

Water has completely different properties from its elements, hydrogen and oxygen.

Sodium is a dangerously reactive metal. It is stored under oil to prevent it reacting with air or water. Chlorine is a very poisonous, reactive gas.

But salt, sodium chloride, is safe to eat in small quantities. Combining the elements to make salt obviously changes them significantly.

A mixture of hydrogen and oxygen gas would explode violently if you held a lighted match to it. Dropping a lighted match into water (a compound of hydrogen and oxygen) doesn't cause a literally earth-shattering explosion.

Reacting the elements to make a compound has again made a huge difference to them.

Covalent Bonding

What is a covalent bond?

In any bond, particles are held together by electrical attractions between something positively charged and something negatively charged. In a covalent bond, a pair of electrons is shared between two atoms. Each of the positively charged nuclei is attracted to the same negatively charged pair of electrons.

In most of the simple examples you will meet at GCSE, each atom in a covalent bond supplies one electron to the pair of electrons. That doesn't have to be the case. Both electrons may come from the same atom.

nucleus of A is attracted to the electron pair

nucleus of B is also attracted to the electron pair

$$A : B$$

pair of electrons (one from each atom)

A and B are held together by this shared attraction.

Covalent bonding in a hydrogen molecule

Covalent bonds are often shown using "dot-and-cross" diagrams. Although the electrons are drawn as dots or as crosses, there is absolutely no difference between them in reality. The dot and the cross simply show that

H ⦾ H

Chapter 2: Bonding

11

Margin boxes
The boxes in the margin give you extra help or information. They might explain something in a little more detail or guide you to linked topics in other parts of the book.

End of Chapter Checklist

If you haven't got a copy of your specification, read the introduction on page vi.

ideas
evidence

You will need to be able to do some or all of the following. Check your Awarding Body's specification (syllabus) to find out exactly what you need to know.

- Explain what is meant by: covalent bond, molecule.
- Draw dot-and-cross diagrams for all the covalent molecules mentioned by your syllabus.
- Explain what is meant by an ionic bond.
- Describe the formation of all the ionic compounds or ions mentioned by your syllabus.
- Describe how atoms are bound together in a metal.
- Explain what is meant by intermolecular forces.

Questions

You will need to use the Periodic Table on page 315. More questions on bonding can be found at the end of Section A on page 53.

1 a) What is meant by a *covalent bond*? How does this bond hold two atoms together?

   b) Draw dot-and-cross diagrams to show the covalent bonding in i) methane, CH_4, ii) hydrogen sulphide, H_2S, iii) phosphine, PH_3, iv) silicon tetrachloride, $SiCl_4$.

2 Draw dot-and-cross diagrams to show the covalent bonding in a) ethene, C_2H_4, b) ethene, C_3H_4 and c) ethanol, CH_3CH_2OH. You will find models of ethene and ethanol on page 14 which might help you.

3 a) What is meant by i) an ion, ii) an ionic bond?

   b) In each of the following cases, write down the electronic structures of the original atoms and then explain (in words or diagrams) what happens when:
      i) sodium bonds with chlorine to make sodium chloride
      ii) lithium bonds with oxygen to make lithium oxide
      iii) magnesium bonds with fluorine to make magnesium fluoride.

4 a) A solid metal is often described as having "an array of positive ions in a sea of electrons". Write down the electronic structure of a magnesium atom and use it to explain what this phrase means.

   b) Metallic bonds are not fully broken until the metal has first melted and then boiled. The boiling points of sodium, magnesium and aluminium are 890°C, 1110°C and 2470°C respectively. What does this suggest about the strengths of the metallic bonds in these three elements?

   c) Find these three metals in the Periodic Table, and suggest why the boiling points show this pattern.

   d) Assuming that an electric current is simply a flow of electrons, suggest why all these elements are good conductors of electricity.

5 The table gives details of the boiling temperatures of some substances made up of covalent molecules. Arrange these substances in increasing order of the strength of their intermolecular attractions.

	Boiling point (°C)
ammonia	−33
ethanamide	221
ethanol	78.5
hydrogen	−253
phosphorus trifluoride	−101
water	100

Don't panic if you don't recognise some of the names. The substances could just as well have been described as A, B, C, D, E and F.

6 Boron and aluminium are both in Group 3 of the Periodic Table. Both form compounds with fluorine (BF_3 and AlF_3). Unusually for elements found in the same group of the Periodic Table, their compounds are bonded differently. BF_3 is covalent whereas AlF_3 is a straightforward ionic compound.

   a) Draw a diagram to show the covalent bonding in BF_3.

   b) Explain, using diagrams or otherwise, the origin of the ionic bonding in AlF_3.

   c) BF_3 is described as an *electron deficient* compound. What do you think that might mean?

20

Ideas and Evidence

ideas
evidence

This icon means that one of the Awarding Bodies has highlighted this topic as useful for learning about Ideas and Evidence. The Ideas and Evidence parts of the specifications help you to understand how scientific ideas have developed over time and how science relates to our everyday lives.

Questions
There are short questions at the end of each chapter. These help you to test your understanding of the material from the chapter. Some of them may also be research questions – you will need to use the internet and other books to answer these.
There are also questions at the end of each section. The end of section questions are written in an exam style and cover topics from all the chapters in the section.

Chapter 1: Atomic Structure

This chapter explores the nature of atoms, and how they differ from element to element. The 100 or so elements are the building blocks from which everything is made – from the simplest substance, like carbon, to the most complex, like DNA.

Copper is an element. If you tried to chop it up into smaller and smaller bits, eventually you would end up with the smallest possible piece of copper. At that point you would have an individual copper atom. You can, of course, split that into still smaller pieces (protons, neutrons and electrons), but you would no longer have copper.

Whether things are man-made...

...or natural, they are all made up of combinations of the same elements.

New atoms are produced in stars...

...or in nuclear processes, like nuclear bombs, nuclear reactors or radioactive decay.

You will find the reason that oxygen atoms go around in pairs described in Chapter 2, page 13.

Chemistry just rearranges existing atoms. For example, when propane burns in oxygen, existing carbon, hydrogen and oxygen atoms combine in new ways:

propane   +   oxygen   →   carbon dioxide   +   water

Chapter 1: Atomic Structure

1

The Structure of the Atom

Atoms are made of protons, neutrons and electrons.

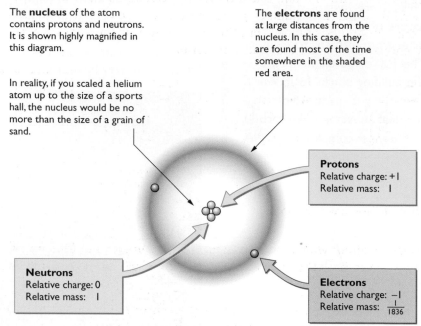

You may have come across diagrams of the atom in which the electrons are drawn orbiting the nucleus, rather like planets around the sun. This is misleading.

It is impossible to know exactly how the electrons are moving in an atom. All you can tell is that they have a particular energy and that they are likely to be found in a certain region of space at some particular distance from the nucleus. Electrons with different energies are found at different distances from the nucleus.

The **nucleus** of the atom contains protons and neutrons. It is shown highly magnified in this diagram.

In reality, if you scaled a helium atom up to the size of a sports hall, the nucleus would be no more than the size of a grain of sand.

The **electrons** are found at large distances from the nucleus. In this case, they are found most of the time somewhere in the shaded red area.

Protons
Relative charge: +1
Relative mass: 1

Neutrons
Relative charge: 0
Relative mass: 1

Electrons
Relative charge: −1
Relative mass: $\frac{1}{1836}$

A helium atom.

Virtually all the mass of the atom is concentrated in the nucleus, because the electrons weigh hardly anything.

The masses and charges are measured relative to each other because the actual values are incredibly small. For example, it would take about 600,000,000,000,000,000,000,000 protons to weigh 1 gram.

Atomic Number and Mass Number

The number of protons in an atom is called its **atomic number** or **proton number**. Each of the one hundred or so different elements has a different number of protons. For example, if an atom has 8 protons it must be an oxygen atom.

Atomic number = number of protons

The **mass number** is the total number of protons and neutrons in the nucleus of the atom.

Mass number = number of protons + number of neutrons

For any particular atom, this information can be shown simply as, for example:

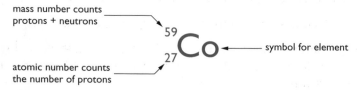

mass number counts
protons + neutrons

$^{59}_{27}\text{Co}$ ← symbol for element

atomic number counts
the number of protons

Warning! When you are writing symbols with two letters in them, the first is a capital letter, and the second must be lower case. If you write CO you are talking about carbon monoxide, not cobalt.

This particular atom of cobalt contains 27 protons. To make the total number of protons and neutrons up to 59, there must also be 32 neutrons.

Isotopes

The number of neutrons in an atom can vary slightly. For example, there are three kinds of carbon atom, called carbon-12, carbon-13 and carbon-14. They all have the same number of protons (because all carbon atoms have 6 protons – its atomic number), but the number of neutrons varies. These different atoms of carbon are called **isotopes**.

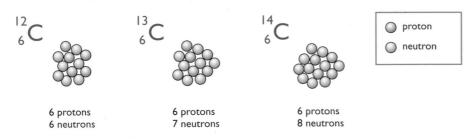

$^{12}_{6}C$	$^{13}_{6}C$	$^{14}_{6}C$
6 protons	6 protons	6 protons
6 neutrons	7 neutrons	8 neutrons

- ○ proton
- ○ neutron

Nuclei of the three isotopes of carbon.

Isotopes are atoms which have the same atomic number, but different mass numbers. They have the same number of protons, but different numbers of neutrons.

The fact that they have varying numbers of neutrons makes no difference whatsoever to their chemical reactions. The chemical properties are governed by the number and arrangement of the electrons and, as you will see shortly, that is identical for all three isotopes.

A radioactive isotope

Carbon-14 is **radioactive**. Its nucleus is unstable and radiation is released as it reorganises into a more stable form. The radiation given off by carbon-14 is used in carbon dating. The nuclei of the carbon-12 and carbon-13 isotopes are perfectly stable, and so these aren't radioactive.

If you are interested you could do an internet search on "Turin shroud" to find out how carbon dating was used to determine its age, and why there is still controversy about it.

Despite carbon dating, there is still considerable controversy about the age of the Turin shroud.

The Electrons

Counting the number of electrons in an atom

Atoms are electrically neutral, and the positiveness of the protons is balanced by the negativeness of the electrons. In a neutral atom it follows that:

Number of electrons = number of protons

So, if an oxygen atom (atomic number = 8) has 8 protons, it must also have 8 electrons; if a chlorine atom (atomic number = 17) has 17 protons, it must also have 17 electrons.

You will see that the key feature in this is knowing the atomic number. You can find that from the Periodic Table.

Remember that the number of protons is the same as the atomic number of the element.

Atomic number and the Periodic Table

Atoms are arranged in the Periodic Table in order of increasing atomic number. You will find a full version of the Periodic Table on page 306. Most Periodic Tables have two numbers against each symbol – be careful to choose the right one. *The atomic number will always be the smaller number.* The other number will either be the mass number of the most common isotope of the element, or the relative atomic mass of the element. The Table will tell you which.

You use a Periodic Table to find out the atomic number of an element, and therefore how many protons and electrons there are in its atoms.

The arrangement of the electrons

The electrons are found at considerable distances from the nucleus in a series of levels called **energy levels** or **shells**. Each energy level can only hold a certain number of electrons. Low energy levels are always filled before higher ones.

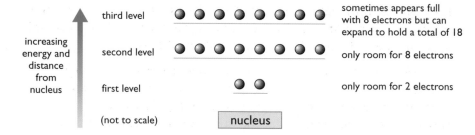

How to work out the arrangement of the electrons

We will use chlorine as an example.

- Look up the atomic number in the Periodic Table. (Make sure that you don't use the wrong number if you have a choice. The atomic number will always be the *smaller* one.)

    The Periodic Table tells you that chlorine's atomic number is 17.

- This tells you the number of protons, and hence the number of electrons.

    There are 17 protons, and so 17 electrons in a neutral chlorine atom.

- Arrange the electrons in levels, always filling up an inner (lower energy) level before you go to an outer one.

    *These will be arranged as 2 in the first level, 8 in the second level, and 7 in the third level. This is written as **2,8,7**. When you have finished, always check to make sure that the electrons add up to the right number – in this case 17.*

The electronic arrangements of the first 20 elements in the Periodic Table

group 1	group 2		group 3	group 4	group 5	group 6	group 7	group 0
								He 2
Li 2,1	**Be** 2,2	10 more elements	**B** 2,3	**C** 2,4	**N** 2,5	**O** 2,6	**F** 2,7	**Ne** 2,8
Na 2,8,1	**Mg** 2,8,2	↓	**Al** 2,8,3	**Si** 2,8,4	**P** 2,8,5	**S** 2,8,6	**Cl** 2,8,7	**Ar** 2,8,8
K 2,8,8,1	**Ca** 2,8,8,2							

H 1

Chapter 10 (page 99) deals in detail with what you need to know about the Periodic Table for GCSE purposes.

Relative atomic mass is explained in Chapter 25 on page 261.

The diagram shows the maximum number of electrons that each energy level can hold.

The third level can expand to hold a total of 18 electrons, but knowledge of this is not required at GCSE.

Don't just accept this diagram! Use the Periodic Table on page 306 and work out each of these electronic structures for yourself (preferably in a random order to make it more difficult). Check your answers when you have finished.

Vertical columns in the Periodic Table are called **groups**. Groups contain elements with similar properties.

There are two important generalisations you can make from this:

- *The number of electrons in the outer level is the same as the group number for groups 1 to 7.*

  This pattern extends right down the Periodic Table for these groups.

  So if you know that barium is in group 2, you know it has 2 electrons in its outer level; iodine (group 7) has 7 electrons in its outer level; lead (group 4) has 4 electrons in its outer level. Working out what is in the inner levels is much more difficult. The simple patterns we have described don't work beyond calcium.

- *The elements in group 0 have 8 electrons in their outer levels (apart from helium, which has 2).*

  These are often thought of as being "full" levels. This is true for helium and neon, but not for the elements from argon downwards. For example, the third energy level will eventually contain 18 electrons.

  The group 0 elements are known as the **noble gases** because they are almost completely unreactive – in fact the three at the top of the group from helium to argon don't react with anything. This lack of reactivity is associated with their electronic structures – often described as **noble gas structures**.

> This idea of "full" levels is best avoided. If you carry it through to A-level, you will give yourself real problems.

Drawing diagrams of electronic arrangements

The electrons in their various energy levels can be shown by drawing circles with dots or crosses on them showing the electrons. It doesn't matter whether you draw dots or crosses.

Hydrogen has one electron and helium has two in the first level. The helium electrons are sometimes shown as a pair (as here), and sometimes as two separate electrons on opposite sides of the circle. Either form is acceptable.

The next four atoms are drawn like this:

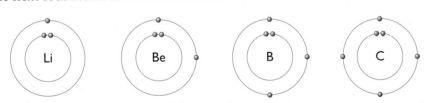

The electrons in the second energy level are drawn singly, up to a maximum of 4. After that, pair them up as necessary. It makes them much easier to count. More importantly, it gives a much better picture of the availability of the electrons in the atom for bonding purposes. This is explored in Chapter 2.

> Drawing circles like this *does not* imply that the electrons are orbiting the nucleus along the circles. The circles represent energy levels. The further the level is from the nucleus, the higher its energy. For theoretical reasons it is impossible to work out exactly how an electron is moving in that energy level.

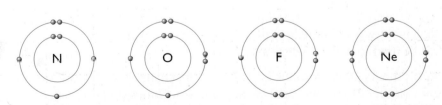

The atoms in the Periodic Table from sodium to argon fill the third level in exactly the same way, and potassium and calcium start to fill the fourth level.

A Brief History of the Atom

The ancient Greeks

Around 2400 years ago, Democritus suggested that you couldn't keep on cutting something into smaller and smaller pieces for ever. Eventually you would get to the smallest possible bit of the substance. The word atom comes from the Greek *atomos* which means "indivisible".

Democritus thought that atoms of everything were all made of the same basic material, but that atoms of different substances had different shapes and sizes.

This is very similar to our modern view of the arrangement of atoms in simple substances like metals.

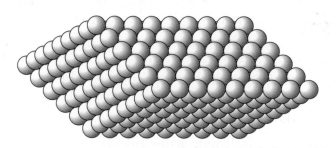

Other Greek thinkers like Plato and Aristotle rejected this idea. They thought that matter was continuous – there was no limit on how finely you could cut it up.

However impressive the Greek philosophers were, they weren't doing science. They were not creating theories which could be tested by experiments. The science had to wait more than 2000 years for the English chemist, John Dalton.

John Dalton (1766–1844)

Dalton took the Greek idea of indivisible atoms and extended it so that he could say useful things about elements and compounds. He suggested that:

- Atoms of a particular element are all the same.
- Atoms of different elements have different masses and different properties.
- Compounds are formed when atoms of different elements combine together in simple whole number ratios like 1:2 or 2:3.
- Chemical changes involve rearranging the atoms.

One of the problems for us is that what Dalton said sounds so obvious. That is because we have been using his ideas for the last 200 years. Dalton's achievement was to come up with a theory that explained known facts and could be used to make predictions.

Dalton's theory could account for two existing chemical laws – the Law of Conservation of Mass and the Law of Constant Composition.

John Dalton.

The Law of Conservation of Mass

A simple version of this says that in a chemical reaction the total mass of everything at the end of the reaction is the same as the total mass at the beginning. This is bound to be the case if all you are doing is rearranging atoms which have got fixed masses.

The diagrams show the reaction between methane (natural gas) and chlorine in sunlight. The atoms have rearranged into new substances. The contents of both boxes must have the same mass because they are both made from exactly the same atoms – even if they are rearranged.

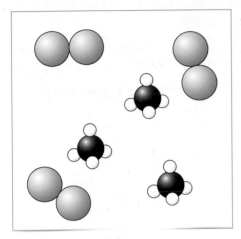 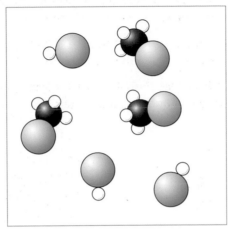

before   after

The Law of Constant Composition

This says that however you make a compound, it always contains the same elements in the same proportions by mass.

Again this follows from Dalton's theory. However you make silver chloride, for example, there is always one silver atom for each chlorine atom. Because silver atoms are about three times heavier than chlorine atoms, the mass ratio is always approximately 3 g of silver to 1 g of chlorine.

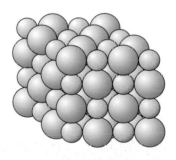

A small part of a silver chloride crystal, AgCl.

Using Dalton's Atomic Theory to make predictions

Not only could Dalton explain existing chemical laws, he could also use his theory to make entirely new predictions which could be tested by doing experiments. That is the basis for modern scientific method.

For example, carbon forms two common oxides, carbon monoxide and carbon dioxide. Look at the simple models. Based on Dalton's theory, you can see that carbon dioxide must contain twice as much oxygen per gram of carbon as carbon monoxide does.

Experiments confirm this. In carbon monoxide, there is 1.33 g of oxygen per gram of carbon. In carbon dioxide, the figure is 2.66 g.

carbon monoxide

carbon dioxide

Refining the atomic theory

Since Dalton's time scientists have gradually got closer and closer to the fine structure of atoms, first with the discovery of electrons, then of protons, and finally of neutrons. More recently, it has been discovered that protons and neutrons are made up of still smaller particles called quarks.

It is now possible to "see" individual atoms and to manipulate them to produce pictures like this, using a scanning tunnelling microscope.

If you are interested in atomic level pictures like this, try an internet search on **scanning tunnelling microscope**. The various IBM sites have some interesting examples.

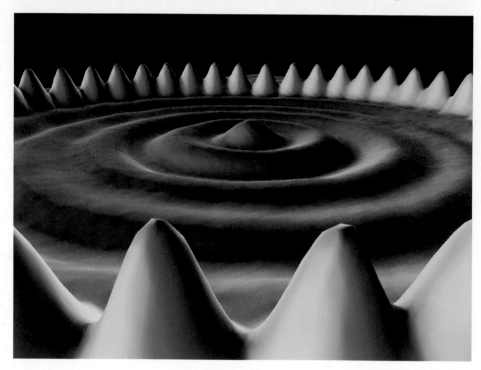

Atomic level art from IBM.

End of Chapter Checklist

If you haven't got a copy of your specification, read the introduction on page vi.

You will need to be able to do some or all of the following.

- State the relative masses and charges of protons, neutrons and electrons.

- Understand what is meant by *atomic (proton) number* and *mass number*.

- Explain the existence of isotopes.

- Know that the nucleus contains protons and neutrons, and that the electrons are found in a series of energy levels.

- Work out the arrangement of the electrons in the first 20 elements in the Periodic Table.

- Know that for elements in groups 1 to 7, the number of electrons in the outer level is the same as the group number.

- Know that noble gases have full (or temporarily full) outer levels.

- Understand the contribution that Dalton made to atomic theory.

Questions

You will need to use the Periodic Table on page 306. More questions on atomic structure can be found at the end of Section A on page 51.

1   Fluorine atoms have a mass number of 19.

   a)   Use the Periodic Table to find the atomic number of fluorine.

   b)   Explain what *mass number* means.

   c)   Write down the number of protons, neutrons and electrons in a fluorine atom.

   d)   Draw a diagram to show the arrangement of the electrons in the fluorine atom.

2   Work out the numbers of protons, neutrons and electrons in each of the following atoms:

   a) $^{56}_{26}Fe$ b) $^{93}_{41}Nb$ c) $^{235}_{92}U$

3   Chlorine has two isotopes, chlorine-35 and chlorine-37.

   a)   What are isotopes?

   b)   Write down the numbers of protons, neutrons and electrons in the two isotopes.

   c)   Write down the arrangement of the electrons in each of the two isotopes.

4   Draw diagrams to show the arrangement of the electrons in a) sodium, b) silicon, c) sulphur.

5   Find each of the following elements in the Periodic Table, and write down the number of electrons in their outer energy level.

   a)   arsenic, As; b) bromine, Br; c) tin, Sn; d) xenon, Xe.

6   The questions refer to the electronic structures below. Don't worry if some of these are unfamiliar to you. All of these are the electronic structures of neutral atoms.

   A   2, 4   (6)

   B   2, 8, 8   (18)

   C   2, 8, 18, 18, 7   (53) 1

   D   2, 8, 18, 18, 8   (54)

   E   2, 8, 8, 2   (20)

   F   2, 8, 18, 32, 18, 4   (8 2)

   a)   Which of these atoms are in group 4 of the Periodic Table?

   b)   Which of these structures represents carbon?

   c)   Which of these structures represents an element in group 7 of the Periodic Table?

   d)   Which of these structures represent noble gases?

   e)   Name element **E**.

   f)   How many protons does element **F** have? Name the element.

Chapter 1: Checklist

9

g) Element **G** has one more electron than element **B**. Draw a diagram to show how the electrons are arranged in an atom of **G**.

7   About 2,400 years ago, Democritus suggested that matter was made of atoms. Dalton reintroduced the idea about 200 years ago. Explain briefly why other scientists took Dalton seriously, whereas Democritus's ideas came to nothing for more than 2000 years.

8   J J Thomson (1856–1940), the discoverer of the electron, thought that atoms were solid, positively charged spheres with electrons embedded in them. Ernest Rutherford (1871–1937) later did an experiment involving alpha particles which showed that atoms are mostly empty space, containing a tiny positively charged nucleus with electrons around it. By doing an internet (or other) search, find out details of this experiment and write a short article, not exceeding one side of A4, describing and explaining it. You should write your article using a computer, and it should contain at least one diagram. Your diagram(s) can be obtained from the internet, generated using a drawing program, or scanned in.

Chapter 2: Bonding

This chapter looks at what happens when atoms combine together – whether into small groups or very large ones.

The elements sodium and chlorine are dangerous. The compound sodium chloride (salt) isn't.

Water has completely different properties from its elements, hydrogen and oxygen.

Sodium is a dangerously reactive metal. It is stored under oil to prevent it reacting with air or water. Chlorine is a very poisonous, reactive gas.

But salt, sodium chloride, is safe to eat in small quantities. Combining the elements to make salt obviously changes them significantly.

A mixture of hydrogen and oxygen gas would explode violently if you held a lighted match to it. Dropping a lighted match into water (a compound of hydrogen and oxygen) doesn't cause a literally earth-shattering explosion.

Reacting the elements to make a compound has again made a huge difference to them.

Covalent Bonding

What is a covalent bond?

In any bond, particles are held together by electrical attractions between something positively charged and something negatively charged. In a covalent bond, a pair of electrons is shared between two atoms. Each of the positively charged nuclei is attracted to the same negatively charged pair of electrons.

In most of the simple examples you will meet at GCSE, each atom in a covalent bond supplies one electron to the pair of electrons. That doesn't have to be the case. Both electrons may come from the same atom.

nucleus of A is attracted to the electron pair

nucleus of B is also attracted to the electron pair

$$A \bullet \bullet B$$

pair of electrons (one from each atom)

A and B are held together by this shared attraction.

Covalent bonding in a hydrogen molecule

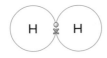

Covalent bonds are often shown using "dot-and-cross" diagrams. Although the electrons are drawn as dots or as crosses, there is absolutely no difference between them in reality. The dot and the cross simply show that the electrons have come from two different atoms. You could equally well use two different colour dots or two different colour crosses.

Both hydrogen nuclei are strongly attracted to the shared pair of electrons.

H——H

The bond can also be shown as a line between the two atoms. Each line represents one pair of shared electrons.

The covalent bond between two hydrogen atoms is very strong. Hydrogen atoms therefore go around in pairs, called a hydrogen **molecule**, with the symbol H_2.

Molecules have a certain fixed number of atoms in them joined together by covalent bonds. Hydrogen molecules are said to be **diatomic** because they contain two atoms. Other sorts of molecules may have as many as thousands of atoms joined together.

Why does hydrogen form molecules?

Whenever a bond is formed (of whatever kind), energy is released; that makes the things involved more stable than they were before. The more bonds an atom can form, the more energy is released and the more stable the system becomes.

In the hydrogen case, each hydrogen atom only has one electron to share, so it can only form one covalent bond. The H_2 molecule is still much more stable than two separate hydrogen atoms.

Covalent bonding in a hydrogen chloride molecule

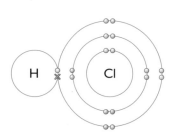

The chlorine atom has one unpaired electron in its outer level which it can share with the hydrogen atom to produce a covalent bond.

Notice that only the electrons in the outer energy level of the chlorine are used in bonding. In the examples you will meet at GCSE, the inner electrons never get used. In fact, the inner electrons are often left out of bonding diagrams. But be careful! In an exam – only leave the inner electrons out if the question tells you to.

The significance of noble gas structures in covalent bonding

If you look at the arrangement of electrons around the chlorine atom in the covalently bonded molecule of HCl, you will see that its structure is now 2,8,8. That is the same as an argon atom. Similarly, the hydrogen now has 2 electrons in its outer level – the same as helium.

> **Warning!** At GCSE people frequently talk about atoms "wanting" to form noble gas structures. This is nonsense! Avoid thinking about it in this way.

At GCSE this is quite common. When atoms bond covalently, they often produce outer electronic structures the same as noble gases – in other words, with four pairs of electrons (or one pair in the case of hydrogen). There are, however, lots of examples where different numbers of pairs are formed, producing structures which are quite unlike noble gases.

Covalent bonding in a chlorine molecule

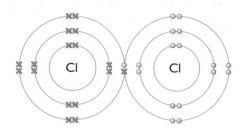

Each chlorine has one unpaired electron in its outer energy level. These are shared between the two atoms to produce a covalent bond.

Chlorine is another diatomic molecule, Cl_2.

Covalent bonding in an oxygen molecule – double bonding

When atoms bond covalently, they do so in a way that forms the maximum number of bonds. That makes the final molecule more stable.

The left-hand diagram shows that forming a single covalent bond between the two oxygen atoms still leaves unpaired electrons. If these are shared as well (as in the right-hand diagram), a more stable molecule is formed.

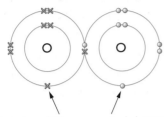

Each atom still has an unpaired electron. If these were shared as well, even more energy would be released.

Sharing 2 electron pairs maximises the bonding and makes the system as stable as possible.

Oxygen is a diatomic molecule, O_2. The double bond can be shown as $O{=}O$.

Covalent bonding in methane, ammonia and water

In methane, the carbon atom has four unpaired electrons. Each of these forms a covalent bond by sharing with the electron from a hydrogen atom. Methane has the formula CH_4

In ammonia, the nitrogen only has three unpaired electrons and so can only form bonds with three hydrogen atoms to give NH_3.

In water, there are two unpaired electrons on the oxygen atom which can bond with hydrogen atoms to give H_2O.

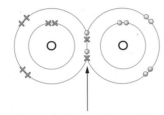

methane

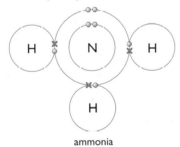

ammonia

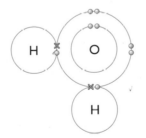

water

Why is water more stable than a mixture of hydrogen and oxygen?

When hydrogen, H_2, and oxygen, O_2, combine to make water, covalent bonds have to be broken in the hydrogen and oxygen molecules, and new ones are made in water.

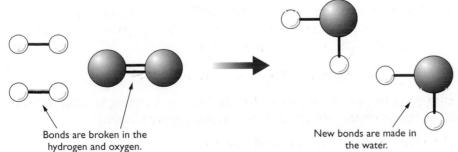

Bonds are broken in the hydrogen and oxygen.

New bonds are made in the water.

It is important to realise that when most compounds are formed, you don't actually make them from atoms of their elements. For example, you never make a water molecule from individual hydrogen and oxygen atoms. Those atoms are already bonded in some way. A chemical reaction reorganises existing bonds

It costs energy to break bonds, but a lot of energy is released when new ones are made. The reaction between hydrogen and oxygen is explosive. This is because much more energy is released in making the new bonds than was used in breaking the old ones. Because so much energy is released, the water is much more stable than the original hydrogen and oxygen.

Covalent bonding in carbon dioxide, CO_2

The carbon atom has four unpaired electrons and so can form four covalent bonds. Each oxygen has two unpaired electrons and can form two covalent bonds.

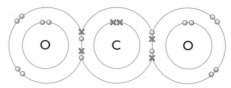

The number of bonds is maximised (and so the system becomes most stable) if the carbon forms two bonds with each oxygen. Using lines to represent pairs of shared electrons, you could draw carbon dioxide as:

$$O = C = O$$

Ways of representing covalent bonds

Apart from full dot-and-cross diagrams, covalent molecules can also be shown in other ways. In models, each link between the atoms represents a covalent bond – a pair of shared electrons.

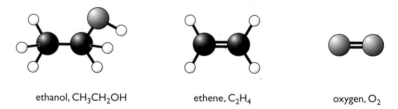

ethanol, CH_3CH_2OH ethene, C_2H_4 oxygen, O_2

On paper, we often simplify the diagrams by leaving out the inner electrons. You might leave out the circles as well, and only show the electrons in the outer energy levels.

Or you might draw each covalent bond as a straight line joining the atoms. Each line means a pair of shared electrons. In diagrams of this sort, sometimes you draw the non-bonding pairs of electrons in the outer level (called **lone pairs**); sometimes you leave them out.

All these diagrams show the covalent bonding in ammonia, NH_3:

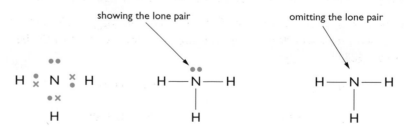

Ionic (Electrovalent) Bonding

In a covalent bond, the electrons are shared between two atoms. Both nuclei are attracted to the same electron pair.

But sometimes it happens that one of the atoms is attracted to the electron pair much more strongly than the other one. The electron pair is then pulled very close to that atom, and away from the other one.

A : B

pair of electrons shared between A and B in a covalent bond

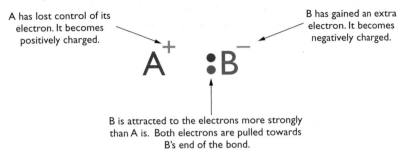

A has lost control of its electron. It becomes positively charged.

B has gained an extra electron. It becomes negatively charged.

B is attracted to the electrons more strongly than A is. Both electrons are pulled towards B's end of the bond.

A has become positively charged because it has effectively lost an electron. It still has the same number of positively charged protons, but now has one less electron to balance them. B is negatively charged because it has gained an extra negative electron.

Atom A has, in effect, given its electron to atom B.

The electrically charged particles are called **ions**. An ion is an atom (or group of atoms) which carries an electrical charge, either positive or negative.

- A positive ion is called a **cation**.

- A negative ion is called an **anion**.

Ionic bonding is bonding in which there has been a transfer of electrons from one atom to another to produce ions. The substance is held together by strong electrical attractions between positive and negative ions.

Ionic bonding in sodium chloride

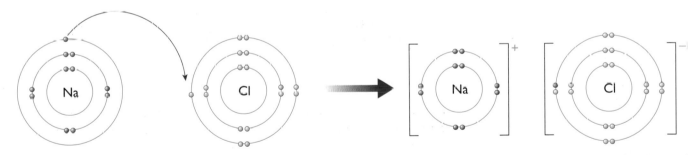

The single electron in the outer energy level of the sodium has been transferred to the chlorine. The sodium chloride is held together by the strong attraction between the sodium ion and the chloride ion. (Notice that it is called a **chloride** ion and not a chlorine ion.)

Overall, lots of energy is given out when this process happens – mainly due to the energy released when the strong bonding between the ions is set up.

You can draw dot-and-cross diagrams to show ionic bonding, but it is much quicker, and takes up much less space, to write electronic structures in the form 2,8,1 or 2,8,7.

This is a simplification! In reality, you don't react sodium with chlorine *atoms*, but with chlorine *molecules*, Cl_2. Before the electron transfer can happen, energy has to be supplied to break the chlorine molecules into individual atoms. You do, in fact, have to heat sodium in chlorine to get it to start to react.

Na   2,8,1 → Na⁺   [2,8]⁺

Cl   2,8,7 → Cl⁻   [2,8,8]⁻

Ionic bonding in magnesium oxide

Mg   2, 8, ② → Mg^{2+} [2, 8]$^{2+}$

O   2, 6 → O^{2-} [2, 8]$^{2-}$

In this case, two electrons are transferred from the magnesium to the oxygen.

The two electrons in the outer energy level of the magnesium are relatively easy to remove, and the oxygen has enough space in its outer level to receive them. More energy is given out this time, mainly due to the very, very strong attractions between the 2+ and 2– ions – the higher the number of charges, the stronger the attractions.

The significance of noble gas structures in ionic bonding

If you look at the structures of the ions formed in the two examples above, each of them has a noble gas structure – 2,8 (the neon structure), or 2,8,8 (the argon structure). You might therefore say that atoms lose or gain electrons so that they achieve a noble gas structure. This is true of the elements in Groups 1 and 2 of the Periodic Table (forming 1+ and 2+ ions), and for those in Groups 6 and 7 when they form 2– and 1– ions – as in all these examples.

But there are lots of common ions which don't have noble gas structures. Fe^{2+}, Fe^{3+}, Cu^{2+}, Zn^{2+}, Ag^+ and Pb^{2+} are all ions that you will come across during a GCSE course – although you won't have to write their electronic structures. Not one of them has a noble gas structure.

Other examples of ionic bonding

GCSE tends to restrict ionic bonding to elements from Groups 1, 2, 6 and 7 because that's where the simple examples are to be found.

Ionic bonds are usually only formed if small numbers of electrons need to be transferred – typically 1 or 2. In cases where the ions produced would have, say, a 3+ charge, the situation is rarely as simple as it might appear at first sight.

The following all involve elements from Groups 1 and 2 combining with those from Groups 6 and 7.

Lithium fluoride

Li   2, ① → Li$^+$ [2]$^+$

F   2, 7 → F$^-$ [2, 8]$^-$

The lithium atom has one electron in its outer energy level that is easily lost, and the fluorine has space to receive one. Lithium fluoride is held together by the strong attractions between lithium and fluoride ions.

Calcium chloride

Cl   2, 8, 7 → Cl$^-$ [2, 8, 8]$^-$

Ca   2, 8, 8, ② → Ca^{2+} [2, 8, 8]$^{2+}$

Cl   2, 8, 7 → Cl$^-$ [2, 8, 8]$^-$

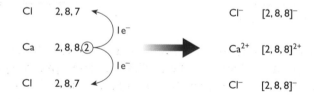

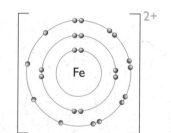

An Fe^{2+} ion – definitely not a noble gas structure!

The calcium has two electrons in its outer energy level that are relatively easy to give away, but each chlorine atom only has room in its outer level to take one of them. You need two chlorines for every one calcium. The formula for calcium chloride is therefore $CaCl_2$. There will be very strong attractions holding the ions together because of the 2+ charge on the calcium ions.

Potassium oxide

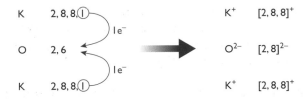

This time the oxygen has room for two electrons in its outer level, but each potassium can only supply one. Potassium oxide's formula is therefore K_2O. The presence of the double negative charge on the O^{2-} ion will help to strengthen the attractions between the ions.

Metallic Bonding

Most metals are hard and have high melting points. That suggests that the forces which hold the particles in the metal together are very strong.

The diagram below shows what happens when sodium atoms bond together to form the solid metal. The outer electron on each sodium atom becomes free to move throughout the whole structure. The electrons are said to be **delocalised**.

If a sodium atom loses its outer electron, that leaves behind a sodium ion. The attraction of each positive ion to the delocalised electrons holds the structure together.

Metallic bonding is sometimes described as an array of positive ions in a "sea of electrons".

Metals are hard and have high melting points.

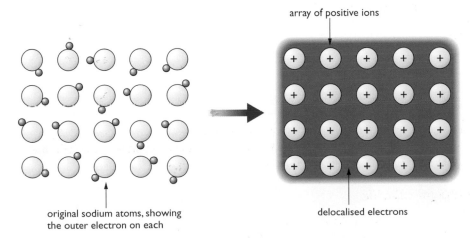

array of positive ions

original sodium atoms, showing the outer electron on each

delocalised electrons

In the case of sodium only one electron per atom is delocalised, leaving ions with only one positive charge on them. The ions don't pack very efficiently either. The effect of all this is that the bonding in sodium is quite weak as

Warning! When they come to write the symbol for a metal like sodium in equations, students who know about metallic bonding sometimes worry whether they should write it as Na or Na^+.

You write it as *atoms* — as Na. Thinking about the structure as a *whole*, the number of electrons exactly balances the number of positive charges. The metal as a *whole* carries no charge.

metals go, which is why sodium is fairly soft with a melting point which is low for a metal.

By contrast, magnesium has two outer electrons, both of which are delocalised into the "sea", leaving behind ions which carry a charge of 2+. It also packs more efficiently. There is a much stronger attraction between the more negative "sea" and the doubly-charged ions, and so the bonding is stronger and the melting point is greater.

Metals like iron have even more outer electrons to delocalise and so the bonding is stronger still.

You can find out more about metallic structures in the next chapter.

Intermolecular Forces

You will remember that water, H_2O, is a molecule with strong covalent bonds between the hydrogen and oxygen. In liquid water, or in ice, there must also be attractions between one molecule and its neighbours – otherwise they wouldn't stick together to make a liquid or a solid.

These forces of attraction between separate molecules are called **intermolecular forces** or **intermolecular attractions**. They are a lot weaker than covalent or ionic bonds, and vary in strength from substance to substance.

For example, the intermolecular forces between hydrogen molecules, H_2, are very, very weak. You have to cool hydrogen to –253°C before the molecules are travelling slowly enough for the intermolecular attractions to be able to hold them together as a liquid.

By contrast, sugar (also a covalent compound) is a solid which doesn't melt until 185°C. The intermolecular forces between sugar molecules must be quite strong.

Intermolecular forces arise from slight electrical distortions in molecules.

In the diagram, you read "δ" as "delta". So "δ+" reads as "delta plus". "δ" is used to mean "slightly", so "δ+" means "slightly positive".

Breaking the intermolecular attractions in water to produce steam.

A hint at how these distortions arise in water molecules is given in the next chapter on page 24.

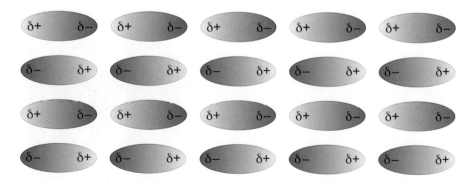

You can see that the slightly positive end of one molecule attracts the slightly negative end of a neighbouring molecule. Heating will supply enough energy to break these intermolecular attractions and cause the substance to either melt or boil.

In melting, some but not all of the intermolecular forces are broken. In boiling, the attractions are totally disrupted and the molecules become free to move around as a gas. It is very important that you realise that melting or boiling a substance made of molecules breaks intermolecular forces – *not* covalent bonds. When you boil water, you get steam – not a mixture of hydrogen and oxygen atoms.

If you boiled a teaspoonful of water (about 5 cm³) in the bottom of an average bucket, enough steam would be produced to fill the bucket. That shows how spread out the water molecules become once you have broken the intermolecular attractions.

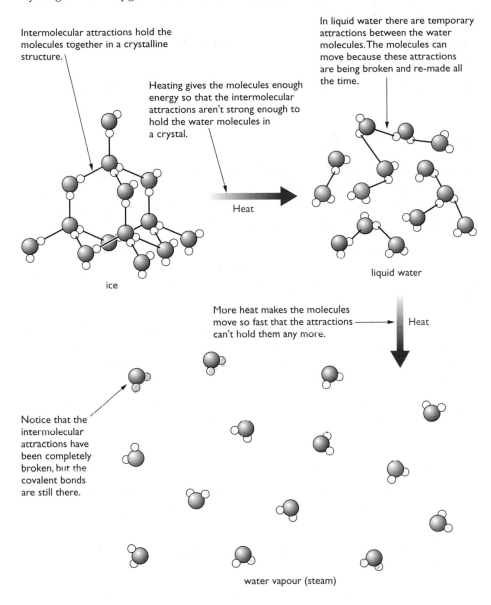

Intermolecular attractions hold the molecules together in a crystalline structure.

Heating gives the molecules enough energy so that the intermolecular attractions aren't strong enough to hold the water molecules in a crystal.

In liquid water there are temporary attractions between the water molecules. The molecules can move because these attractions are being broken and re-made all the time.

Heat

ice

liquid water

More heat makes the molecules move so fast that the attractions can't hold them any more.

Heat

Notice that the intermolecular attractions have been completely broken, but the covalent bonds are still there.

water vapour (steam)

End of Chapter Checklist

If you haven't got a copy of your specification, read the introduction on page vi.

You will need to be able to do some or all of the following. Check your Awarding Body's specification (syllabus) to find out exactly what you need to know.

● Explain what is meant by: covalent bond, molecule.

● Draw dot-and-cross diagrams for all the covalent molecules mentioned by your specification.

● Explain what is meant by an ionic bond.

● Describe the formation of all the ionic compounds or ions mentioned by your specification.

● Describe how atoms are bound together in a metal.

● Explain what is meant by intermolecular forces.

Questions

You will need to use the Periodic Table on page 306. More questions on bonding can be found at the end of Section A on page 51.

1   a)   What is meant by a *covalent bond*? How does this bond hold two atoms together?

   b)   Draw dot-and-cross diagrams to show the covalent bonding in i) methane, CH_4, ii) hydrogen sulphide, H_2S, iii) phosphine, PH_3, iv) silicon tetrachloride, $SiCl_4$.

2   Draw dot-and-cross diagrams to show the covalent bonding in a) ethane, C_2H_6, b) ethene, C_2H_4 and c) ethanol, CH_3CH_2OH. You will find models of ethene and ethanol on page 14 which might help you.

3   a)   What is meant by i) an ion, ii) an ionic bond?

   b)   In each of the following cases, write down the electronic structures of the original atoms and then explain (in words or diagrams) what happens when:

     i)   sodium bonds with chlorine to make sodium chloride

     ii)   lithium bonds with oxygen to make lithium oxide

     iii)   magnesium bonds with fluorine to make magnesium fluoride.

4   a)   A solid metal is often described as having "an array of positive ions in a sea of electrons". Write down the electronic structure of a magnesium atom and use it to explain what this phrase means.

   b)   Metallic bonds are not fully broken until the metal has first melted and then boiled. The boiling points of sodium, magnesium and aluminium are 890°C, 1110°C and 2470°C respectively. What does this suggest about the strengths of the metallic bonds in these three elements?

   c)   Find these three metals in the Periodic Table, and suggest why the boiling points show this pattern.

   d)   Assuming that an electric current is simply a flow of electrons, suggest why all these elements are good conductors of electricity.

5   The table gives details of the boiling temperatures of some substances made up of covalent molecules. Arrange these substances in increasing order of the strength of their intermolecular attractions.

	Boiling point (°C)
ammonia	−33
ethanamide	221
ethanol	78.5
hydrogen	−253
phosphorus trifluoride	−101
water	100

Don't panic if you don't recognise some of the names. The substances could just as well have been described as A, B, C, D, E and F.

6   Boron and aluminium are both in Group 3 of the Periodic Table. Both form compounds with fluorine (BF_3 and AlF_3). Unusually for elements found in the same group of the Periodic Table, their compounds are bonded differently. BF_3 is covalent whereas AlF_3 is a straightforward ionic compound.

   a)   Draw a diagram to show the covalent bonding in BF_3.

   b)   Explain, using diagrams or otherwise, the origin of the ionic bonding in AlF_3.

   c) BF_3 is described as an *electron deficient* compound. What do you think that might mean?

Chapter 3: Structure

The photographs show some substances with quite different physical properties – hardness, melting point and solubility, for example. This chapter explores some of the reasons for these differences, based on the bonding in the substances. It assumes that you are already familiar with the topic of bonding – see Chapter 2.

Metals are strong and easily shaped.

Many substances form brittle crystals which dissolve easily in water.

Diamond (a form of carbon) is obviously crystalline, and is the hardest naturally occurring substance.

Ice is also crystalline, but melts easily to form water.

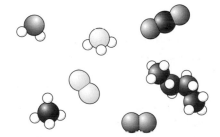

Some simple molecules.

Giant Structures

You can divide substances into two quite different types – giant structures and molecular structures.

You will remember that molecules are made up of *fixed* numbers of atoms joined together by covalent bonds. The number of atoms per molecule is usually fairly small, but can run into thousands in the case of big molecules like plastics, proteins or DNA.

By contrast, giant structures contain huge numbers of either atoms or ions arranged in some regular way, but the number of particles isn't fixed.

Examples will make this clear.

If you aren't sure about this, read page 17 before you go on.

Giant metallic structures

Remember that metals consist of a regular array of positive ions in a "sea of electrons". The metal is held together by the attractions between the positive ions and the delocalised electrons.

The simple physical properties of metals

Metals tend to be strong, with high melting and boiling points because of the powerful attractions involved.

Metals conduct electricity. This is because the delocalised electrons are free to move throughout the structure. Imagine what happens if a piece of metal is attached to an electrical power source.

If you compare this diagram with a similar picture of metallic bonding (see Chapter 2, page 17), you will find that the ions are arranged differently. This diagram shows the staggered rows typical of efficiently packed metals.

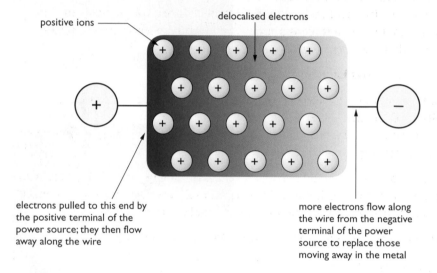

positive ions

delocalised electrons

electrons pulled to this end by the positive terminal of the power source; they then flow away along the wire

more electrons flow along the wire from the negative terminal of the power source to replace those moving away in the metal

Metals are good conductors of heat. This is again due to the mobile delocalised electrons. If you heat one end of a piece of metal, the energy is picked up by the electrons. As the electrons move around in the metal, the heat energy is transferred throughout the structure.

The workability of metals

If a metal is subjected to just a small force, it will stretch and then return to its original shape when the force is released. The metal is described as being **elastic**.

But if a large force is applied, the particles slide over each other and stay in their new positions.

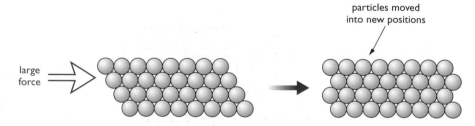

particles moved into new positions

large force

Metals are usually easy to shape because their regular packing makes it simple for the atoms to slide over each other. Metals are said to be **malleable** and **ductile**. Malleable means that it is easily beaten into shape. Ductile means that it is easily pulled out into wires.

Steel being rolled into strips.

Alloys

Metals can be made harder by **alloying** them with other metals. An alloy is a mixture of metals – for example, **brass** is a mixture of copper and zinc.

In an alloy the different metals have slightly differently sized atoms. This breaks up the regular arrangement and makes it more difficult for the layers to slide.

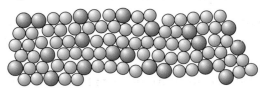

The diagram shows how mixing atoms of only slightly different sizes disrupts the regular packing, and makes it much more difficult for particles to slide over each other when a force is applied. This tends to make alloys harder than the individual metals which make them up.

In some cases, alloys have unexpected properties.

For example, solder – an alloy of tin and lead – melts at a lower temperature than either of the metals individually. Its low melting point and the fact that it is a good conductor of electricity makes it useful for joining components in electrical circuits.

Other common alloys include bronze (a mixture of copper and tin), stainless steel (an alloy of iron with chromium and nickel) and the mixture of copper and nickel ("cupronickel") which is used to make "silver" coins.

You can read more about alloys on pages 148, 151–2 and 156.

Giant ionic structures

An ion is an atom or group of atoms which carries an electrical charge – either positive or negative. If you aren't sure about ionic bonding, you should read pages 15–17 before you go on.

All ionic compounds consist of huge lattices of positive and negative ions packed together in a regular way. A **lattice** is a regular array of particles. The lattice is held together by the strong attractions between the positively and negatively charged ions.

The structure of sodium chloride

In a diagram, the ions are usually drawn in an "exploded" view. The photograph shows how they actually occupy the space.

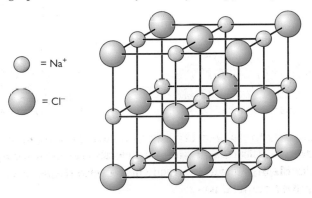

= Na⁺

= Cl⁻

An oil lamp made of both pure copper and its alloy, brass.

Solder is an alloy of lead and tin. Solder has a melting point lower than that of either of the individual metals.

A model of a small part of a sodium chloride crystal.

Warning! The lines in this diagram are *not* covalent bonds. There are just there to help to show the arrangement of the ions. Those ions joined by lines are touching each other. Compare the diagram with the model in the photograph.

Chapter 3: Structure

23

Only ions joined by lines in the diagram are actually touching. Each sodium ion is touched by six chloride ions. In turn, each chloride ion is touched by six sodium ions.

You have to remember that this structure repeats itself over vast numbers of ions.

The structure of magnesium oxide

Magnesium oxide, MgO, contains magnesium ions, Mg^{2+}, and oxide ions, O^{2-}. It has exactly the same structure as sodium chloride.

The only difference is that the magnesium oxide lattice is held together by stronger forces of attraction. This is because in magnesium oxide, 2+ ions are attracting 2– ions. In sodium chloride, the attractions are weaker because they are only between 1+ and 1– ions.

The simple physical properties of ionic substances

Ionic compounds have high melting points and boiling points because of the strong forces holding the lattices together. Magnesium oxide has much higher melting and boiling points than sodium chloride because the attractive forces are much stronger.

Ionic compounds tend to be crystalline. This reflects the regular arrangement of ions in the lattice. Sometimes the crystals are too small to be seen except under powerful microscopes. Magnesium oxide, for example, is always seen as a white powder because the individual crystals are too small to be seen with the naked eye.

Ionic crystals tend to be brittle. This is because any small distortion of a crystal will bring ions with the same charge alongside each other. Like charges repel and so the crystal splits itself apart.

The shape of the sodium chloride crystal reflects the arrangement of the ions.

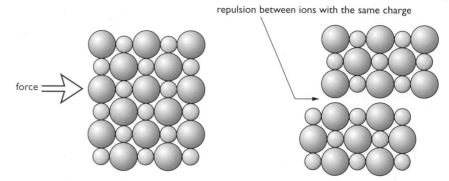

repulsion between ions with the same charge

force

Ionic substances tend to be soluble in water. Although water is a covalent molecule, the electrons in the bonds are attracted towards the oxygen end of the bond. This makes the oxygen slightly negative. It leaves the hydrogen slightly short of electrons and therefore slightly positive.

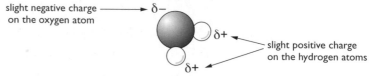

slight negative charge on the oxygen atom → $\delta-$

$\delta+$

slight positive charge on the hydrogen atoms

$\delta+$

For the use of the symbol "δ", see page 18.

Because of this electrical distortion, water is described as a **polar** molecule. There are quite strong attractions between the polar water molecules and the ions in the lattice.

The slightly positive hydrogens in the water molecules cluster around the negative ions, and the slightly negative oxygens are attracted to the positive ions.

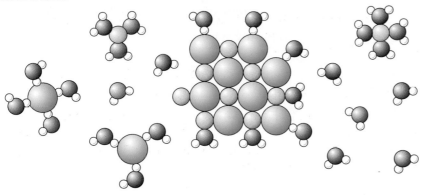

The water molecules then literally pull the crystal apart.

Magnesium oxide isn't soluble in water because the attractions between the water molecules and the ions aren't strong enough to break the very powerful ionic bonds between magnesium and oxide ions.

Ionic compounds tend to be insoluble in organic solvents. Organic solvents contain molecules which have much less electrical distortion than there is in water – their molecules are less polar. There isn't enough attraction between these molecules and the ions in the crystal to break the strong forces holding the lattice together.

The electrical behaviour of ionic substances

Ionic compounds don't conduct electricity when they are solid, because they don't contain any mobile electrons. They do, however, conduct electricity when they melt, or if they are dissolved in water. This happens because the ions then become free to move around. How this enables the compound to conduct electricity is explained in Chapter 11: Electrolysis.

Giant covalent structures

Diamond

Diamond is a form of pure carbon.

Each carbon atom has four unpaired electrons in its outer energy level (shell) and it uses these to form four covalent bonds. In diamond, each carbon bonds strongly to four other carbon atoms. The diagram shows enough of the structure to see what it is happening.

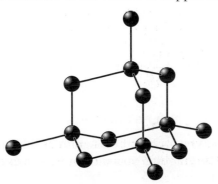

Organic solvents include things like alcohol (ethanol) and hydrocarbons, such as are found in petrol. If you are interested in these, you could explore the organic chemistry section of this book.

In the diagram some carbon atoms only seem to be forming two bonds (or even one bond), but that's not really the case. We are only showing a small bit of the whole structure. The lines in this diagram each represents a covalent bond.

This is a giant covalent structure – it continues on and on in three dimensions. It is not a molecule, because the number of atoms joined up in a real diamond is completely variable, depending on the size of the crystal. Molecules always contain fixed numbers of atoms joined by covalent bonds.

Draw this structure in the following stages:

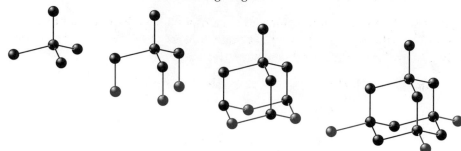

This is a very easy structure to draw as long as you practice it. You should be able to produce a reasonable sketch in 30 seconds.

Diamond is very hard, with a very high melting and boiling point. This is because of the very strong carbon–carbon covalent bonds which extend throughout the whole crystal in three dimensions.

Diamond doesn't conduct electricity. All the electrons in the outer levels of the carbon atoms are tightly held in covalent bonds between the atoms. None are free to move around.

Diamond doesn't dissolve in water or any other solvent. This is again because of the powerful covalent bonds between the carbon atoms. If the diamond dissolved, these would have to be broken.

Graphite

Graphite is also a form of carbon, but the atoms are arranged differently. Graphite has a layer structure rather like a pack of cards. In a pack of cards each card is strong, but the individual cards are easily separated. The same is true in graphite.

atoms in a layer of graphite

Graphite is a soft material with a slimy feel. Although the forces holding the atoms together in each layer are very strong, the attractions between the layers are much weaker. Layers can easily be flaked off.

Graphite (mixed with clay to make it harder) is used in pencils. When you write with a pencil, you are leaving a trail of graphite layers behind on the paper.

Pure graphite is so slippery that it is used as a dry lubricant – for example, to lubricate locks.

Layers of graphite will flake off in the same way that an individual card separates easily from the pack.

Graphite has a high melting and boiling point and is insoluble in any solvents. To melt or dissolve the graphite, you don't just have to break the layers apart – you have to break up the whole structure, including the covalent bonds. That needs very large amounts of energy because the bonds are so strong.

Graphite is less dense than diamond, because the layers in graphite are relatively far apart. The distance between the graphite layers is more than twice the distance between atoms in each layer. In a sense, a graphite crystal contains a lot of wasted space, which isn't there in a diamond crystal.

Graphite conducts electricity. If you look at the diagram of the arrangement of the atoms in each layer in the graphite, you will see that each carbon atom is only joined to three others.

Rubbing layers of graphite off on paper.

Each carbon atom uses three of its electrons to form these simple covalent bonds. The fourth electron in the outer layer of each atom is free to move around throughout the whole of the layer. The movement of these electrons allows the graphite to conduct electricity.

Silicon dioxide, SiO_2

Quartz is a commonly found form of silicon dioxide (silica). It is a hard crystalline solid with a high melting point; it is insoluble in water and other solvents, and it doesn't conduct electricity.

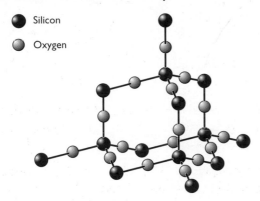

Quartz crystals.

As with all the giant structures, remember that this diagram only shows a small part of the whole thing. This arrangement just goes on and on. Each silicon atom is connected to four oxygen atoms in three dimensions, and each oxygen is attached to two silicons.

The structure looks complicated, but if you look at it closely you will see that the silicon atoms are arranged in a diamond structure, with an oxygen atom slotted in between each silicon atom.

The silicon–oxygen covalent bonds are strong. That means that silicon dioxide has a high melting point and doesn't dissolve in anything. There aren't any mobile electrons in the structure, and so silicon dioxide doesn't conduct electricity.

Simple Molecular Structures

Remember that molecules contain fixed numbers of atoms joined by strong covalent bonds. The forces of attraction between one molecule and its neighbours are much weaker than the covalent bonds, and vary in strength from compound to compound.

Because these intermolecular attractions are relatively weak, *simple molecular compounds tend to be gases, liquids or low melting point solids.*

Molecular substances tend to be insoluble in water unless they react with it. In order for a substance to dissolve, quite strong attractions between water molecules have to be broken so that the dissolving molecules can fit between them. Any new attractions between water molecules and the covalent molecules are not usually big enough to compensate for this.

On the other hand, *molecular substances are often soluble in organic solvents.* In this case, the intermolecular attractions between the two different types of molecules are much the same as in the pure substances.

Molecular substances don't conduct electricity, because the molecules don't have any overall electrical charge and there are no electrons mobile enough to move from molecule to molecule.

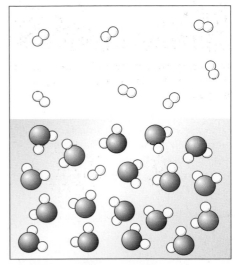

Hydrogen gas is almost insoluble in water.

If you want to find out more, you could do an internet search on *buckminsterfullerene*. If you don't know what a *geodesic dome* is, try a search on that as well.

Buckminsterfullerene

In the 1980s a molecular form of carbon was discovered with the formula C_{60}. This was called buckminsterfullerene after the American inventor of the geodesic dome, Buckminster Fuller. Each molecule of C_{60} looks very much like a football.

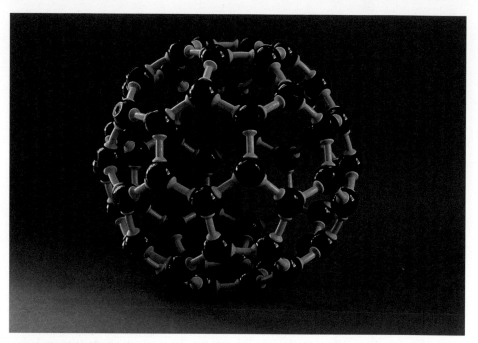

Buckminsterfullerene, C_{60}.

The molecules are packed in a regular way in a crystal lattice, in the same way that you might imagine thousands of footballs packed into a large container – except that you would be talking about far greater numbers of C_{60} molecules.

Elements, Compounds and Mixtures

Elements

Elements are substances which can't be split into anything more simple by chemical means. All the atoms in an element have the same atomic number. You can recognise them in models or diagrams because they consist of atoms of a single colour or size.

It isn't quite true to say that elements consist of only one type of atom. Most elements consist of mixtures of isotopes – with the same atomic number but different numbers of neutrons. When we draw diagrams or make models we aren't usually interested in the differences between the isotopes.

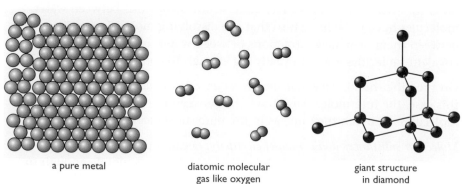

a pure metal

diatomic molecular gas like oxygen

giant structure in diamond

Compounds

All compounds are made from combinations of two or more elements in fixed proportions, and joined by strong bonds.

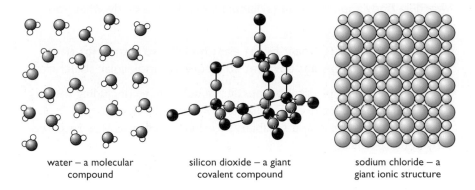

water – a molecular
compound

silicon dioxide – a giant
covalent compound

sodium chloride – a
giant ionic structure

Mixtures

In a mixture, the various components can be in any proportions. An alloy is a mixture rather than a compound because of the totally variable proportions.

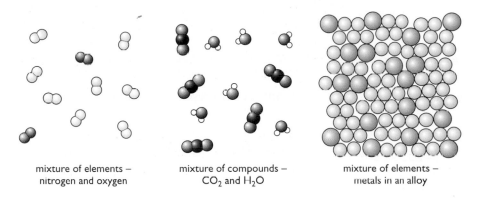

mixture of elements –
nitrogen and oxygen

mixture of compounds –
CO_2 and H_2O

mixture of elements –
metals in an alloy

End of Chapter Checklist

If you haven't got a copy of your specification, read the introduction on page vi.

You will need to be able to do some or all of the following. Check your Awarding Body's specification (syllabus) to find out exactly what you need to know.

- Understand what is meant by (a) giant metallic structure, (b) giant ionic structure, (c) giant covalent structure, (d) molecular structure.

- Relate the simple physical properties of substances (including melting and boiling temperatures, conduction of heat and electricity, workability/hardness, and solubility) to their structures – as required by your specification.

- Recognise and/or draw structures for the compounds mentioned by your specification.

- Understand the differences between the words element, compound and mixture in terms of the way the particles are arranged.

Questions

More questions on structure can be found at the end of Section A on page 51.

1 a) Draw simple diagrams to show the structures of diamond and graphite.

   b) Choose any *one* physical property where diamond and graphite have similar characteristics, and any *two* physical properties where they are different. Use your diagrams to explain the similarity and the differences.

2 a) Most metals are malleable and ductile. Explain what happens to the particles in a metal when it is subjected to a large stress.

   b) State any other physical property of metals and explain how it arises from the metallic structure.

   c) Alloys are mixtures of metals. Explain why an alloy is usually harder than the individual metals that make it up.

3 Explain why sodium chloride *a)* has a high melting point, *b)* has brittle crystals, *c)* is soluble in water.

4 Decide what sort of structure each of the following substances is most likely to have. You can choose between: giant metallic structure, giant covalent structure, giant ionic structure, molecular structure.

   a) Substance A melts at 2300°C. It doesn't conduct electricity even when it is molten. It is insoluble in water.

   b) Substance B is a colourless gas.

   c) Substance C is a yellow solid with a low melting point of 113°C. It doesn't conduct electricity and is insoluble in water.

   d) Substance D forms brittle orange crystals which melt at 398°C. It dissolves freely in water to give an orange solution.

   e) Substance E is a pinkish-brown flexible solid. It conducts electricity.

5 Carbon, silicon and germanium, Ge, are all in Group 4 of the Periodic Table.

   a) Germanium dioxide, GeO_2, has the same structure as SiO_2 in quartz, but the germanium–oxygen bonds are somewhat weaker than silicon–oxygen bonds. Make suggestions about the following physical properties of GeO_2.

      i) What is it likely to look like?

      ii) How hard is it likely to be?

      iii) What will its melting point be relative to SiO_2?

      iv) Will it conduct electricity?

      v) Will it dissolve in water?

   b) Germanium tetrachloride, $GeCl_4$, is a colourless, fuming liquid which produces steamy fumes when it is exposed to damp air. It reacts violently with cold water. What sort of structure do you think $GeCl_4$ has?

6 By doing an internet (or other) search, find out why buckminsterfullerene is named after the American inventor, Buckminster Fuller. Use the results of your search to write a short article to explain the connection. Your article should be produced using a computer, contain at least two pictures or diagrams, and be no longer than one side of A4 paper. You should use your own words, but the artwork can be taken directly from the internet or scanned in from some other source.

Chapter 4: Formulae and Equations

Learning how to understand and write chemical formulae and equations isn't one of the most exciting parts of Chemistry, but once you have the skill to do it, you will find that the subject becomes suddenly much clearer and easier to follow. Don't worry if it sometimes seems to take a long time to work out a formula or write an equation in the early stages. It does take a long time to start with – you just need patience and lots of practice.

Writing Formulae

Formulae for covalent substances

Common everyday examples

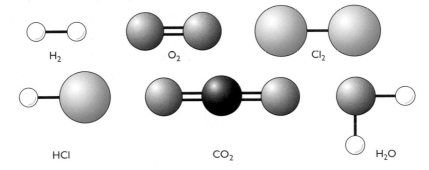

Remember that each line represents a covalent bond – a pair of shared electrons. All of these examples are described in detail on pages 11–14.

The formula simply counts the number of atoms combined to make the compound. You won't usually have to work out the formula of a covalent compound – most of the ones you will meet at GCSE are so simple and so common that you just remember them.

Suppose you had to work one out

Suppose you had to find the formula for phosphine – a simple compound of phosphorus and hydrogen.

Find phosphorus in the Periodic Table. Its atomic number is 15, and so the atom has 15 protons and 15 electrons. The electrons would be arranged 2,8,5. All you are interested in are the electrons in the outer level.

If you aren't happy about this, then you should read the beginning of Chapter 2 on covalent bonding.

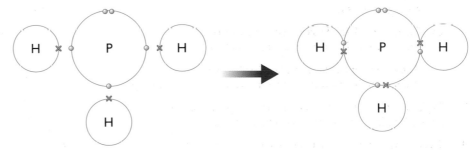

The phosphorus can create three covalent bonds by sharing the single electrons with three hydrogen atoms. That means that the formula for phosphine must be PH_3.

Formulae for ionic compounds

Before you go on, it would be a good idea to remind yourself about ionic bonding by reading pages 15–17.

There are so many different ionic compounds which you might come across at GCSE, that it would be impossible to learn all their formulae. You need a simple way to work them out. You could work a few of them out from first principles using their electronic structures, but that would take ages. Others would be too difficult. You need a simple short-cut method.

The need for equal numbers of "pluses" and "minuses"

Ions are atoms or groups of atoms which carry electrical charges, either positive or negative. Compounds are electrically neutral. Therefore in an ionic compound there must be the right number of each sort of ion so that the total positive charge exactly cancels out the total negative charge. Obviously, then, if you are going to work out a formula, you need to know the charges on the ions.

(1) Cases where you can work out the number of charges on an ion

Any element in Group 2 has two outer electrons which it will lose to form a 2+ ion. Any element in Group 6 has six outer electrons, and it has room to gain two more. This leads to a 2– ion. Similar arguments apply in the other groups shown in the table below.

Element in Periodic Table Group	Charge on ion
1	+1
2	+2
3	+3
6	–2
7	–1

(2) Cases where the name tells you the charge

All metals form positive ions. Names like lead(II) oxide, or iron(III) chloride or copper(II) sulphate tell you directly about the charge on the metal ion. The number after the metal tells you the number of charges. So...

Lead(II) oxide contains a Pb^{2+} ion. Iron(III) chloride contains an Fe^{3+} ion. Copper(II) sulphate contains a Cu^{2+} ion. Find the symbols for the metals from a Periodic Table if you need to.

(3) Ions that need to be learnt

Positive ions		
	zinc	Zn^{2+}
	silver	Ag^+
	hydrogen	H^+
	ammonium	NH_4^+
Negative ions	nitrate	NO_3^-
	hydroxide	OH^-
	hydrogencarbonate	HCO_3^-
	carbonate	CO_3^{2-}
	sulphate	SO_4^{2-}

You will come across other ions during the course, but these are the important ones for now. The ions in this list are the tricky ones – be sure to learn both the formula and the charge for each ion.

Confusing endings!

Don't confuse ions like sulph**ate** with sulph**ide**. A name like copper(II) sulphide means that it contains copper and sulphur *only*. Any "**ide**" ending means that there isn't anything complicated there. Sodium chloride, for example, is just sodium and chlorine combined together.

Once you have an "**ate**" ending, it means that there is something else there as well – often, but not always, oxygen. So, for example, copper(II) sulphate contains copper, sulphur and oxygen.

Not looking carefully at word endings is one of the commonest mistakes students make when they start to write formulae.

Working out the formula for an ionic compound

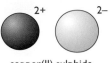

copper(II) sulphide

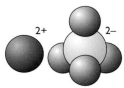

copper(II) sulphate

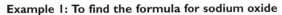

worked
example

Example 1: To find the formula for sodium oxide

Sodium is in Group 1, so the ion is Na^+.

Oxygen is in Group 6, so the ion is O^{2-}.

To have equal numbers of positive and negative charges, you would need 2 sodium ions to provide the 2 positive charges to cancel the 2 negative charges on one oxide ion. In other words, you need:

Na^+ Na^+ O^{2-}

The formula is therefore **Na_2O**.

Example 2: To find the formula for barium nitrate

Barium is in Group 2, so the ion is Ba^{2+}.

Nitrate ions are NO_3^-. You will have to remember this.

To have equal numbers of positive and negative charges, you would need 2 nitrate ions for each barium ion.

	positive ions	negative ions
	Ba^{2+}	NO_3^-
		NO_3^-
total charges	2+	2–

The formula is **$Ba(NO_3)_2$**.

Notice the brackets around the nitrate group. *Brackets must be written if you have more than one of these complex ions (ions containing more than one type of atom).* In any other situation, they are completely unnecessary.

Example 3: To find the formula for iron(III) sulphate

Iron(III) tells you that the metal ion is Fe^{3+}.

Sulphate ions are SO_4^{2-}. You will have to remember this.

To have equal numbers of positive and negative charges, you would need 2 iron(III) ions for every 3 sulphate ions – to give 6+ and 6– in total.

The formula is **$Fe_2(SO_4)_3$**.

> If you didn't write the brackets, the formula would look like this: $BaNO_{32}$. That would read as 1 barium, 1 nitrogen and 32 oxygens. That's not what you mean!

Why aren't ion charges shown in formulae?

Actually, they can be shown. For example, the formula for sodium chloride is NaCl. It is sometimes written Na^+Cl^- if you are trying to make a particular point, but for most purposes the charges are left out. In an ionic compound, the charges are there – whether you write them or not.

Writing Equations

What all the numbers mean

When you write equations it is important to be able to count up how many of each sort of atom you have got. In particular you must understand the difference between big numbers written in front of formulae such as the **2** in 2HCl, and the smaller, subscripted (written slightly lower on the line) numbers such as the **4** in CH_4.

What, for example, is the difference between **2Cl** and $\mathbf{Cl_2}$? The position of the 2 shows whether or not the atoms are joined together.

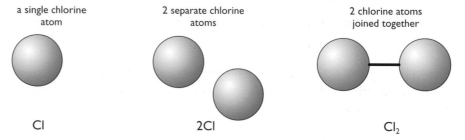

a single chlorine atom

2 separate chlorine atoms

2 chlorine atoms joined together

Cl   2Cl Cl_2

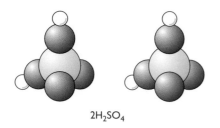

$2H_2SO_4$

Look at the way the numbers work in $2H_2SO_4$. The big number in front tells you that you have 2 sulphuric acid molecules. The little 4, for example, tells you that you have 4 oxygen atoms in each molecule. A small number in a formula only applies to the atom immediately before it in the formula.

If you count the atoms in $2H_2SO_4$, you will find 4 hydrogens, 2 sulphurs and 8 oxygens.

If you have brackets in a formula, a small number refers to everything inside the brackets. For example, in the formula $Ca(OH)_2$, the **2** applies to both the oxygen and the hydrogen. The formula shows 1 calcium, 2 oxygens and 2 hydrogens.

Balancing equations

Chemical reactions involve taking elements or compounds and shuffling their atoms around into new combinations. It follows that you must always end up with the same number of atoms that you started with.

Suppose you had to write an equation for the reaction between methane, CH_4, and oxygen, O_2. Methane burns in oxygen to form carbon dioxide and water. Think of this in terms of rearranging the atoms in some models.

methane   oxygen   carbon dioxide   water

This can't be right! During the rearrangement you seem to have gained an oxygen atom and lost two hydrogens. The reaction must be more complicated than this. Since the substances are all correct, the proportions must be wrong.

Try again:

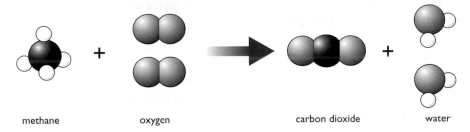

methane   oxygen   carbon dioxide   water

There are now the same number of each sort of atom before and after. This is called **balancing the equation**.

In symbols, this equation would be:

$$CH_4 + 2O_2 \rightarrow CO_2 + 2H_2O$$

Think of each symbol (C or H or O) as representing one atom of that element. Count them up in the equation, and check that there are the same number on both sides.

How to balance equations

Balancing equations isn't a hit-and-miss affair. If you are patient and work systematically, the equations you will meet at GCSE almost balance themselves.

- Work across the equation from left to right, checking one element after another, *except* if an element appears in several places in the equation. In that case, leave it until the end – you will often find that it has sorted itself out.

- If you have a group of atoms (like a sulphate group, for example) which is unchanged from one side of the equation to the other, there is no reason why you can't just count that up as a whole group, rather than counting individual sulphurs and oxygens. It saves time.

- Check everything at the end to make sure that you haven't changed something that you have already counted.

worked
 example

Example 1: Zinc + hydrochloric acid → zinc chloride + hydrogen

Balance the equation:

$$Zn + HCl \rightarrow ZnCl_2 + H_2$$

Work from left to right. Count the zinc atoms. 1 on each side – no problem!

Count the hydrogen atoms. 1 on the left; 2 on the right. If you end up with 2 you must have started with 2. The only way of achieving this is to have 2HCl. (You *must not* change the formula to H_2Cl – there's no such substance.)

$$Zn + 2HCl \rightarrow ZnCl_2 + H_2$$

Now count the chlorines. There are 2 on each side. Good! Finally check everything again to make sure – and that's it.

This is really important! You must *never, never* change a formula in balancing an equation. All you are allowed to do is to write big numbers in front of the formula.

Example 2: Silver nitrate solution + calcium chloride solution → calcium nitrate solution + silver chloride

Balance the equation:

$$AgNO_3 + CaCl_2 \rightarrow Ca(NO_3)_2 + AgCl$$

Working from left to right: 1 silver atom on both sides. That's fine.

The nitrate group is unchanged as it goes from left to right, so save time by counting it as a whole rather than splitting it into individual elements. There's 1 NO_3 group on the left, 2 on the right. That needs correcting. You must have started with $2AgNO_3$.

$$2AgNO_3 + CaCl_2 \rightarrow Ca(NO_3)_2 + AgCl$$

Now check the calcium. 1 on each side.

Now the chlorine. There are 2 on the left, but only 1 on the right. You need 2AgCl.

$$2AgNO_3 + CaCl_2 \rightarrow Ca(NO_3)_2 + 2AgCl$$

Finally, check everything again. It's all OK – but it might not have been. You actually changed the numbers of silver atoms on the left-hand side after you checked them at the beginning. It so happens that the problem corrected itself when you put the 2 in front of the AgCl on the right – but that won't always happen.

Example 3: Ethane + oxygen → carbon dioxide + water

Balance the equation:

$$C_2H_6 + O_2 \rightarrow CO_2 + H_2O$$

Starting from the left, balance the carbons:

$$C_2H_6 + O_2 \rightarrow 2CO_2 + H_2O$$

Now the hydrogens:

$$C_2H_6 + O_2 \rightarrow 2CO_2 + 3H_2O$$

And finally the oxygens. There are 7 oxygens (4 + 3) on the right-hand side, but only 2 on the left. The problem is that the oxygens have to go around in pairs – so how can you get an odd number (7) of them on the left-hand side?

The trick with this is to allow yourself to have halves in your equation; 7 oxygen atoms, O, is the same as $3\frac{1}{2}$ oxygen molecules, O_2.

$$C_2H_6 + 3\frac{1}{2}O_2 \rightarrow 2CO_2 + 3H_2O$$

You might reasonably argue that you can't have half an oxygen molecule, but to get rid of that problem you only have to double everything.

$$2C_2H_6 + 7O_2 \rightarrow 4CO_2 + 6H_2O$$

> Don't go on until you are sure you understand why there are 7 oxygen atoms on the right-hand side.

> In fact, it is acceptable to have halves in equations, but you don't usually come across them at GCSE.

State symbols

State symbols are often, but not always, written after the formulae of the various substances in an equation to show what physical state everything is in. You need to know four different state symbols:

(s)   solid
(l)   liquid
(g)   gas
(aq) in aqueous solution (solution in water)

If those were written into the equations we've just worked out they would look like this:

> Don't worry if this chemistry is unfamiliar to you, or if you wouldn't know at this stage what the state symbols ought to be. That doesn't matter in the least for now.

$$Zn_{(s)} + 2HCl_{(aq)} \rightarrow ZnCl_{2(aq)} + H_{2(g)}$$

$$2AgNO_{3(aq)} + CaCl_{2(aq)} \rightarrow Ca(NO_3)_{2(aq)} + 2AgCl_{(s)}$$

$$2C_2H_{6(g)} + 7O_{2(g)} \rightarrow 4CO_{2(g)} + 6H_2O_{(l)}$$

End of Chapter Checklist

If you haven't got a copy of your specification, read the introduction on page vi.

You will need to be able to do the following.

- Remember the formulae for some simple covalent compounds like water and carbon dioxide.

- Work out the formula of a simple covalent compound from its electronic structure.

- Work out formulae for simple ionic compounds from the symbols and charges of their component ions.

- Balance simple equations for familiar reactions, or reactions where you are given the names of everything involved.

- Understand and use the state symbols (s), (l), (g) and (aq).

Questions

You will need to use the Periodic Table on page 306. More questions on formulae and equations can be found at the end of Section A on page 51.

1   Work out the formulae of the following compounds:

lead(II) oxide	sodium bromide
magnesium sulphate	zinc chloride
potassium carbonate	ammonium sulphide
calcium nitrate	iron(III) hydroxide
iron(II) sulphate	copper(II) carbonate
aluminium sulphate	calcium hydroxide
cobalt(II) chloride	calcium oxide
silver nitrate	iron(III) fluoride
ammonium nitrate	rubidium iodide
sodium sulphate	chromium(III) oxide.

2   a)   Hydrogen sulphide is a simple covalent compound of hydrogen and sulphur.

   i)   Write down the electronic structures of hydrogen and sulphur.

   ii)   Draw a dot-and-cross diagram to show the bonding in hydrogen sulphide.

   iii)   What is the formula for hydrogen sulphide?

   b)   Silane is the simplest compound of silicon and hydrogen. Work out the formula of silane by drawing a dot-and-cross diagram of it.

3   Balance the following equations:

   a) $Ca + H_2O \rightarrow Ca(OH)_2 + H_2$

   b) $Al + Cr_2O_3 \rightarrow Al_2O_3 + Cr$

   c) $Fe_2O_3 + CO \rightarrow Fe + CO_2$

   d) $NaHCO_3 + H_2SO_4 \rightarrow Na_2SO_4 + CO_2 + H_2O$

   e) $C_8H_{18} + O_2 \rightarrow CO_2 + H_2O$

   f) $Fe + HCl \rightarrow FeCl_2 + H_2$

   g) $Zn + H_2SO_4 \rightarrow ZnSO_4 + H_2$

   h) $Fe_3O_4 + H_2 \rightarrow Fe + H_2O$

   i) $Mg + O_2 \rightarrow MgO$

   j) $Pb + AgNO_3 \rightarrow Pb(NO_3)_2 + Ag$

   k) $AgNO_3 + MgCl_2 \rightarrow Mg(NO_3)_2 + AgCl$

   l) $C_3H_8 + O_2 \rightarrow CO_2 + H_2O$

   m) $Fe_2O_3 + C \rightarrow Fe + CO$

4   Rewrite the following equations as balanced symbol equations:

   a)   sodium carbonate + hydrochloric acid (HCl) → sodium chloride + carbon dioxide + water

   b)   sodium hydroxide + sulphuric acid (H_2SO_4) → sodium sulphate + water

   c)   sodium + water → sodium hydroxide + hydrogen (H_2)

   d)   sodium + chlorine (Cl_2) → sodium chloride

   e)   iron(III) oxide + nitric acid (HNO_3) → iron(III) nitrate + water

   f)   zinc + oxygen (O_2) → zinc oxide

   g)   copper(II) oxide + hydrochloric acid → copper(II) chloride + water

   h)   barium chloride + sodium sulphate → barium sulphate + sodium chloride

   i)   zinc + lead(II) nitrate → lead + zinc nitrate

   j)   copper(II) sulphate + potassium hydroxide → copper(II) hydroxide + potassium sulphate

   k)   magnesium + copper(II) oxide → magnesium oxide + copper *l)* sodium + oxygen (O_2) → sodium oxide

m) iron + chlorine (Cl_2) → iron(III) chloride

5 Write balanced symbol equations from the following descriptions. Everything must have a state symbol attached.

a) Solid calcium carbonate reacts with a dilute solution of hydrochloric acid (HCl) to give a solution of calcium chloride and carbon dioxide gas. Water is also formed.

b) Zinc metal reacts with copper(II) sulphate solution to give solid copper and a solution of zinc sulphate.

c) Magnesium reacts with dilute sulphuric acid to give magnesium sulphate solution and hydrogen.

d) Iron(III) sulphate solution and sodium hydroxide solution react to give solid iron(III) hydroxide and a solution of sodium sulphate.

e) Solid aluminium reacts with a dilute solution of hydrochloric acid (HCl) to give a solution of aluminium chloride and hydrogen (H_2) gas.

f) Solid iron(III) oxide reacts with a dilute solution of sulphuric acid to give iron(III) sulphate solution and water.

g) Solid lead(II) carbonate reacts with a dilute solution of nitric acid (HNO_3) to give a solution of lead(II) nitrate, carbon dioxide and water.

h) Magnesium reacts if heated in steam to produce white solid magnesium oxide and hydrogen (H_2).

i) A mixture of carbon and copper(II) oxide heated together produces copper and carbon dioxide.

Chapter 5: Rates of Reaction

Reactions can vary in speed between those which happen within fractions of a second – explosions, for example – and those which never happen at all. Gold can be exposed to the air for thousands of years and not react in any way. This chapter looks at the factors controlling the speeds of chemical reactions.

Some reactions are very fast.

Some reactions happen over several minutes.

Rusting takes days or weeks.

The weathering of limestone and the formation of stalagmites and stalactites takes a very long time.

An Investigation of the Reaction between Marble Chips and Dilute Hydrochloric Acid

Marble chips are made of calcium carbonate and react with hydrochloric acid to produce carbon dioxide gas. Calcium chloride solution is also formed.

$$CaCO_{3(s)} + 2HCl_{(aq)} \rightarrow CaCl_{2(aq)} + H_2O_{(l)} + CO_{2(g)}$$

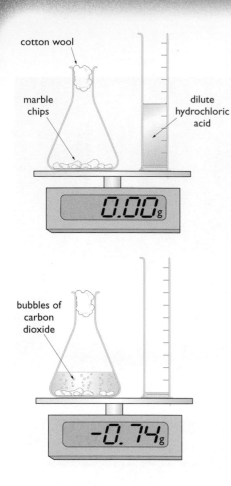

The first diagram shows some apparatus which can be used to measure how the mass of carbon dioxide produced changes with time. The apparatus is drawn as it would look before the reaction starts.

The flask is stoppered with cotton wool and contains marble chips. The cotton wool is to allow the carbon dioxide to escape, but to stop any acid spraying out. The measuring cylinder contains dilute hydrochloric acid. The marble is in large excess – most of it will be left over when the acid is all used up. Everything is placed on a top pan balance which is reset to zero.

The second diagram shows what happens during the reaction. The acid has been poured into the flask and everything has been replaced on the balance.

Notice that once the reaction starts, the balance shows a negative mass. This measures the carbon dioxide escaping through the cotton wool.

The mass of carbon dioxide lost is measured at intervals, and a graph is plotted.

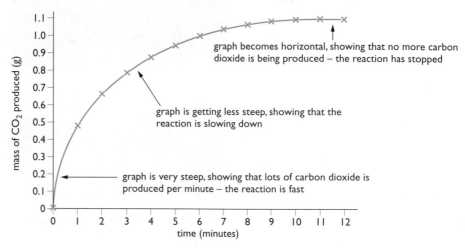

The steeper the slope of the line, the faster the reaction. You can see from the graph that just under 0.5 g of carbon dioxide is produced in the first minute in this example. Less than 0.2 g of extra carbon dioxide is produced in the second minute – the reaction is slowing down.

The reaction is fastest at the beginning. It then slows down until it eventually stops because all the hydrochloric acid has been used up. There will still be unreacted marble chips in the flask.

You can measure how fast the reaction is going at any point by finding the slope of the line at that point. This is called the **rate of the reaction** at that point. You might, for example, find that at a particular time, the carbon dioxide was being lost at the rate of, say, 0.12 g per minute.

Explaining what is happening

We can explain the shape of the curve by thinking about the particles present and how they interact. This is called the **collision theory**.

Reactions can only happen when particles collide. In this case, particles in the acid have to collide with the surface of the marble chips. As the acid particles get used up, the collision rate decreases, and so the reaction slows down.

> You can use a graph to find actual values for the rate of reaction at any particular time by drawing a tangent to the line at the time you are interested in and finding its slope. If the maths of this makes you uneasy, don't worry about it. You are unlikely to be asked to do it in an exam.

early in the reaction
lots of acid particles and lots of collisions

marble chip

later in the reaction
fewer acid particles left

marble chip

The marble is in such large excess that its shape doesn't change much during the reaction.

A different form of graph

At GCSE, you normally plot graphs showing the mass or volume of product formed during a reaction. It is possible, however, that you may come across graphs showing the fall in the concentration of one of the reactants – in this case, the concentration of the dilute hydrochloric acid.

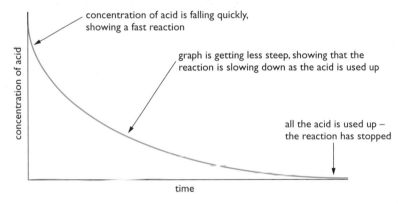

concentration of acid is falling quickly, showing a fast reaction

graph is getting less steep, showing that the reaction is slowing down as the acid is used up

all the acid is used up – the reaction has stopped

concentration of acid

time

Where the graph is falling most quickly, it shows that the reaction is fastest.

Eventually, the graph becomes horizontal because the reaction has stopped.

Changing the conditions in the experiment

Using smaller marble chips

You can easily repeat the experiment using exactly the same quantities of everything, but using much smaller marble chips. The reaction with the small chips happens faster.

Both sets of results are plotted on the same graph. Notice that the same mass of carbon dioxide is produced because you are using the same quantities of everything in both experiments. However, the reaction with the smaller chips starts off much more quickly and finishes sooner.

If you are going to investigate the effect of changing the size of the marble chips, it is important that everything else stays exactly the same.

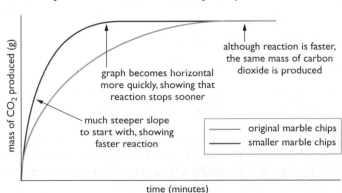

mass of CO_2 produced (g)

although reaction is faster, the same mass of carbon dioxide is produced

graph becomes horizontal more quickly, showing that reaction stops sooner

much steeper slope to start with, showing faster reaction

—— original marble chips
—— smaller marble chips

time (minutes)

Reactions between solids and liquids (or solids and gases) are faster if the solids are present as lots of small bits rather than a few big ones. The more finely divided the solid, the faster the reaction, because the surface area in contact with the gas or liquid is much greater.

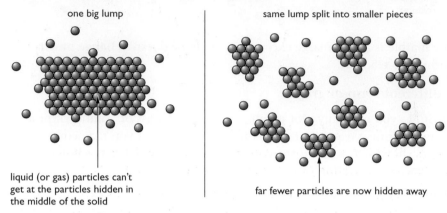

one big lump   same lump split into smaller pieces

liquid (or gas) particles can't get at the particles hidden in the middle of the solid

far fewer particles are now hidden away

High surface areas are frequently used to speed up reactions outside the lab. For example, a **catalytic converter** for a car uses expensive metals like platinum, palladium and rhodium coated onto a honeycomb structure in a very thin layer to keep costs down.

In the presence of these metals harmful substances like carbon monoxide and nitrogen oxides are converted into relatively harmless carbon dioxide and nitrogen. The high surface area means the reaction is very rapid. This is important because the gases in the exhaust system are only in contact with the catalytic converter for a very short time.

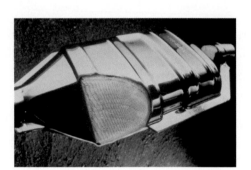

In a catalytic converter, the honeycomb structure gives a very large surface area.

Changing the concentration of the acid

You could repeat the experiment using the original marble chips, but using hydrochloric acid which is only half as concentrated as before. Everything else would be the same – the mass of marble, and the volume of the acid.

You would find that the reaction would be slower, and you would also only get half as much gas given off.

The reason you get half the mass of carbon dioxide is because you only have half the amount of acid present. (You have the same volume of acid, but it is only half as concentrated.) The amount of carbon dioxide is controlled by the amount of acid because that's what runs out first.

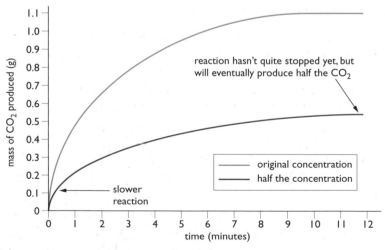

reaction hasn't quite stopped yet, but will eventually produce half the CO_2

original concentration
half the concentration

slower reaction

If you go on to do A-level chemistry, you will come across a few reactions in which increasing the concentration of *one* of the reactants has no effect on how fast the reaction happens. You can ignore that problem for now.

In general terms, if you increase the concentration of the reactants, the reaction becomes faster. Increasing the concentration increases the chances of particles hitting each other.

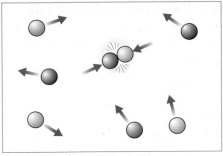

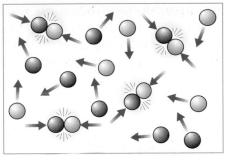

| lower concentration | higher concentration |

Changing the temperature of the reaction

You could do the original experiment again, but this time at a higher temperature. Everything else would be exactly the same as before.

Reactions get faster as the temperature is increased. In this case the same mass of gas would be given off, because you still have the same quantities of everything in the mixture.

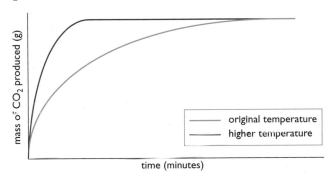

original temperature
higher temperature

time (minutes)

As a rough approximation, a 10°C increase in temperature doubles the rate of a reaction.

There are two factors at work here.

● Increasing the temperature means that the particles are moving faster and so hit each other more often. That will make the reaction go faster, but it only accounts for a small part of the increase in rate.

● Not all collisions end up in a reaction. Many particles just bounce off each other. In order for anything interesting to happen, the particles have to collide with a minimum amount of energy called **activation energy**. A relatively small increase in temperature produces a very large increase in the number of collisions which have enough energy for a reaction to occur.

lower temperature

higher temperature

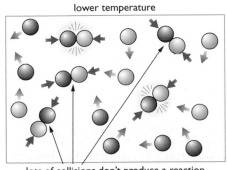

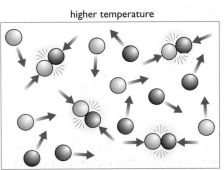

| lots of collisions don't produce a reaction | higher number of useful collisions |

Changing the pressure on the reaction

Changing the pressure on a reaction where the reactants are only solids or liquids makes virtually no difference to the rate of reaction – so in this case, the graphs would be unchanged. But increasing the pressure on a reaction where the reactants are gases does speed the reaction up.

If you have a fixed amount of a gas, you increase the pressure by squeezing it into a smaller volume.

This forces the particles closer together and so they are more likely to hit each other. This is exactly the same as increasing the concentration of the gas.

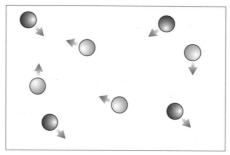

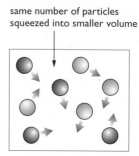

same number of particles squeezed into smaller volume

lower pressure

higher pressure

Catalysts

What are catalysts?

Catalysts are substances which speed up chemical reactions, but aren't used up in the process. They are still there chemically unchanged at the end of the reaction. Because they don't get used up, small amounts of catalyst can be used to process lots and lots of reactant particles – whether atoms or molecules or ions. Different reactions need different catalysts.

Think of a catalyst as being rather like a machine tool in a factory. Because the tool doesn't get used up, one tool can process huge amounts of stainless steel into teaspoons. A different tool could turn virtually endless quantities of plastic into yoghurt pots.

The catalytic decomposition of hydrogen peroxide

Bombardier beetles defend themselves by spraying a hot, unpleasant liquid at their attackers. Part of the reaction involves splitting hydrogen peroxide into water and oxygen using the enzyme catalase. This reaction happens almost explosively, and produces a lot of heat.

Enzymes are biological catalysts and are described in more detail in Chapter 24 on page 255. There are lots of other things which also catalyse this reaction. One of these is manganese(IV) oxide, MnO_2 – also called manganese dioxide. This is what is normally used in the lab to speed up the decomposition of hydrogen peroxide.

The reaction happening with the hydrogen peroxide is:

hydrogen peroxide $\rightarrow$ water + oxygen

$$2H_2O_{2(aq)} \rightarrow 2H_2O_{(l)} + O_{2(g)}$$

Notice that you don't write catalysts into the equation because they are

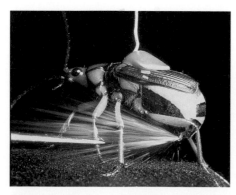

Bombardier beetles use hydrogen peroxide as part of their defence mechanism.

chemically unchanged at the end of the reaction. If you like, you can write their name over the top of the arrow.

Measuring the volume of oxygen evolved

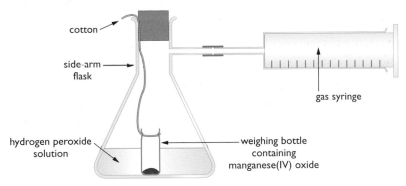

cotton

side-arm flask

gas syringe

hydrogen peroxide solution

weighing bottle containing manganese(IV) oxide

> Using a weighing bottle like this is a simple way of mixing the chemicals together without losing any oxygen before you can get the bung in. It is also impossible to get the bung in quickly without forcing a bit of air into the gas syringe. That would give a misleading reading on the syringe.

When you are ready to start the reaction, shake the flask so that the weighing bottle falls over and the manganese(IV) oxide comes into contact with the hydrogen peroxide. You need to keep shaking so that an even mixture is formed.

You could use this apparatus to find out what happens to the rate of reaction if you:

- change the mass of the catalyst
- change how "lumpy" the catalyst is
- use a different catalyst
- change the concentration of the hydrogen peroxide solution
- change the temperature of the solution.

In each case you could measure the volume of oxygen produced at regular intervals and produce graphs just like the ones earlier in the chapter. However, if you wanted to look in a more detailed way at how a change in concentration or temperature, for example, affects the rate of the reaction, there is a much easier way of doing it.

Exploring the very beginning of the reaction

An easy way of comparing rates under different conditions is to time how long it takes to produce a small, but constant, volume of gas – say, 5 cm^3 – as you vary the conditions. You take measurements of the rate at the beginning of the reaction – the so-called **initial rate**.

This experiment uses an upturned measuring cylinder to measure the volume of the gas. The cylinder is initially full of water.

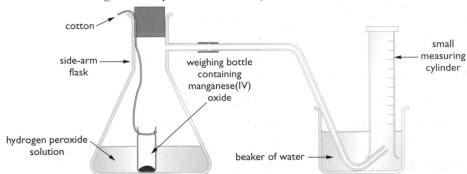

cotton

side-arm flask

weighing bottle containing manganese(IV) oxide

small measuring cylinder

hydrogen peroxide solution

beaker of water

You would shake the flask exactly as before to mix the hydrogen peroxide and manganese(IV) oxide.

This time you record how long it takes for 5 cm³ of oxygen to be collected in the measuring cylinder.

Then you set up the experiment again, changing one of the conditions (for example, the concentration of the hydrogen peroxide) and find out how long it takes to produce the same 5 cm³.

Varying the concentration – some sample results

concentration (mol/dm³)	2.00	1.00	0.50	0.25
time to collect 5 cm³ of oxygen (sec)	10	20	40	80

You will see that every time you halve the concentration, it takes twice as long to produce the 5 cm³ of gas. That means that the rate of reaction has also been halved. You can see this more easily if you work out the initial rate for each reaction.

The rate of the reaction would be worked out in terms of the volume of oxygen produced per second. If it takes 10 seconds to produce 5 cm³ at the beginning of the reaction, then:

$$\text{initial rate} = \frac{5}{10} = 0.5 \text{ cm}^3/\text{sec}$$

If you do this for all the experiments, and then redraw the table, you get these figures:

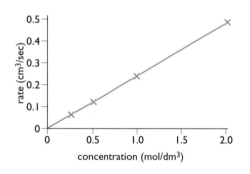

concentration (mol/dm³)	2.00	1.00	0.50	0.25
initial rate (cm³/sec)	0.5	0.25	0.125	0.0625

The graph shows that the rate of the reaction is proportional to the concentration – whatever you do to the concentration also happens to the rate. If you double the concentration, the rate doubles – and so on.

Varying the temperature – a sample graph

You could repeat the experiment, starting with the hydrogen peroxide solution at a range of different temperatures from room temperature up to about 50°C.

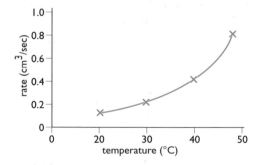

This time, the graph isn't a straight line.

As a rough approximation, the rate of a reaction doubles for every 10°C temperature rise.

Showing that a substance is a catalyst

It isn't difficult to show that manganese(IV) oxide speeds up the decomposition of hydrogen peroxide to produce oxygen. The photograph shows two flasks, both of which contain hydrogen peroxide solution. Without the catalyst, there is only a trace of bubbles in the solution. With it, oxygen is given off quickly.

How can you show that the manganese(IV) oxide is chemically unchanged by the reaction? It still looks the same, but has any been used up? You can only find out by weighing it before you add it to the hydrogen peroxide solution, and then reweighing it at the end.

You can separate it from the liquid by filtering through a weighed filter paper, allowing the paper and residue to dry and then reweighing it to work out the mass of the remaining manganese(IV) oxide. You should find that the mass hasn't changed.

Manganese(IV) oxide speeds up the production of oxygen from hydrogen peroxide solution.

How does a catalyst work?

You will remember that not all collisions result in a reaction happening. Collisions have to involve at least a certain minimum amount of energy, called **activation energy**.

You can show this on an energy diagram. In order for anything interesting to happen, the reactants have to gain enough energy to overcome the activation energy barrier.

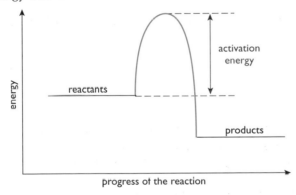

If a reaction is slow, it means that very few collisions have this amount of energy.

Adding a catalyst gives the reaction an alternative and easier way for it to happen – involving a lower activation energy.

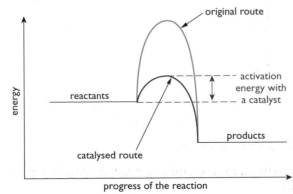

Traffic passes easily through a road tunnel under a mountain.

If the activation energy is lower, many more collisions are likely to be successful. The reaction happens faster because the alternative route is easier.

Catalysts work by providing an alternative route for the reaction, involving a lower activation energy.

You can illustrate this with a simple everyday example. Suppose you have a mountain between two valleys. Only a few very energetic people will climb over the mountain from one valley to the next.

Now imagine building a road tunnel through the mountain. Lots of people will be able to travel easily from one valley to the next.

Catalysts in industry

Catalysts are especially important in industrial reactions because they help substances to react quickly at lower temperatures and pressures than would otherwise be needed. This saves money.

You will meet several examples of industrial use of catalysts later in the course.

End of Chapter Checklist

If you haven't got a copy of your specification, read the introduction on page vi.

You will need to be able to do some or all of the following. Check your Awarding Body's specification (syllabus) to find out exactly what you need to know.

- Suggest examples of reactions which happen at different rates, from very fast to very slow.
- Know and explain (in terms of collision theory) the effect of changing surface area, concentration, pressure and temperature on the rate of a reaction.
- Define what a catalyst is and explain how it speeds up a reaction. Know why catalysts are important in industry.
- Describe simple experiments to investigate rates of reaction.
- Be able to draw and interpret simple graphs showing how the amount of either reactant or product varies with time.

Questions

More questions on rates of reaction can be found at the end of Section A on page 51.

1   A student carried out an experiment to investigate the rate of reaction between an excess of dolomite (magnesium carbonate) and 50 cm³ of dilute hydrochloric acid. The dolomite was in small pieces. The reaction is:

$$MgCO_{3(s)} + 2HCl_{(aq)} \rightarrow MgCl_{2(aq)} + H_2O_{(l)} + CO_{2(g)}$$

He measured the volume of carbon dioxide given off at regular intervals, with these results:

Time (sec)	0	30	60	90	120	150	180	210	240	270	300	330	360
Vol (cm³)	0	27	45	59	70	78	85	90	94	97	99	100	100

a)   Draw a diagram of the apparatus you would use for this experiment, and explain briefly what you would do.

b)   Plot these results on graph paper, with time on the x-axis and volume of gas on the y-axis.

c)   At what time is the gas being given off most quickly? Explain why the reaction is fastest at that time.

d)   Use your graph to find out how long it took to produce 50 cm³ of gas.

e)   In each of the following questions, decide what would happen to the initial rate of the reaction and to the total volume of gas given off if various changes were made to the experiment.

   i)   The mass of dolomite and the volume and concentration of acid were kept constant, but the dolomite was in one big lump instead of small bits.

   ii)   The mass of dolomite was unchanged and it was still in small pieces. 50 cm³ of hydrochloric acid was used which had half the original concentration.

   iii) The dolomite was unchanged again. This time 25 cm³ of the original acid was used instead of 50 cm³.

   iv)   The acid was heated to 40°C before the dolomite was added to it.

2   The effect of concentration and temperature on the rate of a reaction can be explored using the reaction between magnesium ribbon and dilute sulphuric acid.

$$Mg_{(s)} + H_2SO_{4(aq)} \rightarrow MgSO_{4(aq)} + H_{2(g)}$$

A student dropped a 2 cm length of magnesium ribbon into 25 cm³ of dilute sulphuric acid in a boiling tube (a large excess of acid). She stirred the contents of the tube continuously and timed how long it took for the magnesium to disappear.

a)   What would you expect to happen to the time taken for the reaction if she repeated the experiment using the same length of magnesium with a mixture of 20 cm³ of acid and 5 cm³ of water? Explain your answer in terms of the collision theory.

b)   What would you expect to happen to the time taken for the reaction if she repeated the experiment using the original quantities of magnesium and acid, but first heated the acid to 50°C? Explain your answer in terms of the collision theory.

c)   Why is it important to keep the reaction mixture stirred continuously?

3 Catalysts speed up reactions but can be recovered chemically unchanged at the end of the reaction.

a) Explain briefly how a catalyst has this effect on a reaction.

b) Describe how you would find out whether copper(II) oxide was a catalyst for the decomposition of hydrogen peroxide solution. You need to show not only that it speeds the reaction up, but that it is chemically unchanged at the end.

4 By doing an internet (or other) search, find an industrial process which uses a catalyst. Write a short article on the process, using a computer. You should use your own words – pages copied directly from the internet are not acceptable. Your article must fit on a single side of A4 paper, and must not exceed 300 words. It should include at least two pictures or diagrams. You can produce your artwork yourself on the computer, scan it in, or take pictures directly from the internet.

End of Section Questions

You may need to refer to the Periodic Table on page 306.

1 The diagrams show an atom and an ion.

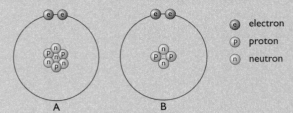

a) Which of the two structures represents an atom? Explain your choice. *(2 marks)*

b) Use the Periodic Table to help you to write the symbol (including the charge) for the structure representing an ion. *(2 marks)*

c) Complete the following table showing the relative masses and charges of the various particles.

Particle	Relative charge	Relative mass
proton	+1	
neutron	0	
electron		1/1836

(3 marks)

d) Find the elements strontium, Sr, and bromine, Br, in the Periodic Table. How many electrons are there in the outer level of each of these atoms? *(2 marks)*

e) Strontium combines with bromine to form strontium bromide. What happens to the electrons in the outer levels when strontium atoms combine with bromine atoms? *(2 marks)*

f) What is the formula for strontium bromide? *(1 mark)*

g) Would you expect strontium bromide to have a high or a low melting point. Explain your answer. *(3 marks)*

Total 15 marks

2 a) Draw dot-and-cross diagrams to show the arrangement of the electrons in *i)* a chlorine atom, *ii)* a chloride ion, *iii)* a chlorine molecule. *(4 marks)*

b) Dichloromethane, CH_2Cl_2, is a liquid with a low boiling point used in paint strippers. Draw a dot-and-cross diagram to show the bonding in dichloromethane. You need only show the electrons in the outer levels of the atoms. *(3 marks)*

c) Dichloromethane contains strong carbon–hydrogen and carbon–chlorine bonds. Despite the presence of these strong bonds, dichloromethane is a liquid. Explain why. *(2 marks)*

Total 9 marks

3

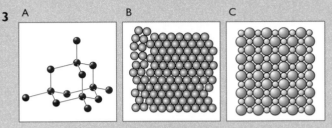

a) Which of the diagrams represents the arrangement of the particles in *i)* magnesium metal, *ii)* solid sodium chloride, *iii)* diamond? *(2 marks)*

b) Explain why:

   i) magnesium can be stretched to form magnesium ribbon; *(1 mark)*

   ii) magnesium conducts electricity; *(2 marks)*

   iii) sodium chloride crystals are brittle; *(1 mark)*

   iv) diamond is extremely hard. *(2 marks)*

c) *i)* State any *one* physical property of graphite which is different from diamond. *(1 mark)*

   ii) Explain how the difference arises from the arrangement of the atoms in the two substances. *(3 marks)*

Total 12 marks

4 In the nineteenth century, John Dalton put forward an atomic theory in which he suggested that atoms of a given element were all alike, but differed from the atoms of other elements. He thought that elements combined in small whole number ratios like 1:2 or 2:3, and that chemical reactions involved rearranging existing atoms into different compounds.

a) Choose a compound whose atoms are arranged in the ratio 1:2 and write its formula. *(1 mark)*

b) The Law of Conservation of Mass states that in a chemical reaction, matter is neither created nor destroyed. Explain how that statement is consistent with Dalton's theory. *(2 marks)*

c) In the twentieth century, a flaw was discovered in Dalton's theory when it was found that there were two different kinds of neon atoms, Ne, one with a mass number of 20 and the other a mass number of 22.

  i) What name is given to these two different kinds of neon atoms? *(1 mark)*

  ii) Write down the numbers of protons, neutrons and electrons in each of these atoms. *(2 marks)*

  iii) Would you expect there to be any differences between the chemical properties of the two sorts of neon atoms? Explain your answer. *(2 marks)*

Total 8 marks

5 In an experiment to investigate the rate of decomposition of hydrogen peroxide solution in the presence of manganese(IV) oxide, 10 cm^3 of hydrogen peroxide solution was mixed with 30 cm^3 of water, and 0.2 g of manganese(IV) oxide was added. The volume of oxygen evolved was measured at 60 second intervals. The results of this experiment are recorded in the table below.

Time (sec)	0	60	120	180	240	300
Vol (cm^3)	0	30	48	57	60	60

a) Balance the equation for the decomposition of hydrogen peroxide, including all the appropriate state symbols.

$$H_2O_2 \rightarrow H_2O + O_2$$

(2 marks)

b) Copy and complete the diagram to show how the volume of oxygen might have been measured.

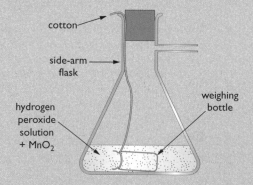

cotton

side-arm flask

hydrogen peroxide solution + MnO_2

weighing bottle

(2 marks)

c) Plot a graph of the results on a piece of graph paper, with time on the horizontal axis and volume of oxygen on the vertical axis. *(4 marks)*

d) Use your graph to find out how long it took to produce 50 cm^3 of oxygen. *(1 mark)*

e) Explain why the graph becomes horizontal after 240 seconds. *(2 marks)*

f) Suppose the experiment had been repeated using the same quantities of everything but with the reaction flask immersed in ice. Sketch the graph you would expect to get. Use the same grid as in part **c)**. Label the new graph **F**. *(2 marks)*

g) On the same grid as in **c)** and **f)**, sketch the graph you would expect to get if you repeated the experiment at the original temperature using 5 cm^3 of hydrogen peroxide solution, 35 cm^3 of water and 0.2 g of manganese(IV) oxide. Label this graph **G**. *(2 marks)*

Total 15 marks

6 During the manufacture of nitric acid from ammonia, the ammonia is oxidised to nitrogen monoxide, NO, by oxygen in the air.

$$4NH_{3(g)} + 5O_{2(g)} \rightarrow 4NO_{(g)} + 6H_2O_{(g)}$$

The ammonia is mixed with air and passed through a stack of large circular gauzes made of platinum–rhodium alloy at red heat (about 900°C). The platinum–rhodium gauzes act as a catalyst for the reaction.

a) Gas particles have to collide before they can react. Use this collision theory to help you to answer the following questions.

  i) Because the gases are only in contact with the catalyst for a very short time, it is important that the reaction happens as quickly as possible. Explain why increasing the temperature to 900°C makes the reaction very fast. *(3 marks)*

  ii) Explain why the reaction rate can also be increased by increasing the pressure. *(2 marks)*

  iii) Explain why the platinum–rhodium alloy is used as gauzes rather than as pellets. *(2 marks)*

b) Platinum and rhodium are extremely expensive metals. Explain why the manufacturer can justify their initial cost. *(2 marks)*

Total 9 marks

Chapter 6: The Reactivity Series

The reactivity series lists elements (mainly metals) in order of decreasing reactivity. This chapter reminds you about some simple reactions involving these metals and their compounds, and then uses these reactions to introduce some new ideas.

potassium

sodium

calcium

magnesium

aluminium

(carbon)

zinc

iron

tin

lead

(hydrogen)

copper

silver

gold

platinum

decreasing reactivity

Rails are welded together using molten iron produced by a reaction between aluminium and iron(III) oxide.

Lead crystals made from the reaction between lead(II) nitrate solution and zinc.

Copper is used in whisky stills because it doesn't react with water or alcohol.

Gold is so unreactive that it will remain chemically unchanged in contact with the air or water for ever.

Displacement Reactions Involving Metal Oxides

The reaction between magnesium and copper(II) oxide

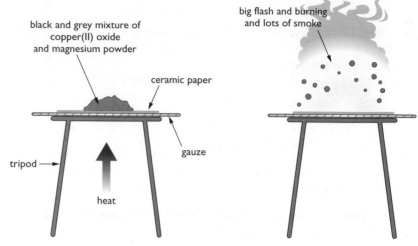

black and grey mixture of copper(II) oxide and magnesium powder

ceramic paper

gauze

tripod

heat

big flash and burning and lots of smoke

The reaction between magnesium and zinc oxide.

At the end, traces of brown copper are left on the ceramic paper.

magnesium + copper(II) oxide → magnesium oxide + copper

$$Mg_{(s)} + CuO_{(s)} \rightarrow MgO_{(s)} + Cu_{(s)}$$

This is an example of a **displacement reaction**. The less reactive metal, copper, has been pushed out of its compound by the more reactive magnesium. Any metal higher in the series will displace one lower down from a compound. If you heated copper with magnesium oxide, nothing would happen because copper is less reactive than magnesium.

The reaction between magnesium and zinc oxide

Heating magnesium with zinc oxide produces zinc metal. This time, though, because the zinc is very hot, it immediately burns in air to form zinc oxide again! This second reaction hasn't been included in the equations below.

magnesium + zinc oxide → magnesium oxide + zinc

$$Mg_{(s)} + ZnO_{(s)} \rightarrow MgO_{(s)} + Zn_{(s)}$$

The reaction between carbon and copper(II) oxide

The mixture is heated in a test tube to avoid the air getting at the hot copper produced and turning it back to copper(II) oxide. The carbon dioxide that is also formed escapes from the tube as a gas. The photograph clearly shows the brown copper formed from the black mixture.

$$C_{(s)} + 2CuO_{(s)} \rightarrow CO_{2(g)} + 2Cu_{(s)}$$

Oxidation and Reduction

Oxidation and reduction – oxygen transfer

A substance has been **oxidised** if it gains oxygen. **Oxidation** is gain of oxygen.

A substance has been **reduced** if it loses oxygen. **Reduction** is loss of oxygen.

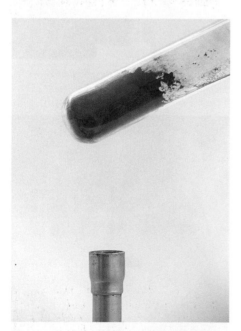

The reaction between carbon and copper(II) oxide. Notice the brown copper formed at the end of the reaction.

In the case of magnesium reacting with copper(II) oxide:

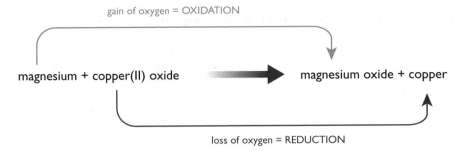

gain of oxygen = OXIDATION

magnesium + copper(II) oxide ➡ magnesium oxide + copper

loss of oxygen = REDUCTION

A **redox** reaction is one in which both **red**uction and **ox**idation are occurring. Oxidation and reduction always go hand-in-hand.

A **reducing agent** is a substance which reduces something else. In this case, the magnesium is the reducing agent.

An **oxidising agent** is a substance which oxidises something else. The copper(II) oxide is the oxidising agent in this reaction.

Oxidation and reduction – electron transfer

We are going to look very closely at what happens in the reaction between magnesium and copper(II) oxide in terms of the various particles involved.

$$Mg_{(s)} + CuO_{(s)} \rightarrow MgO_{(s)} + Cu_{(s)}$$

The magnesium and the copper are metals and are made of metal atoms, but the copper(II) oxide and the magnesium oxide are both ionic compounds.

The copper(II) oxide contains Cu^{2+} and O^{2-} ions, and the magnesium oxide contains Mg^{2+} and O^{2-} ions. Writing those into the equation (not forgetting the state symbols) gives:

$$Mg_{(s)} + Cu^{2+}{}_{(s)} + O^{2-}{}_{(s)} \rightarrow Mg^{2+}{}_{(s)} + O^{2-}{}_{(s)} + Cu_{(s)}$$

Look very carefully at this equation to see what is being changed. Notice that something odd is going on – the oxide ion (O^{2-}) is completely unaffected by the reaction. It ends up with a different partner but is totally unchanged itself. An ion like this is described as a **spectator ion**.

You don't write the spectator ions into the equation because they aren't changed in the reaction. The equation showing just those things being changed looks like this:

$$Mg_{(s)} + Cu^{2+}{}_{(s)} \rightarrow Mg^{2+}{}_{(s)} + Cu_{(s)}$$

This is known as an **ionic equation** and shows the reaction in a quite different light. It shows that the reaction has nothing to do with the oxygen.

What is actually happening is that magnesium *atoms* are turning into magnesium *ions*. The magnesium atoms lose electrons to form magnesium ions.

$$Mg_{(s)} \rightarrow Mg^{2+}{}_{(s)} + 2e^-$$

Those electrons have been gained by the copper(II) ions to make the atoms present in metallic copper.

$$Cu^{2+}{}_{(s)} + 2e^- \rightarrow Cu_{(s)}$$

How do you know whether a substance contains ions or not? As a rough guide for GCSE purposes only, if it is a metal (or ammonium) *compound*, or an acid in solution, it will be ionic – otherwise it's not. If you aren't sure about writing symbols for ions, read pages 31–33.

These equations are called **half equations** or **electron half equations**. They show just part of a reaction from the point of view of one of the substances present.

Remember that we are talking about the reaction between copper(II) oxide and magnesium. We've already described this as a redox reaction, but the equations no longer have any oxygen in them! We now need a wider definition of oxidation and reduction.

Oxidation Is Loss of electrons; Reduction Is Gain of electrons

OILRIG

In this case:

$$Mg_{(s)} \rightarrow Mg^{2+}_{(s)} + 2e^- \qquad \text{Mg is oxidised.}$$
$$Cu^{2+}_{(s)} + 2e^- \rightarrow Cu_{(s)} \qquad \text{Cu}^{2+} \text{ is reduced.}$$

Displacement Reactions Involving Solutions of Salts

Salts are compounds like copper(II) sulphate, silver nitrate, or sodium chloride. You will find a proper definition of what a salt is on pages 64–65. This section explores some reactions between metals and solutions of salts in water.

The reaction between zinc and copper(II) sulphate solution

The more reactive zinc displaces the less reactive copper. The blue colour of the copper(II) sulphate solution fades as colourless zinc sulphate solution is formed.

$$Zn_{(s)} + CuSO_{4(aq)} \rightarrow ZnSO_{4(aq)} + Cu_{(s)}$$

The zinc and the copper are simply metals consisting of atoms, but the copper(II) sulphate and the zinc sulphate are metal compounds and so are ionic.

If you rewrite the equation showing the ions, you will find that the sulphate ions are spectator ions.

$$Zn_{(s)} + Cu^{2+}_{(aq)} + SO_4{}^{2-}_{(aq)} \rightarrow Zn^{2+}_{(aq)} + SO_4{}^{2-}_{(aq)} + Cu_{(s)}$$

Removing the spectator ions (because they aren't changed during the reaction) leaves you with:

$$Zn_{(s)} + Cu^{2+}_{(aq)} \rightarrow Zn^{2+}_{(aq)} + Cu_{(s)}$$

This is another redox reaction:

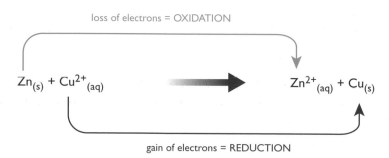

loss of electrons = OXIDATION

$$Zn_{(s)} + Cu^{2+}_{(aq)} \qquad \longrightarrow \qquad Zn^{2+}_{(aq)} + Cu_{(s)}$$

gain of electrons = REDUCTION

Zinc displaces copper from copper(II) sulphate solution.

It wouldn't matter which copper(II) salt you started with, as long as it was soluble in water. Copper(II) chloride or copper(II) nitrate would react in exactly the same way with zinc, because the chloride ions or the nitrate ions would once again be spectator ions, taking no part in the reaction.

56

The reaction between copper and silver nitrate solution

Silver is below copper in the reactivity series, and so a coil of copper wire in silver nitrate solution will produce metallic silver. The photograph shows the silver being produced as a mixture of grey "fur" and delicate crystals.

Notice the solution becoming blue as copper(II) nitrate is produced.

$$Cu_{(s)} + 2AgNO_{3(aq)} \rightarrow Cu(NO_3)_{2(aq)} + 2Ag_{(s)}$$

This time the nitrate ions are spectator ions, and the final version of the ionic equation looks like this:

$$Cu_{(s)} + 2Ag^+_{(aq)} \rightarrow Cu^{2+}_{(aq)} + 2Ag_{(s)}$$

This is another redox reaction:

loss of electrons = OXIDATION

$$Cu_{(s)} + 2Ag^+_{(aq)} \longrightarrow Cu^{2+}_{(aq)} + 2Ag_{(s)}$$

gain of electrons = REDUCTION

Displacing silver from silver nitrate solution.

Reactions of Metals with Water

A general summary

Metals above hydrogen in the reactivity series

Metals above hydrogen in the reactivity series react with water (or steam) to produce hydrogen.

If the metal reacts with cold water, the metal *hydroxide* and hydrogen are formed.

metal + cold water → metal hydroxide + hydrogen

If the metal reacts with steam, the metal *oxide* and hydrogen are formed.

metal + steam → metal oxide + hydrogen

As you go down the reactivity series, the reactions become less and less vigorous.

Metals below hydrogen in the reactivity series

Metals below hydrogen in the reactivity series (such as copper) don't react with water or steam. That is why copper can be used for both hot and cold water pipes.

Potassium or sodium and cold water

These reactions are described in detail in Chapter 10: The Periodic Table (pages 105–107). Both are very vigorous reactions.

$$2K_{(s)} + 2H_2O_{(l)} \rightarrow 2KOH_{(aq)} + H_{2(g)}$$

$$2Na_{(s)} + 2H_2O_{(l)} \rightarrow 2NaOH_{(aq)} + H_{2(g)}$$

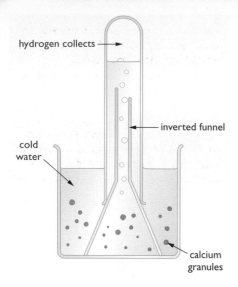

Calcium and cold water

Calcium reacts gently with cold water.

The grey granules sink, but are carried back to the surface again as bubbles of hydrogen are formed around them. The mixture becomes warm as heat is produced.

Calcium hydroxide is formed. This isn't very soluble in water. Some of it dissolves to give a colourless solution, but most of it is left as a white, insoluble solid.

$$Ca_{(s)} + 2H_2O_{(l)} \rightarrow Ca(OH)_{2(aq \ and \ s)} + H_{2(g)}$$

Magnesium and cold water

There is almost no reaction. If the magnesium is very clean, a few bubbles of hydrogen form on it, but the reaction soon stops again. This is because the magnesium gets coated with insoluble magnesium hydroxide which prevents water coming into contact with the magnesium.

Magnesium and steam

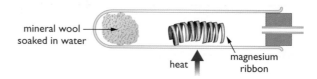

The mineral wool isn't heated directly. Enough heat spreads back along the test tube to vaporise the water.

The magnesium burns with a bright white flame in the steam, producing hydrogen which can be ignited at the end of the delivery tube. White magnesium oxide is formed.

In fact you also get a lot of black product in the tube. This is where the magnesium has reacted with the glass. Ignore this for exam purposes.

$$Mg_{(s)} + H_2O_{(g)} \rightarrow MgO_{(s)} + H_{2(g)}$$

Zinc or iron and steam

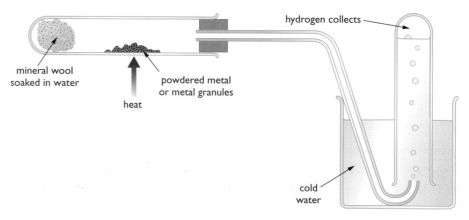

With both zinc and iron, the hydrogen comes off slowly enough to be collected. Neither metal burns.

Care has to be taken during this experiment to avoid "suck back". If you stop heating while the delivery tube is still under the surface of the water, water is sucked back into the hot tube which usually results in it cracking.

With zinc:

Zinc oxide is formed – this is yellow when it is hot, but white on cooling.

$$Zn_{(s)} + H_2O_{(g)} \rightarrow ZnO_{(s)} + H_{2(g)}$$

With iron:

The iron becomes slightly darker grey. A complicated oxide is formed – called triiron tetroxide, Fe_3O_4.

$$3Fe_{(s)} + 4H_2O_{(g)} \rightarrow Fe_3O_{4(s)} + 4H_{2(g)}$$

Notice that in these equations water now has a state symbol (g) because we are talking about it as steam.

Don't worry about the exact name or formula of the iron oxide. You won't be expected to remember it in an exam.

Using Hydrogen as a Reducing Agent

The reduction of copper(II) oxide to copper

Copper won't react with water because copper is below hydrogen in the reactivity series, but that means that you *can* get a reaction between hydrogen and copper(II) oxide.

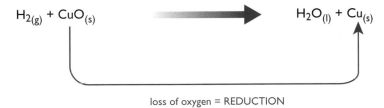

$$H_{2(g)} + CuO_{(s)} \longrightarrow H_2O_{(l)} + Cu_{(s)}$$

loss of oxygen = REDUCTION

The hydrogen removes the oxygen from the copper(II) oxide, so the hydrogen is a reducing agent.

In the experiment in the diagram, hydrogen is passed over hot copper(II) oxide. The oxide glows red hot and continues glowing, even if you remove the Bunsen burner. Lots of heat is released during the reaction.

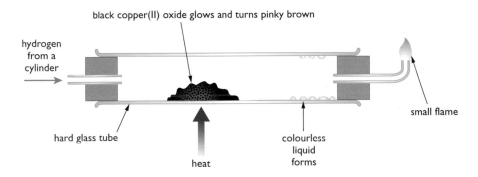

black copper(II) oxide glows and turns pinky brown

hydrogen from a cylinder

small flame

hard glass tube

colourless liquid forms

heat

Warning! This is potentially a very dangerous experiment unless it is carried out by someone knowing exactly what they are doing. Things have to be done in a particular order to avoid explosion, and the hydrogen must be taken from a cylinder and not chemically produced.

The colourless liquid is, of course, water, and the pinky brown solid is copper. The small flame at the end of the apparatus is excess hydrogen being burnt off.

Chapter 6: The Reactivity Series

Making Predictions Using the Reactivity Series

You can make predictions about the reactions of unfamiliar metals if you know their positions in the reactivity series.

A problem involving manganese

This example shows you how you can work out what will happen in reactions that you have never come across before as long you know the position of a metal in the reactivity series.

Suppose you had a question which said this:

Manganese, Mn, lies between aluminium and zinc in the reactivity series and forms a 2+ ion. Solutions of manganese(II) salts are very, very pale pink (almost colourless).

(a) *Use the reactivity series to predict whether manganese will react with copper(II) sulphate solution. If it will react, describe what you would see, name the products and write an equation for the reaction.*

(b) *Explain why you would expect manganese to react with steam. Name the products of the reaction and write the equation.*

(a) The reaction between manganese and copper(II) sulphate solution

Manganese is above copper in the reactivity series and so will displace it from the copper(II) sulphate.

A brown deposit of copper will be formed. The colour of the solution will fade from blue and leave a very pale pink (virtually colourless) solution of manganese(II) sulphate.

$$Mn_{(s)} + CuSO_{4(aq)} \rightarrow MnSO_{4(aq)} + Cu_{(s)}$$

(b) The reaction between manganese and steam

Manganese is above hydrogen in the reactivity series and so reacts with steam to give hydrogen and the metal oxide – in this case, manganese(II) oxide.

You couldn't predict the colour of the manganese(II) oxide and the question doesn't ask you to do it.

$$Mn_{(s)} + H_2O_{(g)} \rightarrow MnO_{(s)} + H_{2(g)}$$

You need to know the charge on the ion so that you can work out the formulae of the manganese compounds. If you aren't sure about working out the formula of an ionic compound, look again at Chapter 4.

potassium

sodium

calcium

magnesium

aluminium

manganese

zinc

iron

lead

(hydrogen)

copper

End of Chapter Checklist

If you haven't got a copy of your specification, read the introduction on page vi.

You will need to be able to do some or all of the following. You could check your Awarding Body's specification (syllabus) to find out exactly what you need to know, but you will probably find that there is little direct reference to the reactivity series. You will, however, need to understand the ideas in this chapter so that you can use them later in the course.

- Know that the reactivity series lists elements (mainly metals) in order of decreasing reactivity and know the positions of the common metals, and of hydrogen and carbon.

- Understand the term *displacement reaction*, and be able to describe some simple displacement reactions involving metals with metal oxides, salt solutions and water/steam.

- Understand the terms oxidation, reduction and redox in terms of both oxygen transfer and electron transfer.

- Understand how to work out simple ionic equations, omitting spectator ions.

- Know that hydrogen can be used to reduce the oxides of metals beneath it in the reactivity series.

- Make simple predictions about the reactions of a metal from its position in the reactivity series.

Questions

More questions on the reactivity series can be found at the end of Section B on page 142.

1 a) List the following metals in order of decreasing reactivity: aluminium, copper, iron, sodium.

b) Some magnesium powder was mixed with some copper(II) oxide and heated strongly. There was a vigorous reaction, producing lots of sparks and a bright flash of light.

    i) Name the products of the reaction.

    ii) Write a balanced symbol equation for the reaction.

    iii) Which substance in the reaction has been reduced?

    iv) Which substance is the oxidising agent?

c) If a mixture of zinc powder and cobalt(II) oxide is heated, the following reaction occurs:

$$Zn_{(s)} + CoO_{(s)} \rightarrow ZnO_{(s)} + Co_{(s)}$$

    i) Which metal is higher in the reactivity series?

    ii) The zinc can be described as a reducing agent. Using this example, explain what is meant by the term *reducing agent*.

    iii) Which substance in this reaction has been oxidised?

d) Aluminium, chromium and manganese are all moderately reactive metals. (Care! We are talking about manganese, *not magnesium*.) Use the following information to arrange them in the correct reactivity series order, starting with the most reactive one.

- Chromium is manufactured by heating chromium(III) oxide with aluminium.

- If manganese is heated with aluminium oxide there is no reaction.

- If manganese is heated with chromium(III) oxide, chromium is produced.

2 Study the following equations and, in each case, decide whether the substance in **bold type** has been oxidised or reduced. Explain your choice in terms of either oxygen transfer or electron transfer as appropriate.

    a) $\mathbf{Zn_{(s)}} + CuO_{(s)} \rightarrow ZnO_{(s)} + Cu_{(s)}$

    b) $\mathbf{Fe_2O_{3(s)}} + 3C_{(s)} \rightarrow 2Fe_{(s)} + 3CO_{(g)}$

    c) $\mathbf{Mg_{(s)}} + Zn^{2+}_{(s)} \rightarrow Mg^{2+}_{(s)} + Zn_{(s)}$

    d) $Zn_{(s)} + \mathbf{Cu^{2+}_{(s)}} \rightarrow Zn^{2+}_{(s)} + Cu_{(s)}$

3 The equation for the reaction when solid magnesium and solid lead(II) oxide are heated together is:

$$Mg_{(s)} + PbO_{(s)} \rightarrow MgO_{(s)} + Pb_{(s)}$$

a) Write down any two things that you might expect to see during this reaction.

b) Rewrite the equation as an ionic equation.

4 Some iron filings were shaken with some copper(II) sulphate solution. The ionic equation for the reaction is:

$$Fe_{(s)} + Cu^{2+}_{(aq)} \rightarrow Fe^{2+}_{(aq)} + Cu_{(s)}$$

a) Write down any one change that you would observe during this reaction.

b) Which substance has been oxidised in this reaction?

c) Write down the full (not ionic) equation for this reaction.

5 Some experiments were carried out to place the metals copper, nickel and silver in reactivity series order.

Experiment 1: A piece of copper was placed in some green nickel(II) sulphate solution. There was no change to either the copper or the solution.

Experiment 2: A coil of copper wire was suspended in some silver nitrate solution. A furry grey growth appeared on the copper wire, out of which grew spiky silvery crystals. The solution gradually turned from colourless to blue.

a) Use this information to place copper, nickel and silver in reactivity series order, starting with the most reactive one.

b) In another experiment, a piece of nickel was placed in some copper(II) sulphate solution.

i) Write down any one change that you would observe during this reaction.

ii) Write the full balanced equation for this reaction. (Assume that nickel(II) sulphate solution is formed.)

iii) Write the ionic equation for this reaction, and use it to explain which substance has been oxidised during the reaction.

6 **a)** Look carefully at the following equations and then decide what you can say about the position of the metal X in the reactivity series. Explain your reasoning.

$$X_{(s)} + H_2O_{(g)} \rightarrow XO_{(s)} + H_{2(g)}$$
$$X_{(s)} + CuSO_{4(aq)} \rightarrow XSO_{4(aq)} + Cu_{(s)}$$
$$X_{(s)} + FeSO_{4(aq)}: \text{ no reaction}$$

b) Decide whether X will react with the following substances. If it will react, write down the names of the products.

i) silver nitrate solution, *ii)* zinc oxide, *iii)* cold water, *iv)* copper(II) chloride solution.

7 Given some small bits of the metal titanium, and any simple apparatus that you might need, describe how you would find out the approximate position of titanium in the reactivity series using only water (or steam). You need only find out that the reactivity is "similar to iron" or "similar to magnesium" or whatever. Your experiments should be done in an order that guarantees maximum safety – for example, if its reactivity turned out to be similar to potassium, heating it in steam wouldn't be a good idea!

Chapter 7: Acids

This chapter explores what acids are, and some simple patterns in their chemistry in the lab. Wider issues relating to acids – such as the production and uses of sulphuric acid, and problems concerning acid rain – are dealt with in later chapters.

John George Haigh (the "acid bath murderer") killed six people and disposed of their bodies by dissolving them in drums of concentrated sulphuric acid...

...so it's no wonder protective clothing has to be worn to clean up spills of the acid!

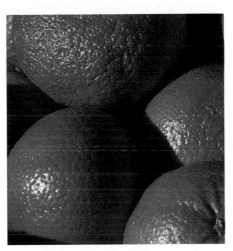

But not all acids are dangerous – oranges contain citric acid...

...and slightly acidic soils favour the growth of plants like rhododendrons.

pH and Indicators

The pH scale

The pH scale ranges from about 0 to about 14 and tells you how acidic or how alkaline a solution is.

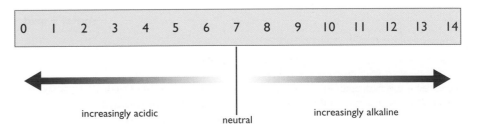

| 0 | 1 | 2 | 3 | 4 | 5 | 6 | 7 | 8 | 9 | 10 | 11 | 12 | 13 | 14 |

increasingly acidic

neutral

increasingly alkaline

Using universal indicator solution to measure the pH of various solutions.

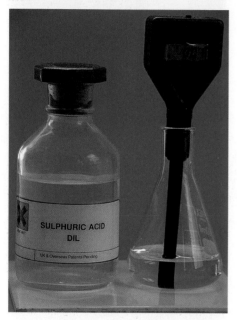

Measuring the pH of dilute sulphuric acid.

One of the common acids in the lab, nitric acid, has much more complex reactions with metals. You won't be asked about this at GCSE.

Measuring pH

Using universal indicator

Universal indicator is made from a mixture of dyes which change colour in a gradual way over a range of pHs. It can be used as a solution or as a paper. The commonest form is known as *wide range* universal indicator. It changes through a variety of colours from pH 1 right up to pH 14, but isn't very accurate.

Using a pH meter

You can measure pH much more accurately using a pH meter. Before you can use a pH meter, you have to adjust it to make sure that it is reading accurately. To do this you put it into a solution whose pH is known and adjust the reading so that it gives exactly that value.

Simple indicators

Any substance which has more than one colour form, depending on the pH, can be used as an indicator. One of the commonest ones is **litmus**.

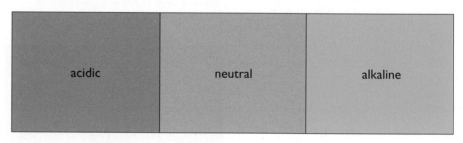

| acidic | neutral | alkaline |

Litmus is red in acidic solutions and blue in alkaline ones. The neutral purple colour is an equal mixture of the red and blue forms.

Reacting Acids with Metals

Simple dilute acids react with metals depending on their positions in the reactivity series.

- Metals below hydrogen in the series don't react with dilute acids.

- Metals above hydrogen in the series react to produce hydrogen gas.

- The higher the metal is in the reactivity series, the more vigorous the reaction. You would never mix metals like sodium or potassium with acids, because their reactions are too violent.

A summary equation for metals above hydrogen in the reactivity series

metal + acid → salt + hydrogen

Salts

Common salt is sodium chloride. This is produced if the hydrogen in hydrochloric acid is replaced by sodium.

All simple acids contain hydrogen. When that hydrogen is replaced by a metal, the compound formed is called a **salt**. Magnesium sulphate is a salt, so is zinc chloride, and so is lead(II) nitrate.

Sulphuric acid can be thought of as the **parent acid** of all the sulphates.

Parent acid	Salts
sulphuric acid	sulphates
hydrochloric acid	chlorides
nitric acid	nitrates

It doesn't matter if the replacement can't be done directly. For example, you can't make copper(II) sulphate from copper and dilute sulphuric acid because they don't react. There are, however, other ways of making it from sulphuric acid. Copper(II) sulphate is still a salt.

Reactions involving magnesium and acids

With dilute sulphuric acid

There is rapid fizzing and a colourless gas is evolved which pops with a lighted splint (the test for hydrogen). The reaction mixture gets very warm as heat is produced. The magnesium gradually disappears to leave a colourless solution of magnesium sulphate.

$$Mg_{(s)} + H_2SO_{4(aq)} \rightarrow MgSO_{4(aq)} + H_{2(g)}$$

This is a displacement reaction. The more reactive magnesium has displaced the less reactive hydrogen.

With dilute hydrochloric acid

The reaction looks exactly the same. The only difference is that this time a solution of magnesium chloride is formed.

$$Mg_{(s)} + 2HCl_{(aq)} \rightarrow MgCl_{2(aq)} + H_{2(g)}$$

Why are the reactions so similar?

Acids in solution are ionic. Dilute sulphuric acid contains hydrogen ions and sulphate ions. Dilute hydrochloric acid contains hydrogen ions and chloride ions.

You can rewrite the equations as ionic equations. In the sulphuric acid case:

$$Mg_{(s)} + 2H^+_{(aq)} + SO_4^{2-}{}_{(aq)} \rightarrow Mg^{2+}{}_{(aq)} + SO_4^{2-}{}_{(aq)} + H_{2(g)}$$

You can see that the sulphate ion hasn't been changed by the reaction. It is a spectator ion, and so we leave it out of the ionic equation:

$$Mg_{(s)} + 2H^+_{(aq)} \rightarrow Mg^{2+}{}_{(aq)} + H_{2(g)}$$

Repeating this with hydrochloric acid, you find that the chloride ions are also spectator ions.

$$Mg_{(s)} + 2H^+_{(aq)} + 2Cl^-_{(aq)} \rightarrow Mg^{2+}{}_{(aq)} + 2Cl^-_{(aq)} + H_{2(g)}$$

Leaving the spectator ions out produces the ionic equation:

$$Mg_{(s)} + 2H^+_{(aq)} \rightarrow Mg^{2+}{}_{(aq)} + H_{2(g)}$$

The reactions look the same because they are the same. All acids in solution contain hydrogen ions. That means that magnesium will react with any simple dilute acid in the same way.

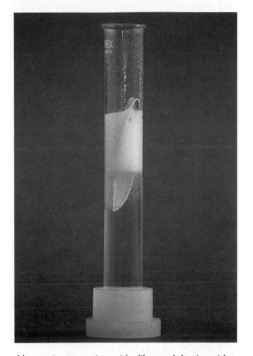

Magnesium reacting with dilute sulphuric acid.

You need to have read about ionic equations in Chapter 6 (pages 55–57) in order to understand this section.

Making hydrogen from the reaction between zinc and dilute sulphuric acid.

Reactions involving zinc and acids

Again the reactions between zinc and the two acids look exactly the same. The reactions are slower because zinc is lower down the reactivity series than magnesium. The reaction can be speeded up if it is heated or if the zinc is impure. A little copper(II) sulphate solution is often added to these reactions to make the zinc impure.

The full equations are:

$$Zn_{(s)} + H_2SO_{4(aq)} \rightarrow ZnSO_{4(aq)} + H_{2(g)}$$

$$Zn_{(s)} + 2HCl_{(aq)} \rightarrow ZnCl_{2(aq)} + H_{2(g)}$$

The ionic equations turn out to be identical because the sulphate ions and chloride ions are spectator ions:

$$Zn_{(s)} + 2H^+{}_{(aq)} \rightarrow Zn^{2+}{}_{(aq)} + H_{2(g)}$$

Reacting Acids with Metal Oxides

The metal magnesium reacts with dilute sulphuric acid; the metal copper doesn't. However, both magnesium oxide and copper(II) oxide react similarly with acids. All the metal oxide and acid combinations that you will meet at GCSE behave in exactly the same way.

A summary equation for acids and metal oxides

metal oxide + acid → salt + water

Reacting dilute sulphuric acid with copper(II) oxide

Copper(II) oxide reacting with hot dilute sulphuric acid.

> Remember that for GCSE purposes, acids in solution and metal (and ammonium) compounds are ionic.

The black powder reacts with hot dilute sulphuric acid to produce a blue solution of copper(II) sulphate.

$$CuO_{(s)} + H_2SO_{4(aq)} \rightarrow CuSO_{4(aq)} + H_2O_{(l)}$$

Most metal oxide and acid combinations need to be heated to get the reaction started.

The ionic equation for an acid/metal oxide reaction

In the reaction between copper(II) oxide and dilute sulphuric acid, everything in the equation is ionic apart from the water.

$$Cu^{2+}{}_{(s)} + O^{2-}{}_{(s)} + 2H^+{}_{(aq)} + SO_4{}^{2-}{}_{(aq)} \rightarrow Cu^{2+}{}_{(aq)} + SO_4{}^{2-}{}_{(aq)} + H_2O_{(l)}$$

Look carefully to find the spectator ions. The sulphate ion isn't changed at all, and the Cu^{2+} ion has only changed to the extent that it started as solid and ends up in solution. In this particular reaction we count that as unchanged. Leaving the spectator ions out gives:

$$O^{2-}{}_{(s)} + 2H^+{}_{(aq)} \rightarrow H_2O_{(l)}$$

This would be equally true of any simple metal oxide reacting with any acid. Oxide ions combine with hydrogen ions to make water. This is a good example of a **neutralisation reaction**. The presence of the hydrogen ions is what makes the sulphuric acid acidic. If something combines with these and removes them from solution, then obviously the acid has been neutralised.

Bases

Bases are defined as substances which combine with hydrogen ions. In the ionic equation above, an oxide ion is acting as a base because it combines with hydrogen ions to make water.

The simple metal oxides you will meet at GCSE are described as **basic oxides**.

> The bases you will meet at GCSE include: metal oxides (because they contain oxide ions), metal hydroxides (containing hydroxide ions), metal carbonates (containing carbonate ions) and ammonia. All of these have the ability to combine with hydrogen ions.

Reacting Acids with Metal Hydroxides

All metal hydroxides react with acids, but the ones most commonly used in the lab are the soluble hydroxides – usually sodium, potassium or calcium hydroxide solutions.

A summary equation for acids and metal hydroxides

metal hydroxide + acid → salt + water

Reacting dilute hydrochloric acid with sodium hydroxide solution

Mixing sodium hydroxide solution and dilute hydrochloric acid produces a colourless solution – not much seems to have happened. But if you repeat the reaction with a thermometer in the tube, the temperature rises several degrees, showing that there has been a chemical change.

Sodium chloride solution has been formed.

$$NaOH_{(aq)} + HCl_{(aq)} \rightarrow NaCl_{(aq)} + H_2O_{(l)}$$

The ionic equation for this shows that the underlying reaction is between hydroxide ions and hydrogen ions in solution to produce water.

$$OH^-_{(aq)} + H^+_{(aq)} \rightarrow H_2O_{(l)}$$

This is another good example of a neutralisation reaction. The hydroxide ion is a base because it combines with hydrogen ions. Sodium hydroxide is a soluble base.

Soluble bases are called **alkalis** and their solutions have pHs greater than 7.

Following the course of a neutralisation reaction

If everything involved in a neutralisation reaction is a colourless solution, how can you tell when exactly enough acid has been added to an alkali to produce a neutral solution?

Using an indicator

Some indicators will change colour when you have added even one drop too much acid. You normally avoid litmus because its colour change isn't very sharp and distinct. A common alternative is **methyl orange**.

Methyl orange is yellow in alkaline solutions and red in acids. You run acid in from the burette, swirling the flask all the time. The alkali is neutralised when the solution shows the first trace of orange. If it turns red, you have added too much acid. You will find more about carrying out this experiment (called a **titration**) on pages 80–81 and 289.

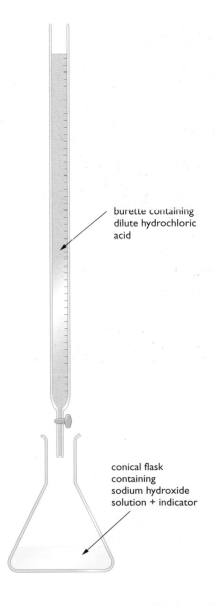

burette containing dilute hydrochloric acid

conical flask containing sodium hydroxide solution + indicator

Using a pH meter

Instead of using an indicator, you could use a pH meter.

The sodium hydroxide solution would be put in a beaker rather than a flask so that there is room to fit a pH meter in as well. The acid would be added a little at a time and the pH recorded after each addition. The mixture would have to be well stirred. The results can be drawn on a graph.

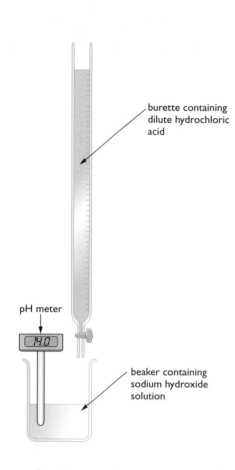

burette containing dilute hydrochloric acid

pH meter

14.0

beaker containing sodium hydroxide solution

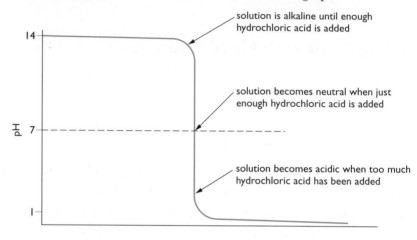

solution is alkaline until enough hydrochloric acid is added

solution becomes neutral when just enough hydrochloric acid is added

solution becomes acidic when too much hydrochloric acid has been added

volume of acid added to the sodium hydroxide solution

Notice how fast the pH changes around the neutral point.

Using a thermometer

You could also plot the temperature as you added the acid.

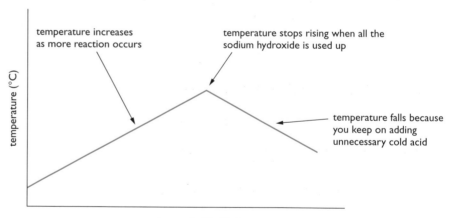

temperature increases as more reaction occurs

temperature stops rising when all the sodium hydroxide is used up

temperature falls because you keep on adding unnecessary cold acid

volume of acid added

The reaction gives out heat and so the temperature rises as you add more and more acid. Eventually all the sodium hydroxide is used up and so no more heat is given out. As you add more cold acid, the temperature falls again.

Using a conductivity meter

The **conductivity** of a solution is a measure of how well it conducts electricity. Solutions conduct electricity because of the ions they contain. How well they conduct depends on the number and type of ions present. Small ions move faster and so help conduct electricity better than bigger ones.

> The reason that the presence of ions allows solutions to conduct electricity is dealt with in Chapter 11: Electrolysis.

The apparatus would be similar to the pH measurement except that a conductivity meter would be used instead of a pH meter.

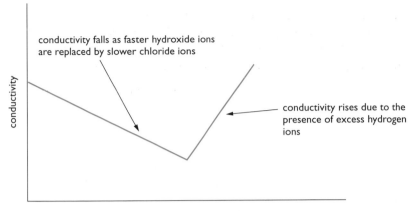

At the start, the solution contains $Na^+_{(aq)}$ and $OH^-_{(aq)}$. The hydroxide ions move quickly and this means the solution conducts electricity well.

As you add hydrochloric acid, the hydroxide ions react with hydrogen ions to make water – so they are not available to help conduct the electricity.

$$OH^-_{(aq)} + H^+_{(aq)} \rightarrow H_2O_{(l)}$$

The chloride ions being added from the acid are slower moving than the hydroxide ions, so don't make up for their loss.

After the solution has been fully neutralised, if you go on adding more acid, you are adding extra hydrogen ions, $H^+_{(aq)}$. These are fast moving, so the conductivity of the solution increases rapidly.

Reacting Acids with Carbonates

Carbonates react with cold acids to produce carbon dioxide gas.

A summary equation for acids and carbonates

carbonate + acid → salt + carbon dioxide + water

Reacting dilute sulphuric acid with copper(II) carbonate

A colourless gas is given off which turns lime water milky. This is the test for carbon dioxide. The green copper(II) carbonate reacts to give a blue solution of copper(II) sulphate.

$$CuCO_{3(s)} + H_2SO_{4(aq)} \rightarrow CuSO_{4(aq)} + CO_{2(g)} + H_2O_{(l)}$$

The ionic equation for carbonate/acid reactions

$$CO_3{}^{2-}{}_{(s)} + 2H^+_{(aq)} \rightarrow CO_{2(g)} + H_2O_{(l)}$$

Because carbonate ions are combining with hydrogen ions, carbonate ions are bases.

The reaction between copper(II) carbonate and dilute sulphuric acid.

Occasionally, you might come across an acid reacting with a carbonate solution. In that case, replace (s) by (aq).

Theories of Acids and Bases

The Arrhenius theory

As a part of his work on the theory of ions, the Swedish chemist Arrhenius suggested in 1887 that acids produced hydrogen ions when they were dissolved in water. That is much the line that we have taken so far in this chapter.

He also thought that bases were solutions containing hydroxide ions. We have taken a wider view than this, defining a base as something which combines with hydrogen ions.

The problem with the Arrhenius theory is that it is very restricted. A simple example of this involves hydrochloric acid and ammonia, NH_3. Ammonia gas is very soluble in water and gives an alkaline solution with a pH of about 11. Hydrochloric acid is a solution of the gas hydrogen chloride in water.

Concentrated hydrochloric acid gives off hydrogen chloride gas, and concentrated ammonia solution releases ammonia gas. If these gases are allowed to mix, solid white ammonium chloride is produced.

$$NH_{3(g)} + HCl_{(g)} \rightarrow NH_4Cl_{(s)}$$

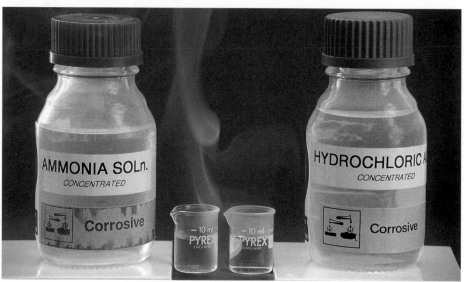

Ammonia and hydrogen chloride gases reacting to give a white smoke of ammonium chloride.

You would get exactly the same product if you neutralised dilute hydrochloric acid with ammonia solution. This time the ammonium chloride would be in solution.

Despite the similarity, the Arrhenius theory wouldn't accept the reaction between the gases as being acid–base, because it doesn't involve hydrogen ions and hydroxide ions in solution in water. There are lots of other similar examples – often involving solvents other than water. A resolution of the problem had to wait until 1923.

The Bronsted–Lowry theory

Bronsted was a Danish chemist; Lowry an English one. They defined acids and bases as follows:

- An **acid** is a proton (hydrogen ion) donor.

- A **base** is a proton (hydrogen ion) acceptor.

In this theory, when hydrogen chloride dissolves in water to give hydrochloric acid, there is a transfer of a proton from the HCl to the water.

> A hydrogen ion is a raw proton – the hydrogen nucleus minus its electron.

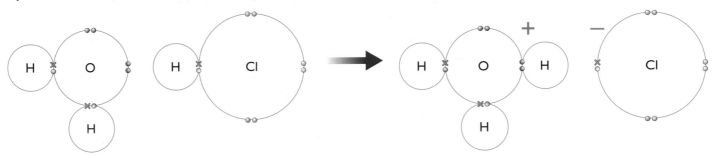

The hydrogen nucleus breaks away from the chlorine, leaving its electron behind.

In symbols:

$$H_2O_{(l)} + HCl_{(g)} \rightarrow H_3O^+_{(aq)} + Cl^-_{(aq)}$$

The $H_3O^+_{(aq)}$ ion is called a **hydroxonium ion**. This is the ion that we normally write simply as $H^+_{(aq)}$ – you can think of it as a hydrogen ion riding on a water molecule.

In this example, according to the Bronsted–Lowry theory, the HCl is an acid because it is giving a proton (a hydrogen ion) to the water. The water is acting as a base because it is accepting the proton.

In a similar way, hydrogen chloride gas reacts with ammonia gas to produce ammonium chloride:

> Only the outer electrons are shown in these diagrams to avoid confusion. Notice the new bond formed between the hydrogen and the water molecule. Both electrons in the bond come from the oxygen. This is described as a **co-ordinate covalent bond** or a **dative covalent bond**. Once the bond has been made, there is absolutely no difference between this bond and the other two normal covalent bonds between the oxygen and the hydrogens.

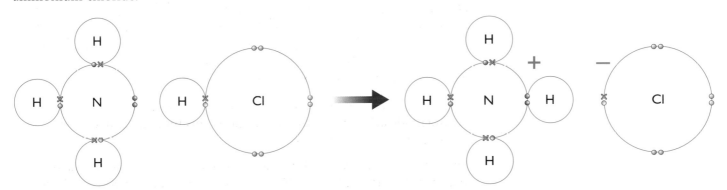

The ammonia acts as a base by accepting the proton; the HCl acts as an acid by donating it. This time an ammonium ion, NH_4^+, is formed. Notice the co-ordinate bond that is formed between the nitrogen and the new hydrogen.

$$NH_{3(g)} + HCl_{(g)} \rightarrow NH_4^+{}_{(s)} + Cl^-{}_{(s)}$$

Acids in solution

Acids in solution are acidic because of the presence of the hydroxonium ion. We would normally write a neutralisation reaction between an acid and a hydroxide, for example, as:

$$H^+_{(aq)} + OH^-_{(aq)} \rightarrow H_2O_{(l)}$$

What actually happens is that the hydroxonium ion donates a proton to the base, OH^-.

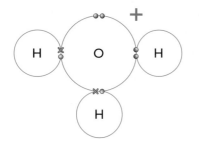 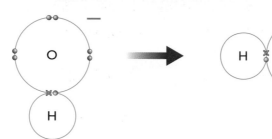

$$H_3O^+_{(aq)} + OH^-_{(aq)} \rightarrow 2H_2O_{(l)}$$

For GCSE purposes, we normally use the simplified version:

$$H^+_{(aq)} + OH^-_{(aq)} \rightarrow H_2O_{(l)}$$

Strong and Weak Acids and Alkalis

Strong acids

Hydrochloric acid is a strong acid. This means that when hydrogen chloride gas comes into contact with water, it reacts completely to give hydroxonium ions and chloride ions.

$$H_2O_{(l)} + HCl_{(g)} \rightarrow H_3O^+_{(aq)} + Cl^-_{(aq)}$$

Sulphuric acid and nitric acid are also strong acids. There are no un-ionised acid molecules left in their solutions. An acid which is 100% ionised in solution is called a **strong acid**.

Weak acids

An acid which is only partially ionised in solution is described as a **weak acid**. Ethanoic acid (in vinegar), citric acid (in oranges and lemons) or carbonic acid (formed when carbon dioxide dissolves is water) are all examples of weak acids.

In dilute ethanoic acid only about 1% of the ethanoic acid molecules have actually formed ions at any one time. As fast as the molecules react with water to produce ions, the ions react back again to give the original molecules. You have a reversible reaction.

$$CH_3COOH_{(aq)} + H_2O_{(l)} \rightleftharpoons CH_3COO^-_{(aq)} + H_3O^+_{(aq)}$$

If you are meeting this for the first time, don't worry too much about the formula for ethanoic acid for now.

Recognising strong and weak acids

If you had a strong and a weak acid of *equal concentration*, the weak acid will have a lower concentration of hydrogen ions in solution. That means:

- Its pH won't be so low. The greater the concentration of hydrogen ions in solution, the lower the pH. Strong acids have lower pHs than weak acids of the same concentration.

- The reactions will be slower. For example, weak acids react with metals or carbonates more slowly than strong acids of the same concentration do.

Be careful to distinguish between the words *strong* and *weak* as opposed to the words *concentrated* and *dilute*.

- *Concentrated* and *dilute* tell you about the amount of acid which has gone into the solution.

- *Strong* and *weak* tell you about the proportion of the acid which has reacted with the water to form ions.

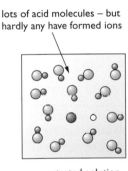

lots of acid molecules – but hardly any have formed ions

concentrated solution
of a weak acid

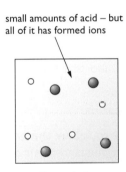

small amounts of acid – but all of it has formed ions

dilute solution
of a strong acid

> For simplicity, we normally talk about "hydrogen ions in solution", even if we should really be calling them hydroxonium ions.

Strong alkalis

A strong alkali is one which is fully ionised in solution. Examples of strong alkalis include sodium and potassium hydroxide. These are ionic compounds containing Na^+ or K^+ ions along with OH^- ions. These ions are present in the solid, but dissolving them in water makes them free to move around. The presence of lots of hydroxide ions makes the solution strongly alkaline with a very high pH.

> Remember that an alkali is a soluble base – a soluble substance which will combine with hydrogen ions.

Weak alkalis

The simplest example of a weak alkali is ammonia, NH_3. When ammonia dissolves in water, a very small proportion of it (typically about 1%) reacts.

$$NH_{3(g)} + H_2O_{(l)} \rightleftharpoons NH_4^+{}_{(aq)} + OH^-{}_{(aq)}$$

The hydroxide ions tend to recapture the H^+ from the ammonium ions, NH_4^+, as fast as they are formed.

The relatively small numbers of hydroxide ions in solution (compared with sodium hydroxide solution of the same concentration) gives the ammonia solution a pH of around 11.

End of Chapter Checklist

If you haven't got a copy of your specification, read the introduction on page vi.

You will need to be able to do some or all of the following. Check your Awarding Body's specification (syllabus) to find out exactly what you need to know.

● Understand the pH scale and the use of universal indicator and simple indicators like litmus.

● Know how the reactions of metals with acids are related to the position of the metal in the reactivity series.

● Describe and write equations (full and ionic) for the reactions between common metals and dilute sulphuric or hydrochloric acid.

● Understand that simple metal oxides are bases, and know that they react with acids to form a salt and water.

● Describe and write equations (full and ionic) for the reaction of copper(II) oxide with dilute sulphuric acid.

● Know that soluble bases are called alkalis, that they contain OH^- ions in solution, and that they react with acids to form a salt and water.

● Write equations (full and ionic) for the reactions between common alkalis and acids.

● Know how to follow the process of neutralisation using an indicator, a pH meter, measurement of heat evolved or measurement of conductivity, and be able to explain the shapes of any graphs produced.

● Know that carbonates react with acids to give a salt, carbon dioxide and water, and be able to describe and write equations (full and ionic) for common examples.

● Understand what is meant by an acid and a base according to the Arrhenius theory and the Bronsted–Lowry theory, and know the limitations of the Arrhenius theory.

● Understand the terms strong, weak, concentrated and dilute as applied to acids and alkalis.

Questions

More questions on acids can be found at the end of Section B on page 142.

1 **a)** Which of the following will react with dilute sulphuric acid? Copper, copper(II) oxide, copper(II) hydroxide, copper(II) carbonate.

b) In the case of each of the substances which does react, write the full equation (including state symbols) for the reaction. All of these substances are insoluble solids.

2 Draw a labelled diagram of the apparatus you would use to collect a few test tubes of hydrogen gas from the reaction between magnesium and dilute sulphuric acid. Write the full equation for the reaction.

3 **a)** Nickel, Ni, is a silvery metal just above hydrogen in the reactivity series. Nickel(II) compounds in solution are green. Describe what you would see if you warmed some nickel with dilute sulphuric acid in a test tube. Include a description of how you would test for any gas given off.

b) Write the full equation for the reaction between nickel and dilute sulphuric acid.

c) Nickel(II) carbonate is a green insoluble powder. Describe what you would see if you added a spatula measure of nickel(II) carbonate to some dilute hydrochloric acid in a test tube. Include a description of how you would test for any gas given off.

d) Write *i)* a full equation, *ii)* the ionic equation, for the reaction between nickel(II) carbonate and dilute hydrochloric acid.

4 Heat is evolved when sodium hydroxide solution and dilute hydrochloric acid react to make sodium chloride and water. Dilute hydrochloric acid was added, a little at a time, to $25\ cm^3$ of sodium hydroxide solution in a conical flask. After each addition, the mixture was stirred with a thermometer and the temperature was recorded.

a) Sketch a graph of the results you would expect to get if you kept on adding acid until it was in excess.

b) Explain the shape of your graph.

c) Describe in detail, with the help of a labelled diagram, exactly how you would carry out the experiment. You should include any important safety factors.

5 a) Which of the following equations represent reactions between acids and bases? For each of the equations which is an acid–base reaction, state which substance is the acid and which the base.

i) $MgO_{(s)} + H_2SO_{4(aq)} \rightarrow MgSO_{4(aq)} + H_2O_{(l)}$

ii) $CO_3^{2-}{}_{(s)} + 2H^+{}_{(aq)} \rightarrow CO_{2(g)} + H_2O_{(l)}$

iii) $2Al_{(s)} + 6HCl_{(aq)} \rightarrow 2AlCl_{3(aq)} + 3H_{2(g)}$

iv) $H_2O_{(l)} + HCl_{(g)} \rightarrow H_3O^+{}_{(aq)} + Cl^-{}_{(aq)}$

v) $Zn_{(s)} + Cu^{2+}{}_{(aq)} \rightarrow Zn^{2+}{}_{(aq)} + Cu_{(s)}$

vi) $NH_{3(g)} + HCl_{(g)} \rightarrow NH_4^+{}_{(s)} + Cl^-{}_{(s)}$

vii) $NH_{3(g)} + H_2O_{(l)} \rightleftharpoons NH_4^+{}_{(aq)} + OH^-{}_{(aq)}$

viii) $NaOH_{(aq)} + HCl_{(aq)} \rightarrow NaCl_{(aq)} + H_2O_{(l)}$

b) Only one of these would count as an acid–base reaction if you restricted yourself strictly to the Arrhenius theory of acids and bases. Which one? Explain your choice.

6 Sodium hydride, NaH, is a white ionic solid in which the hydrogen exists as an H^- ion. The electronic structures of the two ions are:

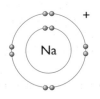

Sodium hydride reacts violently with water. The hydride ion reacts with the water like this:

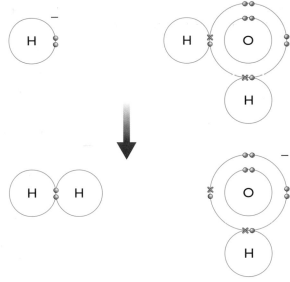

a) At the beginning of the reaction there were sodium ions, hydride ions and water present.

i) Which of these acts as an acid? Explain your reasoning.

ii) Which of these acts as a base? Explain your reasoning.

b) Name the products of this reaction.

c) Describe what you might expect to see during the course of the reaction.

d) Assuming that you used a reasonable quantity of sodium hydride, what would you expect the pH of the final solution to be?

7 Ethanoic acid (found in vinegar), citric acid (found in oranges and lemons) and ascorbic acid (vitamin C) are all weak acids.

a) What do you understand by the term "weak acid"?

b) Assuming you were given solutions of the three acids of the same concentration, and any other simple apparatus or materials that you might want, describe two different experiments that you could carry out with all three acids which would enable you to rank them in order of decreasing acid strength.

8 The following results were obtained when dilute sodium hydroxide solution was run into dilute ethanoic acid of the same concentration. The pH of the mixture was measured at intervals during the experiment.

Volume of NaOH solution added (cm³)	pH
0	2.88
3.0	3.89
6.0	4.26
9.0	4.51
12.0	4.73
15.0	4.94
18.0	5.17
21.0	5.48
24.0	6.14
24.5	6.45
24.9	7.16
25.0	8.73
25.1	11.30
25.5	12.00
27.0	12.59

a) Plot a graph of these results showing the volume of the sodium hydroxide solution on the horizontal (x) axis, and the pH on the vertical (y) axis.

b) What evidence is there from the graph that ethanoic acid is a weak acid?

c) A common indicator used for this titration is phenolphthalein. Phenolphthalein is colourless at pH's less than 8.3, and bright pink at pH's above 10.0. In between 8.3 and 10.0, it's colour gradually deepens. Use your graph to find the volume of sodium hydroxide solution which would have to be added before you observed the first trace of very pale pink in the solution.

9 Hydrogen cyanide ("prussic acid" or "hydrocyanic acid") is a weak acid. By doing an internet (or other) search, write a short article not exceeding 300 words on anything you discover about hydrogen cyanide which you find interesting. Your article should be word processed and include at least one picture or diagram.

Chapter 8: Making Salts

This chapter looks at some of the practical problems in making pure samples of salts in the lab. Remember that a salt is what is formed when the hydrogen in an acid is replaced by a metal. For example, sulphates come from sulphuric acid, chlorides from hydrochloric acid, and nitrates from nitric acid.

Soluble and Insoluble Salts

The importance of knowing whether a salt is soluble or insoluble in water

Remember that acids react with carbonates to give a salt, carbon dioxide and water. In the case of calcium carbonate (for example, marble chips) reacting with dilute hydrochloric acid, calcium chloride solution is produced.

$$CaCO_{3(s)} + 2HCl_{(aq)} \rightarrow CaCl_{2(aq)} + CO_{2(g)} + H_2O_{(l)}$$

If you try the reaction between calcium carbonate and dilute sulphuric acid, nothing much seems to happen if you use large marble chips. You will get a few bubbles when you first add the acid, but the reaction soon stops.

Calcium carbonate reacting with dilute hydrochloric acid.

Calcium carbonate not reacting with dilute sulphuric acid.

How quickly the reaction stops may well depend on the size of the marble chips (because that affects the surface area), the concentration of the acid, the volume of acid added, and the amount the flask is shaken. This would make an interesting investigation – particularly if you could then explain any pattern in your results.

The problem is that the calcium sulphate produced in the reaction is almost insoluble in water. As soon as the reaction starts, a layer of calcium sulphate is formed around the calcium carbonate, stopping any further reaction.

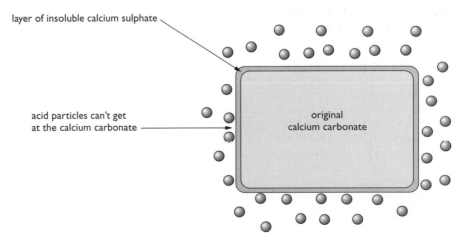

layer of insoluble calcium sulphate

acid particles can't get at the calcium carbonate

original calcium carbonate

Any attempt to produce an insoluble salt from the reaction between a solid and a liquid will fail for this reason.

Solubility patterns

To keep the table simple, it includes one or two compounds (like aluminium carbonate, for example) which don't actually exist. Don't worry about these. The problem won't arise at GCSE.

Hydroxides have been included for the sake of completeness although they are bases and not salts.

The list is in reactivity series order, apart from the ammonium group. Ammonium compounds often have similarities with sodium and potassium compounds, and so are included near them.

Ethanoates are salts made from ethanoic acid.

There is no clear cut-off between "insoluble" and "almost insoluble" compounds. The ones picked out as "almost insoluble" include the more common ones that you will need to know about elsewhere in the course.

	nitrate	chloride	sulphate	ethanoate	carbonate	hydroxide
ammonium	soluble	soluble	soluble	soluble	soluble	soluble
potassium	soluble	soluble	soluble	soluble	soluble	soluble
sodium	soluble	soluble	soluble	soluble	soluble	soluble
barium	soluble	soluble	insoluble	soluble	insoluble	almost insoluble
calcium	soluble	soluble	almost insoluble	soluble	insoluble	almost insoluble
magnesium	soluble	soluble	soluble	soluble	insoluble	insoluble
aluminium	soluble	soluble	soluble	soluble	insoluble	insoluble
zinc	soluble	soluble	soluble	soluble	insoluble	insoluble
iron	soluble	soluble	soluble	soluble	insoluble	insoluble
lead	soluble	insoluble	insoluble	soluble	insoluble	insoluble
copper	soluble	soluble	soluble	soluble	insoluble	insoluble
silver	soluble	insoluble	soluble	soluble	insoluble	insoluble

key:

soluble   insoluble   almost insoluble (slightly soluble)

It can seem a bit daunting to have to remember all this, but it isn't as hard as it looks at first sight. Except for the carbonates and hydroxides, most of these compounds are soluble. Learn the exceptions in the sulphates and chlorides. The reason for the exceptions in the carbonates and hydroxides is that all sodium, potassium and ammonium compounds are soluble.

Note that:

- All sodium, potassium and ammonium compounds are *soluble*.
- All nitrates are *soluble*.
- All common ethanoates are *soluble*.
- Most common chlorides are *soluble*, except lead(II) chloride and silver chloride.
- Most common sulphates are *soluble*, except lead(II) sulphate, barium sulphate and calcium sulphate.
- Most common carbonates are *insoluble*, except sodium, potassium and ammonium carbonates.
- Most metal hydroxides are *insoluble* (or *almost insoluble*), except sodium, potassium and ammonium hydroxides.

Making Soluble Salts (Except Sodium, Potassium and Ammonium Salts)

These all involve reacting a solid with an acid. You can use any of the following mixtures:

- **acid + metal** (but only for the moderately reactive metals from magnesium to iron in the reactivity series)

- **acid + metal oxide** or **hydroxide**

- **acid + carbonate**

Whatever mixture you use, the method is essentially the same.

Making magnesium sulphate crystals

Enough magnesium is added to some dilute sulphuric acid so that there is some left over when the reaction stops bubbling. This is to make sure that there is no acid left in the final mixture.

$$Mg_{(s)} + H_2SO_{4(aq)} \rightarrow MgSO_{4(aq)} + H_{2(g)}$$

The unused magnesium is then filtered off, and the magnesium sulphate solution is concentrated by boiling it until crystals will form when it is cooled. You can test for this by cooling a small sample of the solution quickly. If the sample crystallises, so will the whole solution.

Dilute sulphuric acid with excess magnesium.

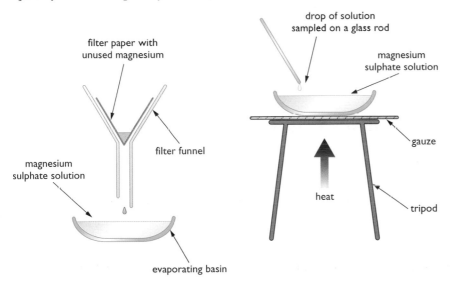

In the diagram, a small drop on the end of a glass rod is cooled rapidly in the air to see whether crystals form.

Finally, the solution is left to form colourless magnesium sulphate crystals. Any uncrystallised solution can be poured off the crystals, and the crystals can be blotted dry with paper tissue.

Why not just evaporate the solution to dryness? Water of crystallisation

It would seem much easier to just boil off all the water rather than crystallising the solution slowly, but evaporating to dryness wouldn't give you magnesium sulphate crystals. Instead, you would produce a white powder of **anhydrous** magnesium sulphate.

Copper(II) sulphate crystals.

"Anhydrous" means "without water". When many salts form their crystals, water from the solution becomes chemically bound up with the salt. This is called **water of crystallisation**. A salt which contains water of crystallisation is said to be **hydrated**.

$$MgSO_{4(aq)} + 7H_2O_{(l)} \rightarrow MgSO_4 \cdot 7H_2O_{(s)}$$

The extra water in the equation comes from the water in the solution.

Making copper(II) sulphate crystals from copper(II) oxide

The method is identical except that you add an excess of black copper(II) oxide to *hot* dilute sulphuric acid. You can easily see when you have an excess because you are left with some unreacted black solid.

$$CuO_{(s)} + H_2SO_{4(aq)} \rightarrow CuSO_{4(aq)} + H_2O_{(l)}$$

$$CuSO_{4(aq)} + 5H_2O_{(l)} \rightarrow CuSO_4 \cdot 5H_2O_{(s)}$$

How do you know whether you need to heat the mixture?

Carbonates react with dilute acids in the cold, and so does magnesium. Most other things that you are likely to come across need to be heated.

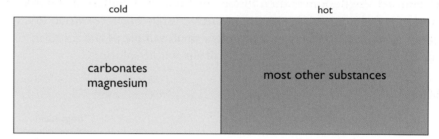

cold	hot
carbonates magnesium	most other substances

Making Sodium, Potassium and Ammonium salts

The need for a different method

In the method we've just been looking at, you add an excess of a solid to an acid, and then filter off the unreacted solid. You do this to make sure that all the acid is used up.

The problem is that all sodium, potassium and ammonium compounds are soluble in water. The solid you added to the acid would not only react with the acid, but any excess would just dissolve in the water present. You wouldn't have any visible excess to filter off. There's no simple way of seeing when you have added just enough of the solid to neutralise the acid.

Solving the problem by doing a titration

Before you go on, it would be a good idea to read page 67 for the background to these reactions.

You normally make these salts from sodium or potassium hydroxide or ammonia solution, but you can also use the carbonates. Fortunately, solutions of all these are alkaline. This means that you can find out when you have a neutral solution by using an indicator.

The method of finding out exactly how much of two solutions you need to neutralise each other is called a titration. The point at which an indicator changes colour during the **titration** is called the **end point** of the titration.

Having found out how much acid and alkali are needed, you can make a pure solution of the salt by mixing these same volumes again but without the indicator.

Making sodium sulphate crystals

$$2NaOH_{(aq)} + H_2SO_{4(aq)} \rightarrow Na_2SO_{4(aq)} + 2H_2O_{(l)}$$

Sodium hydroxide solution (25 cm^3) is transferred to a conical flask using a pipette, and a few drops of methyl orange are added as the indicator.

Dilute sulphuric acid is run in from the burette until the indicator turns from yellow to orange.

not enough acid	just right!	too much acid

Colour changes for methyl orange.

The volume of acid needed is noted, and the same volumes of acid and alkali are mixed together in a clean flask without any indicator. The solution can be crystallised by evaporating it to the point that crystals will form on cooling, and then leaving it for the crystals to form. The crystals are finally separated from any remaining solution and allowed to dry.

$$Na_2SO_{4(aq)} + 10H_2O_{(l)} \rightarrow Na_2SO_4 \cdot 10H_2O_{(s)}$$

Making sodium chloride crystals

$$NaOH_{(aq)} + HCl_{(aq)} \rightarrow NaCl_{(aq)} + H_2O_{(l)}$$

You would need to do the titration using dilute hydrochloric acid rather than dilute sulphuric acid. However, once you have re-mixed the acid and the alkali without the indicator, you can then evaporate the sodium chloride solution to dryness rather than crystallising it slowly. Sodium chloride crystals don't contain any water of crystallisation, so you can save time by evaporating all the water off in one go. The disadvantage is that you end up with either a powder or very tiny crystals.

Making ammonium sulphate crystals

$$2NH_{3(aq)} + H_2SO_{4(aq)} \rightarrow (NH_4)_2SO_{4(aq)}$$

Using ammonia solution rather than sodium hydroxide solution makes no difference to the method. Although simple ammonium salts don't have water of crystallisation, you would still crystallise them slowly rather than evaporating them to dryness. Heating dry ammonium salts tends to break them up.

Making Insoluble Salts

Precipitation reactions

To make an insoluble salt, you do a **precipitation reaction**. A **precipitate** is a fine solid that is formed by a chemical reaction involving liquids or gases. A precipitation reaction is simply a reaction which produces a precipitate.

Apparatus for carrying out a titration.

A precipitate of silver chloride.

For example, if silver chloride is produced from a reaction involving solutions, you get a white precipitate formed, because silver chloride won't dissolve in water – and so is seen as a fine white solid.

The photograph shows the results of this reaction:

$$AgNO_{3(aq)} + NaCl_{(aq)} \rightarrow AgCl_{(s)} + NaNO_{3(aq)}$$

Explaining what's happening

Silver nitrate solution contains silver ions and nitrate ions in solution. The positive and negative ions are attracted to each other, but the attractions aren't strong enough to make them stick together. Similarly, sodium chloride solution contains sodium ions and chloride ions – again, the attractions aren't strong enough for them to stick together.

When you mix the two solutions, the various ions meet each other. When silver ions meet chloride ions, the attractions are so strong that the ions clump together and form a solid. The sodium and nitrate ions remain in solution because they aren't sufficiently attracted to each other.

> The water molecules in the solutions have been left out to avoid cluttering the diagram.

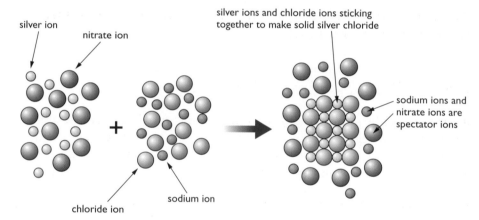

Writing ionic equations for precipitation reactions

> Ionic equations for precipitation reactions are simple to write. Write down the formula for the precipitate on the right-hand side of the equation. Write down the formulae for the ions that have clumped together to produce it on the left-hand side. Don't forget the state symbols.

The ionic equation for a precipitation reaction is much easier to write than the full equation. All that is happening in one of these reactions is that the ions of an insoluble salt are clumping together to form the solid. The ionic equation simply shows that happening. You don't need to worry at all about the spectator ions – they aren't doing anything.

$$Ag^+_{(aq)} + Cl^-_{(aq)} \rightarrow AgCl_{(s)}$$

This means that if you mix any solution containing silver ions with any solution containing chloride ions you will get the same white precipitate of silver chloride.

Making pure barium sulphate

$$Ba^{2+}_{(aq)} + SO_4{}^{2-}_{(aq)} \rightarrow BaSO_{4(s)}$$

You can mix solutions of any soluble barium compound (for example, barium chloride or barium nitrate) with any soluble sulphate. The sulphate doesn't necessarily have to be a salt. Dilute sulphuric acid contains sulphate ions, so you can perfectly well use that.

Suppose you decide to use barium chloride solution and dilute sulphuric acid.

You would mix the solutions to get a white precipitate of barium sulphate. The hydrogen ions from the sulphuric acid and the chloride ions are just spectator ions and aren't involved at all.

The mixture is filtered to get the precipitate. The solid barium sulphate is impure because of the presence of the spectator ions and any excess barium chloride solution or sulphuric acid. It is washed with pure water while it is still on the filter paper and then left to dry.

In an exam, simply use the words "filter, wash and dry the precipitate".

Making pure lead(II) iodide

It doesn't matter if the salt is unfamiliar to you, as long as you are told that it is insoluble in water. For example, to make lead(II) iodide you would have to mix a solution containing lead(II) ions with one containing iodide ions.

$$Pb^{2+}_{(aq)} + 2I^-_{(aq)} \rightarrow PbI_{2(s)}$$

The most common soluble lead(II) salt is lead(II) nitrate. A simple source of iodide ions would be sodium or potassium iodide solution, because all sodium and potassium salts are soluble.

The photograph shows the yellow precipitate of lead(II) iodide. This can now be filtered, washed and dried.

A precipitate of lead(II) iodide.

Making Salts by Direct Combination

A few salts can be made directly from their elements. Aluminium chloride and iron(III) chloride are good examples. Both can be made by passing dry chlorine over the heated metal. Because everything is done in the absence of water, the anhydrous salts are formed.

Making anhydrous aluminium chloride from aluminium and chlorine

Pure dry chlorine is passed over heated aluminium foil in a long glass tube.

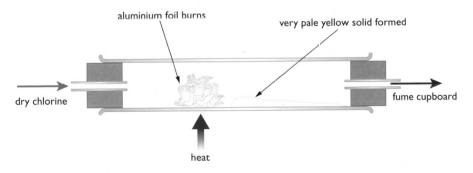

The aluminium burns in the chlorine. The aluminium chloride formed sublimes (turns straight from a solid to a gas), and condenses back to a solid again further along the tube where it is cooler.

$$2Al_{(s)} + 3Cl_{2(g)} \rightarrow 2AlCl_{3(s)}$$

Making anhydrous iron(III) chloride from iron and chlorine

This time, pure dry chlorine is passed over heated iron wool in the long glass tube. The iron burns and can produce a whole range of coloured products.

$$2Fe_{(s)} + 3Cl_{2(g)} \rightarrow 2FeCl_{3(s)}$$

Iron(III) chloride crystals scraped out of the reaction tube.

Pure anhydrous iron(III) chloride is actually black, but it reacts with any trace of water present to produce various orange and red compounds. There is usually enough water present in rubber bungs and tubing to affect the iron(III) chloride while it is still in the tube.

As soon as the iron(III) chloride is exposed to water vapour in the air, it reacts with it to form orange or yellow solutions. You can see this happening in the photograph. Look closely at the end of the tube and at the small droplets on the white tile.

Summarising the methods of making salts

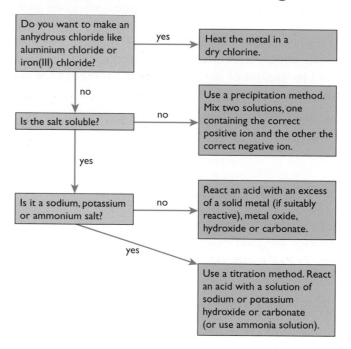

End of Chapter Checklist

If you haven't got a copy of your specification, read the introduction on page vi.

You will need to be able to do some or all of the following. Check your Awarding Body's specification (syllabus) to find out exactly what you need to know.

- Know which salts are soluble and which are insoluble in water.

- Give practical details of how you would make a pure, dry sample of a salt from the reaction between a suitable solid (metal, metal oxide, metal hydroxide or metal carbonate) and an acid. Know which salts can be made by this method.

- Describe how to make pure, dry samples of sodium, potassium or ammonium salts using a titration method.

- Describe how to make pure, dry samples of insoluble salts using precipitation reactions.

- Write ionic equations for precipitation reactions.

- Describe how to make samples of anhydrous aluminium chloride and iron(III) chloride by direct combination of the metal with chlorine.

Questions

More questions on salts can be found at the end of Section B on page 142.

1 Sort the following compounds into two lists – those which are soluble in water, and those which are insoluble.

sodium chloride, lead(II) sulphate, zinc nitrate, calcium carbonate, iron(III) sulphate, lead(II) chloride, potassium sulphate, copper(II) carbonate, silver chloride, aluminium nitrate, barium sulphate, ammonium chloride, magnesium nitrate, calcium sulphate, sodium phosphate, nickel(II) carbonate, chromium(III) hydroxide, potassium dichromate(VI)

2 a) Describe in detail the preparation of a pure, dry sample of copper(II) sulphate crystals, $CuSO_4 \cdot 5H_2O$, starting from copper(II) oxide.

   b) Write full equations for i) the reaction producing copper(II) sulphate solution, ii) the crystallisation reaction.

3 a) Read the following description of a method for making sodium sulphate crystals, $Na_2SO_4 \cdot 10H_2O$, and then explain the reasons for each of the underlined phrases or sentences.

25 cm³ of sodium carbonate solution was transferred to a conical flask using a pipette, and a few drops of methyl orange were added. Dilute sulphuric acid was run in from a burette until the solution became orange. The volume of acid added was noted. That same volume of dilute sulphuric acid was added to a fresh 25 cm³ sample of sodium carbonate solution in a clean flask, but without the methyl orange. The mixture was evaporated until a sample taken on the end of a glass rod crystallised on cooling in the air. The solution was left to cool. The crystals formed were separated from the remaining solution and dried.

   b) Write equations for i) the reaction producing sodium sulphate solution, ii) the crystallisation reaction.

4 Suggest solutions which could be mixed together to make each of the following insoluble salts. In each case, write the ionic equation for the reaction you choose.

   a) silver chloride,   b) calcium carbonate,
   c) lead(II) sulphate,   d) lead(II) chloride.

5 Describe in detail the preparation of a pure, dry sample of barium carbonate. Write the ionic equation for the reaction you use.

6   There are four main methods of making salts:

A   reacting an acid with an excess of a suitable solid

B   using a titration

C   using a precipitation reaction

D   by direct combination

For each of the following salts, write down the letter of the appropriate method, and name the substances you would react together. You should state whether they are used as solids, solutions or gases. Write an equation (full or ionic as appropriate) for each reaction.

a) zinc sulphate

b) barium sulphate

c) potassium nitrate (nitric acid is HNO_3)

d) anhydrous aluminium chloride

e) copper(II) nitrate

f) lead(II) chromate(VI) (a bright yellow insoluble solid; chromate(VI) ions have the formula CrO_4^{2-}).

Chapter 9: Analysis

> This chapter summarises some of the many ways of testing for substances – from methods which are simple enough to use in any chemistry lab to very powerful modern techniques.

Collecting and Identifying Gases

Collecting gases

Gases can be collected:

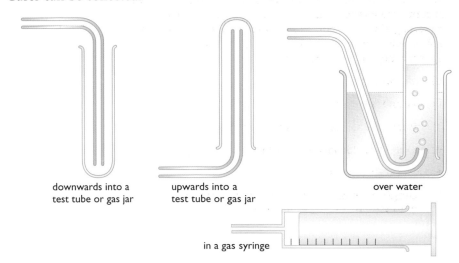

downwards into a
test tube or gas jar

upwards into a
test tube or gas jar

over water

in a gas syringe

Collecting into a gas syringe is fine if you want to measure the volume of the gas. To test the gas, though, you would have to transfer it to a test tube using one of the first three methods.

Hazards associated with collecting the gases

Hydrogen forms explosive mixtures with air. Chlorine, sulphur dioxide, hydrogen chloride and ammonia are all poisonous. Sulphur dioxide can trigger asthma attacks.

How to collect and test individual gases

Hydrogen, H_2

Hydrogen is less dense than air and is almost insoluble in water. Collect it over water or upwards into a test tube or gas jar.

Hydrogen pops when a lighted splint is held to the mouth of the tube. The hydrogen combines explosively with oxygen in the air to make water.

$$2H_{2(g)} + O_{2(g)} \rightarrow 2H_2O_{(l)}$$

Oxygen, O_2

Oxygen has almost the same density as air and is almost insoluble in water. You normally collect it over water.

Oxygen relights a glowing splint.

Carbon dioxide, CO_2

Carbon dioxide is denser than air and can be collected downwards into a test tube or gas jar. It is only slightly soluble in water and so it can be collected over water as well.

Carbon dioxide turns lime water milky (or chalky). Lime water is calcium hydroxide solution. Carbon dioxide reacts with it to form a white precipitate of calcium carbonate.

$$Ca(OH)_{2(aq)} + CO_{2(g)} \rightarrow CaCO_{3(s)} + H_2O_{(l)}$$

Chlorine, Cl_2

Chlorine is denser than air and is usually collected downwards into a test tube or gas jar. Because chlorine is green it is easy to see when the tube or gas jar you are collecting it into is full. Chlorine is too soluble to collect it satisfactorily over water, but you can collect it over concentrated salt solution instead. It is less soluble in the salt solution.

Chlorine is a green gas which bleaches damp litmus paper.

Sulphur dioxide, SO_2

Sulphur dioxide is denser than air, and can be collected downwards into a test tube or gas jar. It is too soluble to collect over cold water, but you can collect it over hot water. Gases become less soluble as the temperature increases.

Sulphur dioxide is an acidic gas, turning blue litmus paper red. You can pick out sulphur dioxide from other acidic gases because it turns potassium dichromate(VI) paper from orange to green.

Hydrogen chloride, HCl

Hydrogen chloride is denser than air and is extremely soluble in water (to make hydrochloric acid). It must be collected downwards into a test tube or gas jar.

Hydrogen chloride is a steamy acidic gas. The steaminess is due to the hydrogen chloride reacting with water vapour in the air to form droplets of concentrated hydrochloric acid. It turns damp blue litmus paper red.

Ammonia, NH_3

Ammonia is less dense than air and is extremely soluble in water. It can only be collected upwards into a test tube or gas jar.

Ammonia is the only alkaline gas that you will meet at GCSE. It turns damp red litmus paper blue.

Testing for Water

Using anhydrous copper(II) sulphate

Water turns white anhydrous copper(II) sulphate blue.

Anhydrous copper(II) sulphate lacks water of crystallisation and is white. Dropping water onto it replaces the water of crystallisation, and turns it blue.

If you aren't sure about water of crystallisation, see pages 79–80.

$$CuSO_{4(s)} + 5H_2O_{(l)} \rightarrow CuSO_4 \cdot 5H_2O_{(s)}$$

This test works for anything which contains water. It does *not* show that the water is pure.

Chlorine bleaches damp litmus paper.

An explanation of the chemistry of the potassium dichromate(VI) test is beyond GCSE.

Testing for water with anhydrous copper(II) sulphate.

Using cobalt chloride paper

Cobalt chloride paper is simply filter paper which has been dipped into cobalt(II) chloride solution and then dried thoroughly in a desiccator. A desiccator is a piece of glassware or a small cabinet which contains a tray of some substance which absorbs water.

In the absence of any water, the paper is blue. Adding water to it turns it pink.

Anything which contains water will turn cobalt chloride paper from blue to pink. The water doesn't have to be pure.

Testing for water with cobalt chloride paper.

Testing for Ions

Flame tests

A flame test is used to show the presence of certain metal ions in a compound. A platinum or nichrome wire is cleaned by dipping it into concentrated hydrochloric acid and then holding it in a hot Bunsen flame. This is repeated until the wire doesn't impart any colour to the flame.

The wire is dipped back into the acid, then into a tiny sample of the solid you are testing, and back into the flame.

> Nichrome (a nickel–chromium alloy) is a cheap alternative to platinum. It does, however, always produce a faint yellow colour in the flame which you have to ignore.

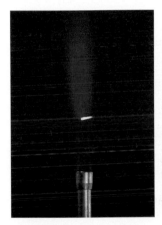

Red shows Li^+ ions.

Strong persistent orange shows Na^+ ions.

Lilac (pink) shows K^+ ions.

Orange-red ("brick red") shows Ca^{2+} ions.

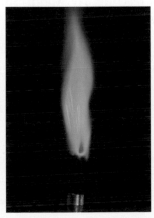

Blue-green shows Cu^{2+} ions.

Pale green shows Ba^{2+} ions.

> The calcium flame test often has so much orange in it that it can be confused with sodium. The orange-red often appears for quite a short time only.

If you aren't sure about precipitation reactions and the ionic equations for them, you must read pages 81–83 before you go on.

Testing for positive ions using sodium hydroxide solution

Of the common hydroxides, only sodium, potassium and ammonium hydroxides dissolve in water. Most metal hydroxides are insoluble. This means that if you add sodium hydroxide solution to a solution containing the metal ions, you will get a precipitate of the metal hydroxide.

The colour of these precipitates can help you to identify the metal ion.

Blue precipitate

This shows the presence of copper(II) ions. The precipitate is copper(II) hydroxide.

$$Cu^{2+}_{(aq)} + 2OH^-_{(aq)} \rightarrow Cu(OH)_{2(s)}$$

Any copper(II) salt in solution will react with sodium hydroxide solution in this way. For example, with copper(II) sulphate solution, the full equation would read

$$CuSO_{4(aq)} + 2NaOH_{(aq)} \rightarrow Cu(OH)_{2(s)} + Na_2SO_{4(aq)}$$

An orange-brown precipitate

This shows the presence of iron(III) ions. The precipitate is iron(III) hydroxide.

$$Fe^{3+}_{(aq)} + 3OH^-_{(aq)} \rightarrow Fe(OH)_{3(s)}$$

Any iron(III) compound in solution will give this precipitate. An example full equation might be

$$FeCl_{3(aq)} + 3NaOH_{(aq)} \rightarrow Fe(OH)_{3(s)} + 3NaCl_{(aq)}$$

Notice how much more complicated the full equations for these reactions are. They also obscure what is going on. Use ionic equations for precipitation reactions wherever possible.

A green precipitate

This shows the presence of iron(II) ions. The precipitate is iron(II) hydroxide.

$$Fe^{2+}_{(aq)} + 2OH^-_{(aq)} \rightarrow Fe(OH)_{2(s)}$$

This could be the result of reacting, say, iron(II) sulphate solution with sodium hydroxide solution:

$$FeSO_{4(aq)} + 2NaOH_{(aq)} \rightarrow Fe(OH)_{2(s)} + Na_2SO_{4(aq)}$$

The green precipitate darkens on standing and turns orange around the top of the tube. This is due to the iron(II) hydroxide being oxidised to iron(III) hydroxide by the air.

White precipitates

White precipitates are more common and therefore less informative. The examples you may need for GCSE are:

- *A white precipitate which doesn't dissolve when you add excess sodium hydroxide solution.*

This shows the presence of either magnesium or calcium ions in solution.

The blue precipitate of copper(II) hydroxide.

The orange-brown precipitate of iron(III) hydroxide.

The green precipitate of iron(II) hydroxide.

Precipitate formed from magnesium sulphate solution and sodium hydroxide solution.

The precipitate may be either magnesium hydroxide or calcium hydroxide.

$$Mg^{2+}_{(aq)} + 2OH^-_{(aq)} \rightarrow Mg(OH)_{2(s)}$$

$$\text{or}\quad Ca^{2+}_{(aq)} + 2OH^-_{(aq)} \rightarrow Ca(OH)_{2(s)}$$

You could tell which you had by doing a flame test on the original compound. Calcium gives an orange-red colour, whereas magnesium has no flame colour.

- *A white precipitate which dissolves when you add excess sodium hydroxide solution.*

This shows the presence of aluminium ions in the solution. The white precipitate is aluminium hydroxide.

$$Al^{3+}_{(aq)} + 3OH^-_{(aq)} \rightarrow Al(OH)_{3(s)}$$

The precipitate dissolves because the aluminium hydroxide reacts with excess hydroxide ions to give $Al(OH)_4^-$ ions – called tetrahydroxoaluminate ions.

$$Al(OH)_{3(s)} + OH^-_{(aq)} \rightarrow Al(OH)_4^-_{(aq)}$$

The full equation for this is:

$$Al(OH)_{3(s)} + NaOH_{(aq)} \rightarrow NaAl(OH)_{4(aq)}$$

The compound formed is called sodium tetrahydroxoaluminate.

> The name looks distinctly worrying, but it simply describes the ion: *tetra* means "four"; *hydroxo* refers to the OH groups; *aluminate* shows the presence of the aluminium in a negative ion.

No precipitate, but a smell of ammonia

This shows the presence of an ammonium salt. Sodium hydroxide solution reacts with ammonium salts (either solid or in solution) to produce ammonia gas. In the cold, there is just enough ammonia gas produced for you to be able to smell it. If you warm it, you can test the gas coming off with a piece of damp red litmus paper. Ammonia is alkaline and turns the litmus paper blue.

$$NH_4^+_{(s\ or\ aq)} + OH^-_{(aq)} \rightarrow NH_{3(g)} + H_2O_{(l)}$$

A typical full equation might be:

$$NH_4Cl_{(s)} + NaOH_{(aq)} \rightarrow NaCl_{(aq)} + NH_{3(g)} + H_2O_{(l)}$$

> Recognising gases by smelling them has to be done with great care. In this case, there is usually so little ammonia present in the cold, that it is safe to do as long as you take the normal precautions. You shouldn't, however, make any attempt to smell the mixture when it is warm.

Testing for carbonates and sulphites

Testing for carbonates by heating them

Most carbonates split up to give the metal oxide and carbon dioxide when you heat them. This is a good example of **thermal decomposition** – breaking something up by heating it. You can test the carbon dioxide given off by passing it through lime water, and there may be helpful colour changes as well.

For example, copper(II) carbonate is a green powder which decomposes on heating to produce black copper(II) oxide.

$$CuCO_{3(s)} \rightarrow CuO_{(s)} + CO_{2(g)}$$

Zinc carbonate is a white powder which decomposes on heating to give zinc oxide, which is yellow when it is hot, but turns back to white on cooling.

$$ZnCO_{3(s)} \rightarrow ZnO_{(s)} + CO_{2(g)}$$

Copper(II) carbonate turns black on heating.

See page 77 for a discussion of the insoluble salt problem.

Testing for carbonates using dilute acids

If you add a dilute acid to a solid carbonate, carbon dioxide is produced in the cold. It is probably best to use dilute nitric acid. Some acid–carbonate combinations can produce an insoluble salt which coats the solid carbonate and stops the reaction, but this doesn't happen if you use nitric acid because all nitrates are soluble.

Add a little dilute nitric acid, look for bubbles of gas produced in the cold, and test the gas with lime water.

The ionic equation shows any carbonate reacting with any acid.

$$CO_3{}^{2-}{}_{(s)} + 2H^+{}_{(aq)} \rightarrow CO_{2(g)} + H_2O_{(l)}$$

For example, using zinc carbonate and dilute nitric acid:

$$ZnCO_{3(s)} + 2HNO_{3(aq)} \rightarrow Zn(NO_3)_{2(aq)} + CO_{2(g)} + H_2O_{(l)}$$

Testing for sulphites using dilute acids

Sulphites contain the ion $SO_3{}^{2-}$, and have formulae just like carbonates. Carbonates react with dilute acids to give off CO_2. Sulphites react with dilute acids to give off SO_2.

$$SO_3{}^{2-}{}_{(s)} + 2H^+{}_{(aq)} \rightarrow SO_{2(g)} + H_2O_{(l)}$$

The only difference is that sulphites usually need warming with the acid before you produce enough sulphur dioxide to test adequately. Remember that sulphur dioxide is an acidic gas which turns potassium dichromate(VI) paper from orange to green.

The full equation would be exactly like the carbonate equation – replacing the carbon by sulphur.

Testing for sulphates

Make a solution of your suspected sulphate, add enough dilute *hydrochloric* acid to make it acidic and then add some barium *chloride* solution. A sulphate will produce a white precipitate of barium sulphate.

$$Ba^{2+}{}_{(aq)} + SO_4{}^{2-}{}_{(aq)} \rightarrow BaSO_{4(s)}$$

You acidify the solution to destroy other compounds which might also produce white precipitates when you add the barium chloride solution. For example, if you didn't add acid, you would also get a white precipitate if there was a carbonate present because barium carbonate is also white and insoluble. The acid reacts with and removes the carbonate ions.

You could equally well use *nitric* acid and barium *nitrate* solution. You must *never* acidify the solution with sulphuric acid because sulphuric acid contains sulphate ions. If you add those, you are bound to get a precipitate of barium sulphate, whatever else is present.

Testing for chlorides, bromides and iodides

This is very similar to the test for sulphates. Make a solution of your suspected chloride, bromide or iodide and add enough dilute *nitric* acid to make it acidic. Then add some silver *nitrate* solution.

The acid is added to react with and remove other substances which might also produce precipitates with silver nitrate solution.

A white precipitate (of silver chloride) shows the presence of chloride ions.

$$Ag^+{}_{(aq)} + Cl^-{}_{(aq)} \rightarrow AgCl_{(s)}$$

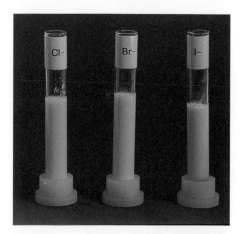

Precipitates of silver chloride, silver bromide and silver iodide.

A pale cream precipitate (of silver bromide) shows the presence of bromide ions.

$$Ag^+{}_{(aq)} + Br^-{}_{(aq)} \rightarrow AgBr_{(s)}$$

A yellow precipitate (of silver iodide) shows the presence of iodide ions.

$$Ag^+{}_{(aq)} + I^-{}_{(aq)} \rightarrow AgI_{(s)}$$

Testing for nitrates

This is a potentially very dangerous reaction, with a good chance of splashing hot corrosive sodium hydroxide solution on yourself or others. Don't try it without full safety protection and supervision.

Mix a small amount of the solid you are testing with some aluminium powder and then add sodium hydroxide solution. Warm *very* gently until a reaction just starts.

The aluminium reacts with the sodium hydroxide solution to produce hydrogen, but if a nitrate is present you also get ammonia gas evolved. You can test for this with moist red litmus paper.

The aluminium and sodium hydroxide mixture is a powerful reducing agent, and reduces the nitrate ions, NO_3^-, to ammonia, NH_3.

Testing for hydrogen ions or hydroxide ions

Hydrogen ions, H⁺

The presence of hydrogen ions makes a solution acidic. You can test this with indicators or by adding a reactive metal like magnesium (looking for hydrogen evolved) or a carbonate like sodium carbonate (looking for carbon dioxide). You will find details of all this in Chapter 7: Acids.

Hydroxide ions, OH⁻

Hydroxide ions in a solution make it alkaline. You can test this with indicators. Alternatively, you can make use of the fact that hydroxide ions react with ammonium ions to produce ammonia gas.

$$NH_4^+{}_{(s)} + OH^-{}_{(aq)} \rightarrow NH_{3(g)} + H_2O_{(l)}$$

Add some solid ammonium chloride to a solution which might contain hydroxide ions, warm it and test for ammonia with moist red litmus paper.

Instrumental Analysis

Modern methods of analysis use machines to recognise even tiny amounts of substances – and not only recognise them, but tell you how much of them is present. Detection techniques get more sensitive all the time and machines get smaller and cheaper. By coupling the detector to a computer, the results can be instantly matched with known patterns. That makes the whole process automatic.

It is now possible to detect and measure pollutants in the air or water, contamination in food, or the presence of drugs in the body in terms of "parts per million" or even "parts per billion".

All of these precipitates tend to discolour towards greys and pale purples on exposure to light. The bromide and iodide colours are quite difficult to distinguish between in practice. There is a follow-on test involving ammonia solution which helps to sort them out, but this is beyond GCSE.

Important! Before you do this test, you must first show that the substance isn't an ammonium salt. An ammonium salt would also produce ammonia in the presence of sodium hydroxide – irrespective of the presence of the aluminium.

There are lots of different instrumental methods. The two described here are chosen because they are easy to understand at GCSE.

Atomic emission spectroscopy

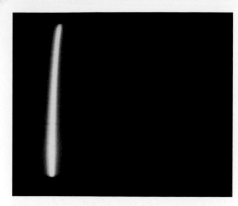

A sodium emission spectrum.

This is nothing more than a sophisticated flame test! It can be used to show the presence of particular elements in a sample. It won't tell you which compounds are present.

A solution of the sample is injected into a flame. The light emitted is analysed by splitting it into its component colours. The photograph shows part of the spectrum produced if the sample contained sodium.

Each element has a unique spectrum. No other element has lines in exactly those positions. The various frequencies of the lines in the spectrum can be measured automatically and the machine's computer can compare them with its database to obtain the correct match.

The concentration of a particular element can be found by measuring the intensity of the lines. The fainter the lines, the less concentrated the solution being tested.

You could use this technique as a means of quality control if you were manufacturing chemicals. For example, potassium compounds often have some sodium contamination. Atomic emission spectroscopy would tell you instantly if you had sodium compounds present, and if they were within acceptable limits.

Infra-red spectroscopy

Warning! In an exam, don't use the abbreviation "IR" spectroscopy – you won't get any marks unless you write the words in full!

Infra-red spectroscopy is used to identify whole compounds – particularly organic compounds like alcohol or drugs.

Visible light is made up of a continuous range of frequencies which we see as individual colours. Infra-red light is also made up of a range of different frequencies. If each of these frequencies is shone in turn through a sample of an organic compound, some frequencies are absorbed by the compound; others pass through unchanged.

Particular frequencies are absorbed by particular bonds in the compounds. An infra-red spectrum shows troughs where these frequencies have been absorbed. The position of the troughs tells you about the type of bond present.

This spectrum is for background interest only. Don't worry about the scales used. "Wavenumber", for example, is simply a measure of frequency. You will not be expected to identify particular bonds in compounds. All you would need to know is that infra-red spectroscopy is used to identify particular compounds.

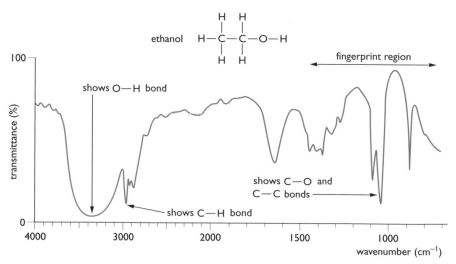

Infra-red spectrum of ethanol.

Any compound containing the marked bonds will have troughs in those positions, but only ethanol will have that particular pattern of troughs in the **fingerprint region** at the right of the spectrum.

Infra-red spectroscopy is used, for example, in breath tests for alcohol (the common name for ethanol), and for testing for drugs.

Breathalysers

Infra-red machines need too much power to be used in small hand-held breathalysers, but can be used in vehicles or in a police station. The machines are computerised and the computer looks for the absorption due to the C–H bond at 2950 cm^{-1}. This is one of the troughs shown in the diagram.

If it finds the 2950 cm^{-1} trough, it then checks the rest of the spectrum to see if it matches ethanol. There will be systems to eliminate signals due to other innocent substances in the breath caused, for example, by diseases like diabetes.

The amount of ethanol in the breath can be found by measuring the strength of the various absorptions. If a breath test is positive, it can be checked by testing a blood sample.

Drug testing

Infra-red spectroscopy can be used to test for the presence of drugs in a similar way. If the result is positive, it can be checked by other methods.

There are problems with drug testing. Methods are now so sensitive that unbelievably small amounts can be detected. This raises questions about possible contamination of samples, and whether such small quantities of the drug might have got into the body in entirely innocent ways.

Infra-red breathalysers are used in the fight against drink-driving.

You might like to do an internet search to read about any recent controversial drug tests (for example, in athletes). Try searching the sites belonging to the main newspapers for recent news items.

End of Chapter Checklist

If you haven't got a copy of your specification, read the introduction on page vi.

You will need to be able to do some or all of the following. Check your Awarding Body's specification (syllabus) to find out exactly what you need to know.

● Choose an appropriate method for collecting a particular gas.

● Know how to test for hydrogen, oxygen, carbon dioxide, chlorine, sulphur dioxide, hydrogen chloride and ammonia.

● Know how to test for the presence of water using anhydrous copper(II) sulphate or cobalt chloride paper.

● Describe how to carry out a flame test, and know the flame colours produced by lithium, sodium, potassium, calcium, barium and copper compounds.

● Know how to use sodium hydroxide solution to test for the presence of Cu^{2+}, Fe^{3+}, Fe^{2+}, Mg^{2+}, Ca^{2+}, Al^{3+} and NH_4^+ ions in solution. Write full or ionic equations for the reactions involved.

● Know how to test for carbonates using a dilute acid, and that the thermal decomposition of copper(II) carbonate and zinc carbonate leads to distinctive colour changes. Write equations for these reactions (full or ionic as appropriate).

● Know how to test for sulphites using dilute acids. Write equations (full or ionic) for these reactions.

● Know the tests for sulphates, chlorides, bromides and iodides. Write equations (full or ionic) for these reactions.

● Know how to test for a nitrate.

● Know how to test for H^+ or OH^- in a solution.

● Name two instrumental methods of analysis (one for elements and one for compounds) and give examples of their use, including advantages and disadvantages where appropriate.

● Understand that improvements in technology have made instrumental methods cheaper, faster, more automatic and more sensitive.

Questions

More questions on analysis can be found at the end of Section B on page 142.

I Four common methods of collecting gases are:

    A   downwards into a gas jar

    B   upwards into a gas jar

    C   over water

    D   in a gas syringe

Which method would you use in each of the following cases?

a) To collect a dry sample of carbon dioxide in order to do a reaction with it.

b) To measure the amount of hydrogen produced in 10 seconds during the reaction between dilute sulphuric acid and magnesium.

c) To collect a sample of carbon monoxide – a colourless, odourless, poisonous gas, insoluble in water and with approximately the same density as air.

d) To collect a sample of ammonia in order to do a reaction with it.

2 Name the gas being described in each of the following cases.

   a) A green gas which bleaches damp litmus paper.

   b) A gas which dissolves readily in water to produce a solution with a pH of about 11.

   c) A gas which turns blue litmus paper red and potassium dichromate(VI) paper from orange to green.

   d) A gas which produces a white precipitate with calcium hydroxide solution.

   e) A gas which pops when a lighted splint is placed in it.

   f) A steamy gas which turns blue litmus paper red.

   g) A gas which relights a glowing splint.

3 Describe fully how you would carry out the following tests. In each case, describe what you would expect to happen.

   a) A flame test for lithium ions in lithium chloride.

   b) A test for iron(II) ions in iron(II) sulphate.

   c) A test for sulphate ions in iron(II) sulphate.

   d) A test for the presence of water.

   e) A test for the carbonate ion in sodium carbonate.

   f) A test for the hydroxide ion in sodium hydroxide.

4 **A** is an orange solid which dissolves in water to give an orange solution. When sodium hydroxide solution is added to a solution of **A**, an orange-brown precipitate, **B**, is formed. Adding dilute nitric acid and silver nitrate solution to a solution of **A** gives a white precipitate, **C**.

   a) Identify **A**, **B** and **C**.

   b) Write equations (full or ionic) for the reactions producing **B** and **C**.

5 **D** is a white crystalline solid which dissolves in water to give a colourless solution. Addition of sodium hydroxide solution to a solution of **D** produces a white precipitate, **E**, insoluble in excess sodium hydroxide solution. A flame test on **D** proves negative. Addition of dilute hydrochloric acid and barium chloride solution to a solution of **D** gives a white precipitate, **F**.

   a) Identify **D**, **E** and **F**.

   b) Write equations (full or ionic) for the reactions producing **E** and **F**.

6 **G** is a colourless crystalline solid which reacts with dilute nitric acid to give a colourless solution, **H**, and a colourless, odourless gas, **I**, which turns lime water milky. **G** has a bright orange flame colour.

   a) Identify **G**, **H** and **I**.

   b) Write an equation (full or ionic) for the reaction between **G** and dilute nitric acid.

7 **a)** Identify the substances lettered from **A** to **L** in the flow scheme below. (Don't worry if you don't know anything about electrolysis. This uses electricity to split compounds up. The cathode is the negative electrode and the anode is the positive one. In this case, the electrodes would be made of carbon and attached to a 6 volt battery.)

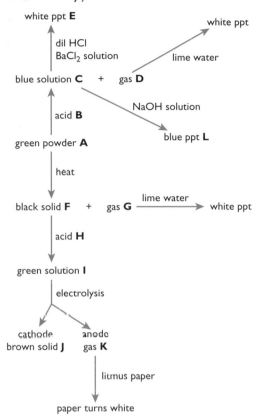

   b) Write equations (full or ionic) for the reactions

   i) between **A** and **B**

   ii) solution **C** and barium chloride solution

   iii) solution **C** and sodium hydroxide solution

   iv) heating **A**

   v) the reaction between **F** and **H**

8 The indigo blue mineral covelline consists of crystals of copper(II) sulphide, CuS. If copper(II) sulphide is heated strongly in a stream of air, it is oxidised to produce black copper(II) oxide and sulphur dioxide.

$$2CuS_{(s)} + 3O_{2(g)} \rightarrow 2CuO_{(s)} + 2SO_{2(g)}$$

You are given a crushed sample of covelline, and have to show that it contains both copper and sulphur. Describe an experiment to carry out the oxidation of covelline so that you can test the gases to show that sulphur dioxide is produced. How would you show that the resulting solid is copper(II) oxide? You should give full experimental details including diagram(s), and explain all the necessary safety precautions you would take.

9   **a)**   Name an instrumental method of analysis which might
           be used to test for:

      i)   the presence of mercury in shellfish

      ii)   the presence of a drug in the urine of an athlete.

   b)   An athlete who tested positive for a banned drug
           claimed that he was entirely innocent. Suggest any two
           factors which might have caused him to be wrongly
           accused. (You might find it helpful to find out about any
           recent cases in the news.)

   c)   Suggest any three reasons why instrumental methods
           of analysis are increasingly commonly used.

Chapter 10: The Periodic Table

Suppose a library containing tens of thousands of books had its books arranged in the sort of heaps that you often find at car-boot sales or jumble sales. Finding what you wanted would be a nightmare. Chemists faced a similar problem. As knowledge of chemical facts grew in the eighteenth and nineteenth centuries, scientists had to find patterns, both to avoid being swamped by the mass of information and also to provide the basis for understanding the facts. This chapter explores some of the patterns that were found.

At this point in the history of Chemistry, atomic weight was measured relative to the mass of a hydrogen atom. For example, a sodium atom weighed 23 times as much as a hydrogen atom.

For the moment we shall persist in using the term "atomic weight" rather than relative atomic mass, because that's what it was called at the time.

The Search for Patterns

ideas
evidence

Arranging books in a well-ordered way makes it easy to find what you want.

Libraries would be impossible to use if they were arranged like this.

Johann Wolfgang Döbereiner

In work beginning in 1817, Döbereiner investigated some groups of three elements with similar chemical properties – for example, lithium, sodium and potassium. He noticed that the atomic weight (what we would now call the relative atomic mass) of the middle one was the average of the other two:

$$\text{lithium: } 7 \qquad \text{sodium: } 23 \qquad \text{potassium: } 39$$
$$23 = (7 + 39) \div 2$$

He found similar patterns with calcium, strontium and barium; with sulphur, selenium and tellurium; and with chlorine, bromine and iodine. These became known as Döbereiner's triads.

At the time, this was no more than a curiosity, but the real advances some 50 years later were again based on patterns in atomic weights.

John Newlands

In 1864–65, Newlands noticed that by arranging the known elements into atomic weight order, elements with similar properties recurred at regular intervals. Every eighth element was similar. He proposed a law which said that "The properties of the elements are a periodic function of their atomic weights".

By drawing an analogy with music, his law became known as the "Law of Octaves."

1	2	3	4	5	6	7
H	Li	Be	B	C	N	O
F	Na	Mg	Al	Si	P	S
Cl	K	Ca	Cr	Ti	Mn	Fe

John Newlands

Dimitri Mendeleev. You may find various spellings of "Mendeleev". Don't worry about this. They are all attempts at converting a Russian name into English.

This works up to a point, but it soon goes wrong. For example, there is no similarity at all between the metal manganese, Mn, and the non-metals nitrogen and phosphorus in Newlands' sixth group.

Newlands' ideas were greeted with ridicule at the time. One of his critics suggested that he might as well have arranged his elements in alphabetical order. With hindsight his theory was bound to fail.

There were lots of undiscovered elements, and some of the known ones had been given the wrong atomic weights. We also know now that the patterns get more complicated when you get beyond calcium.

Dimitri Mendeleev

In 1869, only a few years after Newlands, Mendeleev used the same idea but improved it almost beyond recognition.

He arranged the elements in atomic weight order, but accepted that the pattern was going to be more complicated than Newlands suggested. Where necessary he left gaps in the table to be filled by as yet undiscovered elements. He also had the confidence to challenge existing values for atomic weight where they forced an element into the wrong position in his Periodic Table.

And he was right! One of Mendeleev's early triumphs was to leave a gap underneath silicon in the table for a new element which he called "ekasilicon". He predicted what the properties of this element and its compounds would be. When what we now call germanium was isolated in 1886, it proved to have almost exactly the properties that Mendeleev predicted.

Part of Mendeleev's Table

	Group							
	1	2	3	4	5	6	7	8
Period 1	H							
Period 2	Li	Be	B	C	N	O	F	
Period 3	Na	Mg	Al	Si	P	S	Cl	
Period 4	K	Ca	*	Ti	V	Cr	Mn	Fe Co Ni
	Cu	Zn	*	*	As	Se	Br	
Period 5	Rb	Sr	Y	Zr	Nb	Mo	*	Ru Rh Pd
	Ag	Cd	In	Sn	Sb	Te	I	

You might notice that the noble gases (helium, neon, argon and the rest) are missing – they weren't known at this time. When argon was discovered in 1894, it was obvious that it didn't fit in anywhere in the existing table.

Mendeleev put argon into a new group – Group 0. This suggested that there must be other members of this group, and helium, neon, krypton and xenon were quickly found over the next four years.

There were some residual problems in Mendeleev's Table which had to be left for later. The atomic weight order wasn't *exactly* followed.

For example, tellurium, Te, has a slightly higher atomic weight than iodine – but iodine obviously has to go where it is in the Table because its properties are very similar to chlorine and bromine.

Group 6	Group 7
O	F
S	Cl
Cr / Se	Mn / Br
Mo / I	* / Te

correct order in terms
of atomic weight

Group 6	Group 7
O	F
S	Cl
Cr / Se	Mn / Br
Mo / Te	* / I

Mendeleev swapped these so that their
properties matched the rest of the group

Another similar problem occurs when you come to slot argon into the Table.

Argon's atomic weight is slightly higher than potassium's. Putting the elements in their correct order of increasing atomic weight would put argon into the same group as lithium and sodium, and potassium in with gases like neon. That's obviously silly!

The need for sub-groups (starting in Period 4) is also awkward. For example, lithium, sodium, potassium and rubidium are described as being in Group 1a; copper and silver are in Group 1b. There's not much resemblance between the two sub-groups.

These problems are completely overcome in the modern version of the Periodic Table.

Group 0	Group 1
He	Li
Ne	Na
K	Ar

This is what the table would look like if potassium and argon were put in the order of their atomic weights.

The Modern Periodic Table

In the modern Periodic Table, the elements are arranged in order of atomic number – not of relative atomic mass. That solves the problems associated with the positions of tellurium and iodine, and argon and potassium. Stretching the Table to keep the transition elements separate gets rid of the problem of having sub-groups.

Now that we are back in the modern day, it is appropriate to use the proper term "relative atomic mass". You will find this explained in full on pages 261–262.

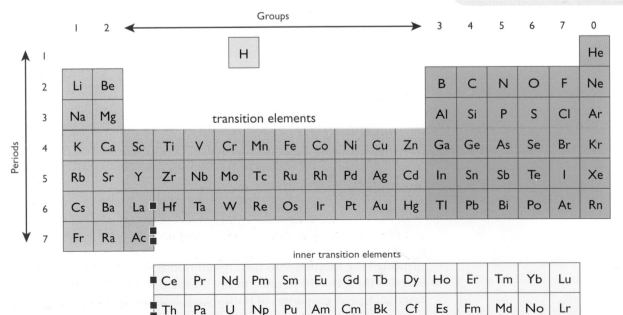

The Periodic Table.

For a larger version of the Periodic Table including atomic numbers and other information, see page 306.

The inner transition elements are usually dropped out of their proper places, and written separately at the bottom of the Periodic Table. The reason for this isn't very subtle. If you put them where they should be, everything has to be drawn slightly smaller to fit on the page. That makes it more difficult to read.

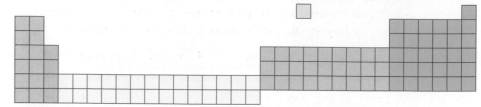

Full width Periodic Table.

Most GCSE Periodic Tables stop at the end of the second inner transition series. Some continue beyond that to include a small number of more recently discovered elements.

New elements are still being discovered, although they are highly radioactive and have incredibly short existences. Knowing where they are in the Periodic Table, however, means that you can make good predictions about what their properties would be.

The Periodic Table and atomic structure

Now would be a good time to revise electronic structures on pages 3-6.

Remember that the atomic number counts the number of protons in the atoms of the element, and therefore the number of electrons in the neutral atom. Elements in the same Group have the same number of electrons in their outer energy levels (shells). That governs how they react and means that they are likely to have similar chemical properties.

There is, however, a change in properties (sometimes gradual, sometimes quite rapid) from the top to the bottom of a Group. You will find examples later in this chapter.

Metals and non-metals in the Periodic Table

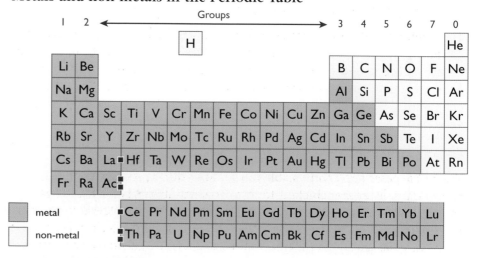

Although the division into metals and non-metals is shown as clear-cut, in practice there is a lot of uncertainty on the dividing line. For example, arsenic, As, has properties of both metals and non-metals. Even tin, Sn, is close enough to the boundary that it is possible to find some non-metallic character hidden away in its properties.

Group 0 – The Noble Gases

Physical properties

The noble gases are all colourless gases. Radon, at the bottom of the Group, is radioactive. Argon makes up almost 1% of the air. Helium is the second lightest gas (after hydrogen). Radon is nearly 8 times denser than air.

All the gases are **monatomic**. That means that their molecules consist of single atoms.

Their densities and boiling points illustrate typical patterns (trends) in physical properties as you go down a Group in the Periodic Table.

He
Ne
Ar
Kr
Xe
Rn

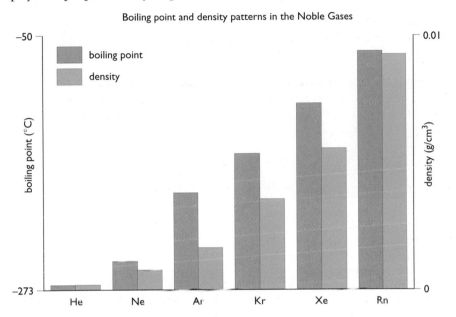

Boiling point and density patterns in the noble gases.

The density increases as the atoms get heavier.

The boiling points also increase as you go down the Group. This is because the attractions between one molecule and its neighbours get stronger as the atoms get bigger. More energy is needed to break the stronger attractions. In helium, these intermolecular attractions are very, very weak. Very little energy is needed to break these attractions, and so helium's boiling point is very low.

Chemical reactivity

The noble gases don't form stable ions, and so don't produce ionic compounds. They are reluctant to form covalent bonds because in most cases it costs too much energy to rearrange the full energy levels to produce the single electrons that an atom needs if it is to form simple covalent bonds by sharing electrons. That means that these gases are generally unreactive.

Until the 1960s scientists thought that the noble gases were completely unreactive. Then they found that they could make xenon combine with fluorine just by heating the two together. After that a number of other compounds of both xenon and krypton were found – mainly combined with fluorine and oxygen.

Summarising the main features of the Group 0 elements

Group 0 elements:

- are colourless, monatomic gases
- become denser as you go down the Group
- have higher boiling points as you go down the Group
- are generally unreactive.

Uses of the noble gases

Helium is used in weather balloons because it is less dense than air, and safer than hydrogen.

Helium is also used in mixtures with oxygen for deep sea diving. For deep diving, this is safer than using oxygen by itself, or using air.

Argon is used to fill light bulbs because the hot filament won't react with it.

Some of the noble gases are used in advertising lighting. Neon, for example, produces a bright red colour.

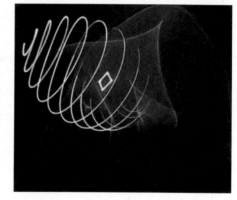

A light show using krypton and argon lasers.

Li
Na
K
Rb
Cs
Fr

Group 1 – The Alkali Metals

This Group contains the familiar reactive metals sodium and potassium, as well as some less common ones.

Francium at the bottom of the Group is radioactive. One of its isotopes is produced during the radioactive decay of uranium-235, but is extremely short-lived. Once you know about the rest of Group 1 you can predict what francium would be like, but you can't realistically observe its properties. We will make those predictions later.

Physical properties

	Melting point (°C)	Boiling point (°C)	Density (g/cm³)
Li	181	1342	0.53
Na	98	883	0.97
K	63	760	0.86
Rb	39	686	1.53
Cs	29	669	1.88

Nobody is expecting you to remember these values! They are here just to show the patterns. You *will* be expected to know those patterns.

Notice that the melting and boiling points of the elements are very low for metals, and get lower as you go down the Group.

Their densities tend to increase – although not as tidily as the noble gases. Lithium, sodium and potassium are all less dense than water, and so will float on it.

The metals are also very soft and are easily cut with a knife, becoming softer as you go down the Group. They are shiny and silver when freshly cut, but tarnish within seconds on exposure to air.

Storage and handling

All these metals are extremely reactive, and get more reactive as you go down the Group. They all react quickly with air to form oxides, and anywhere between rapidly and violently with water to form strongly alkaline solutions of the metal hydroxides.

To stop them reacting with oxygen or water vapour in the air, lithium, sodium and potassium are stored under oil. If you look carefully at the photograph, you will see traces of bubbles in the beaker containing the sodium. There must have been a tiny amount of water present in the oil that the sodium was placed in.

Rubidium and caesium are so reactive that they have to be stored in sealed glass tubes to stop any possibility of oxygen getting at them.

Great care must be taken not to touch any of these metals with bare fingers. There could be enough sweat on your skin to give a reaction producing lots of heat and a very corrosive metal hydroxide.

Lithium, sodium and potassium have to be kept in oil to stop them reacting with oxygen in the air.

The reactions with water

All of these metals react with water to produce a metal hydroxide and hydrogen.

metal + cold water → metal hydroxide + hydrogen

The main difference between the reactions is how fast they happen. The reaction between sodium and water is typical.

Strictly speaking, most of the time the sodium is reacting, it is present as molten sodium – not solid sodium. Writing (l) for the state symbol, though, has the potential for confusing an examiner, and is probably best avoided!

Sodium

$$2Na_{(s)} + 2H_2O_{(l)} \rightarrow 2NaOH_{(aq)} + H_{2(g)}$$

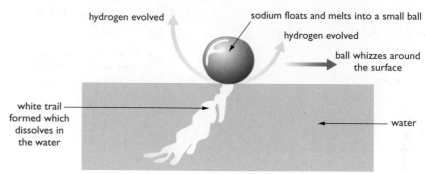

hydrogen evolved

sodium floats and melts into a small ball

hydrogen evolved

ball whizzes around the surface

white trail formed which dissolves in the water

water

The sodium floats because it is less dense than water. It melts because its melting point is low and lots of heat is produced by the reaction. Because the hydrogen isn't given off symmetrically around the ball, the sodium is pushed around the surface of the water – literally like a jet-propelled hovercraft.

The white trail formed is the sodium hydroxide which dissolves to make a strongly alkaline solution.

Lithium

$$2Li_{(s)} + 2H_2O_{(l)} \rightarrow 2LiOH_{(aq)} + H_{2(g)}$$

The reaction is very similar to sodium's reaction, except that it is slower. Lithium's melting point is higher and the heat isn't produced so quickly, so the lithium doesn't melt.

Potassium

$$2K_{(s)} + 2H_2O_{(l)} \rightarrow 2KOH_{(aq)} + H_{2(g)}$$

Potassium's reaction is faster than sodium's. Enough heat is produced to ignite the hydrogen which burns with a lilac flame. The reaction often ends with the potassium spitting around.

The lilac colour is due to contamination of the normally blue hydrogen flame by potassium compounds.

Rubidium and caesium

These react even more violently than potassium. Both of them sink in the water and so they don't have the opportunity to move around the surface, losing excess heat. Caesium tends to react explosively. The heat produced instantly vaporises all the water around it.

As you go down the Group, the metals become more reactive.

Explaining the increase in reactivity

In all of these reactions, the metal atoms are losing electrons and forming metal ions in solution. For example

$$Na_{(s)} \rightarrow Na^+_{(aq)} + e^-$$

The electrons released by the metal are gained by the water molecules, producing hydroxide ions and hydrogen.

$$2H_2O_{(l)} + 2e^- \rightarrow 2OH^-_{(aq)} + H_{2(g)}$$

You are unlikely to need this part of the reaction for GCSE purposes. It is included here just for completeness.

The differences between the reactions depend in part on how easily the outer electron of the metal is lost in each case. That depends on how strongly it is attracted to the nucleus in the original atom. Remember that the nucleus of an atom is positive because it contains protons, and so attracts the negative electrons.

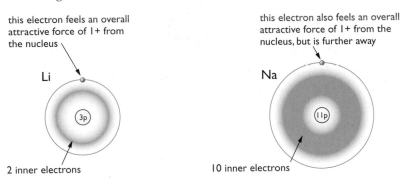

this electron feels an overall attractive force of 1+ from the nucleus

Li

2 inner electrons

this electron also feels an overall attractive force of 1+ from the nucleus, but is further away

Na

10 inner electrons

In every single atom in the elements of this Group, the outer electron will feel an overall attractive force of 1+ from the nucleus, but the effect of the force falls very quickly as distance increases. The bigger the atom, the more easily the outer electron is lost.

Compounds of the alkali metals

All Group 1 metal ions are colourless. That means that their compounds will be colourless or white unless they are combined with a coloured negative ion. Potassium dichromate(VI) is orange, for example, because the dichromate(VI) ion is orange.

The compounds are typical ionic solids and are mostly soluble in water. Some lithium compounds (like lithium carbonate, for example) are rather less soluble.

You don't normally come across bottles containing the oxides of the metals in the lab. They react instantly with water (including water vapour in the air) to form hydroxides. For example:

$$Na_2O_{(s)} + H_2O_{(l)} \rightarrow 2NaOH_{(aq)}$$

The hydroxides all form strongly alkaline solutions. This is the origin of the name "alkali metals".

Summarising the main features of the Group 1 elements

Group 1 elements:

- are metals
- are soft, with melting points and densities which are very low for metals
- react rapidly with air to form coatings of the metal oxide
- react with water to produce the metal hydroxide and hydrogen
- increase in reactivity as you go down the Group
- form compounds in which the metal has a 1+ ion
- have oxides which react with water to produce soluble alkaline hydroxides
- have white compounds which dissolve to produce colourless solutions.

Potassium dichromate(VI) solution is orange because the dichromate(VI) ion is orange.

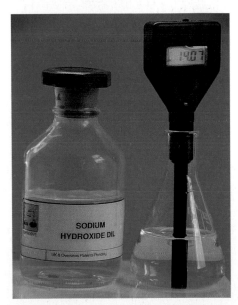

Sodium hydroxide solution is strongly alkaline.

Remember that francium is highly radioactive and short-lived. You couldn't actually observe any of these things.

Using these features to predict the properties of francium

If you followed the trends in Group 1, you could predict that:

- Francium is very soft with a melting point just above room temperature.

- Its density is probably just over 2 g/cm^3.

- Francium will be a silvery metal but will tarnish almost instantly in air.

- It would react violently with water to give francium hydroxide and hydrogen.

- Francium hydroxide solution will be strongly alkaline.

- Francium oxide will react with water to give francium hydroxide.

- Francium compounds are white and dissolve in water to give colourless solutions.

F
Cl
Br
I
At

Group 7 – The Halogens

The name "halogen" means "salt-producing". When they react with metals these elements produce a wide range of salts like calcium fluoride, sodium chloride, potassium iodide and so on.

The halogens are non-metallic elements with diatomic molecules – F_2, Cl_2 and so on. As the molecules get larger towards the bottom of the Group, the melting and boiling points increase. Fluorine and chlorine are gases. Bromine is a liquid which turns to vapour very easily, and iodine is a solid.

Astatine is radioactive and is formed during the radioactive decay of other elements like uranium and thorium. Most of its isotopes are so unstable that their lives can be measured in seconds or fractions of a second.

	State	Colours
F_2	gas	yellow
Cl_2	gas	green
Br_2	liquid	dark red liquid – red/brown vapour
I_2	solid	dark grey solid – purple vapour

Chlorine, bromine and iodine.

Iodine has a purple vapour.

Because the halogens are non-metals, they will be poor conductors of heat and electricity. When they are solid (for example, iodine at room temperature), their crystals will be brittle and crumbly.

Safety

Fluorine is so dangerously reactive that you would never expect to come across it in a school lab.

Apart from any safety problems associated with the reactivity of the elements (especially fluorine and chlorine), all the elements have extremely poisonous vapours and have to be handled in a fume cupboard.

Liquid bromine is also very corrosive and great care has to be taken to keep it off the skin. It is a good idea to have a beaker of dilute sodium thiosulphate solution handy whenever you use bromine. This reacts at once with any bromine you might have got on your skin or the bench, without being particularly harmful itself.

Reactions with hydrogen

The halogens react with hydrogen to form **hydrogen halides** – hydrogen fluoride, hydrogen chloride, hydrogen bromide and hydrogen iodide.
For example:

$$H_{2(g)} + Br_{2(g)} \rightarrow 2HBr_{(g)}$$

The hydrogen halides are all steamy, acidic, poisonous gases. In common with all the compounds formed between the halogens and non-metals, the gases are covalently bonded. They are very soluble in water, reacting with it to produce solutions of acids. For example, hydrochloric acid is a solution of hydrogen chloride in water.

The reactivity falls significantly as you go down the Group.

> Notice that you would react hydrogen with bromine vapour – not liquid bromine.

Halogen	Reaction with hydrogen
F_2	violent explosion, even in the cold and dark
Cl_2	violent explosion if exposed to a flame or to sunlight
Br_2	mild explosion if a bromine vapour/hydrogen mixture is exposed to a flame
I_2	partial reaction to form hydrogen iodide if iodine vapour is heated continuously with hydrogen

Reactions with metals

With iron

If bromine or iodine vapour are passed over hot iron, the reactions look similar to the chlorine reaction, but the reactivity again falls as you go down the Group. For example, the iron burns more brightly with chlorine than it does with bromine vapour. If you used fluorine, the iron would burn even more brightly still. Notice that the less reactive iodine forms iron(II) iodide, FeI_2, rather than iron(III) iodide, FeI_3.

> The reaction between **chlorine** and **iron** is described on pages 83–84. You should refer back to this before you go on.

> For the reason for this, see page 111.

Halogen	Product	Equation
F_2	iron(III) fluoride	$2Fe_{(s)} + 3F_{2(g)} \rightarrow 2FeF_{3(s)}$
Cl_2	iron(III) chloride	$2Fe_{(s)} + 3Cl_{2(g)} \rightarrow 2FeCl_{3(s)}$
Br_2	iron(III) bromide	$2Fe_{(s)} + 3Br_{2(g)} \rightarrow 2FeBr_{3(s)}$
I_2	iron(II) iodide	$Fe_{(s)} + I_{2(g)} \rightarrow FeI_{2(s)}$

The reaction between sodium and chlorine

Sodium burns in chlorine with its typical orange flame to produce white, solid sodium chloride.

$$2Na_{(s)} + Cl_{2(g)} \rightarrow 2NaCl_{(s)}$$

Sodium chloride is, of course, an ionic solid. Typically, when the halogens react with metals from Groups 1 and 2, they form ions.

It is useful to look at this from the point of view of the sodium and of the chlorine separately by writing the ionic equation for the reaction:

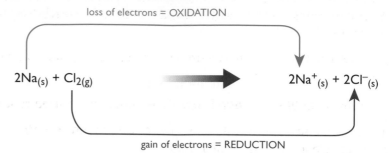

loss of electrons = OXIDATION

$$2Na_{(s)} + Cl_{2(g)} \longrightarrow 2Na^+{}_{(s)} + 2Cl^-{}_{(s)}$$

gain of electrons = REDUCTION

The sodium has lost electrons and so has been oxidised to sodium ions. This means that chlorine is acting as an **oxidising agent**. That is typical of the reactions of chlorine. It is a strong oxidising agent.

Exploring another reaction of chlorine as an oxidising agent

If you add chlorine solution ("chlorine water") to iron(II) chloride solution, the very pale green solution becomes yellowish.

The iron(II) chloride has been converted into iron(III) chloride.

$$2FeCl_{2(aq)} + Cl_{2(aq)} \rightarrow 2FeCl_{3(aq)}$$

Adding some sodium hydroxide solution to a sample of the product gives an orange-brown precipitate, showing that the solution now contains iron(III) ions. If you still had iron(II) ions present, you would have got a green precipitate. This is explored on page 90.

The ionic equation shows that this is another example of chlorine acting as an oxidising agent.

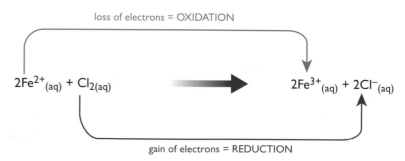

loss of electrons = OXIDATION

$$2Fe^{2+}{}_{(aq)} + Cl_{2(aq)} \longrightarrow 2Fe^{3+}{}_{(aq)} + 2Cl^-{}_{(aq)}$$

gain of electrons = REDUCTION

Any other iron(II) salt would be oxidised in exactly the same way.

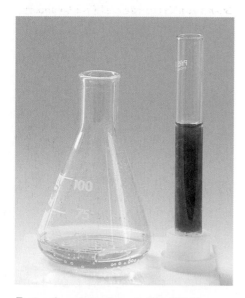

Sodium burning in chlorine to produce sodium chloride.

If you aren't sure about ionic equations and oxidation and reduction, you should read pages 54–56.

Testing the solution formed when chlorine reacts with iron(II) chloride solution.

This is the reason that iron(III) chloride is formed when you pass chlorine over hot iron. Chlorine is a powerful enough oxidising agent to remove three electrons from the iron and oxidise it all the way to iron(III) chloride.

Iodine, on the other hand, is a much weaker oxidising agent. It is powerful enough to remove two electrons from the iron to make iron(II) ions, but won't oxidise iron(II) ions onwards to iron(III) ions. This is why iron(II) iodide is formed when iodine vapour is passed over hot iron.

Displacement reactions involving the halogens

Just as you can use the Reactivity Series of metals to make sense of their displacement reactions, so you can also use a corresponding Reactivity Series for the halogens.

We shall concentrate on the three commonly used halogens, but the trend continues for the rest of the Group.

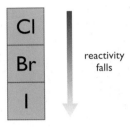

Reacting chlorine with potassium bromide or potassium iodide solutions

If you add chlorine solution to colourless potassium bromide solution, the solution becomes orange as bromine is formed.

$$2KBr_{(aq)} + Cl_{2(aq)} \rightarrow 2KCl_{(aq)} + Br_{2(aq)}$$

The more reactive chlorine has displaced the less reactive bromine from potassium bromide.

Similarly, adding chlorine solution to potassium iodide solution gives a dark red solution of iodine. If an excess of chlorine is used, you may get a dark grey precipitate of iodine.

$$2KI_{(aq)} + Cl_{2(aq)} \rightarrow 2KCl_{(aq)} + I_{2(aq \ or \ s)}$$

Both of these are redox reactions. In each case, the chlorine is acting as an oxidising agent.

Bromine and iodine displaced from potassium bromide and potassium iodide solutions.

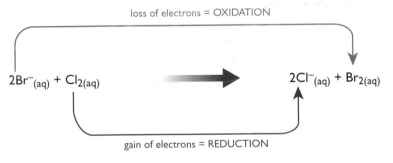

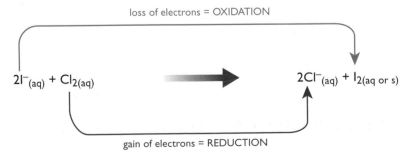

Small amounts of an organic solvent are sometimes added to these reactions. The bromine or iodine produced dissolve in the solvents and give strong colours which can make them easier to recognise.

The potassium ions are spectator ions. The reaction would be the same with any soluble bromide or iodide.

Bromine displaces iodine from potassium iodide solution.

The reaction of bromine with potassium iodide solution

In exactly the same way, the more reactive bromine displaces the less reactive iodine from potassium iodide solution.

Adding bromine solution ("bromine water") to colourless potassium iodide solution gives a dark red solution of iodine (or a dark grey precipitate if an excess of bromine is added).

$$2KI_{(aq)} + Br_{2(aq)} \rightarrow 2KBr_{(aq)} + I_{2(aq\ or\ s)}$$

The ionic equation shows bromine acting as an oxidising agent.

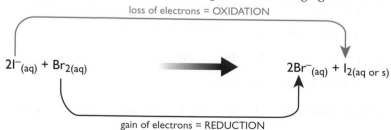

loss of electrons = OXIDATION

$$2I^-_{(aq)} + Br_{2(aq)} \qquad\qquad 2Br^-_{(aq)} + I_{2(aq\ or\ s)}$$

gain of electrons = REDUCTION

Explaining the trend in the reactivity of the halogens

As you go down the Group, the oxidising ability of the halogens falls.

When a halogen oxidises something, it does so by removing electrons from it.

Each halogen has the ability to oxidise the ions of those underneath it in the Group, but not those above it. Chlorine can remove electrons from bromide or iodide ions, and bromine can remove electrons from iodide ions.

The reason has to do with how easily the atoms gain electrons to make their ions.

Chlorine is a strong oxidising agent because its atoms readily attract an extra electron to make chloride ions. Bromine is less successful at attracting electrons, and iodine is less successful still.

You have to consider the amount of attraction the incoming electron feels from the nucleus. In chlorine, there are 17 positively charged protons offset by the 10 negatively charged electrons in the inner energy levels (shells). That means that the new electron feels an overall pull of 7+ from the centre of the atom.

A similar argument with bromine shows that the new electron also feels an overall pull from the nucleus of 7+, but in the bromine case, it is further away.

Warning! This argument falls down if you try to include fluorine . Fluorine atoms don't in fact accept electrons more readily than chlorine atoms do – for reasons which are beyond GCSE. The greater reactivity of fluorine has to be explained in a more complicated way. However, if your examiners ask you to compare fluorine and chlorine you will have to use the explanation described here – even though it would be wrong!

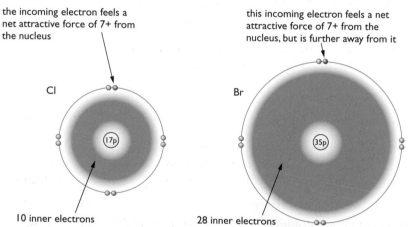

the incoming electron feels a net attractive force of 7+ from the nucleus

this incoming electron feels a net attractive force of 7+ from the nucleus, but is further away from it

Cl

Br

(17p)

(35p)

10 inner electrons

28 inner electrons

The incoming electron is further and further from the nucleus as you go down the Group, and so it is less strongly attracted. That means the ion is less easily formed.

It also means that it is easier to remove the extra electron from, say, a bromide ion than from a chloride ion. Bromide ions are more easily oxidised than chloride ions. Iodide ions hold the extra electron weakly enough that they are very easily oxidised.

Summarising the main features of the Group 7 elements

Group 7 elements:

- have diatomic molecules, X_2
- go from gases to liquid to solid as you go down the Group
- have coloured poisonous vapours
- form ionic salts with metals and molecular compounds with non-metals
- become less reactive towards the bottom of the Group
- are oxidising agents, with oxidising ability decreasing down the Group
- will displace elements lower down the Group from their salts.

Uses of the halogens and their compounds

Fluorine

Fluorides such as sodium fluoride are added to toothpaste or sometimes to drinking water supplies to help prevent tooth decay.

There is a lot of controversy about this. A study by the University of York in September 2000 showed that much of the research in this area wasn't very good. By comparing the results of lots of research projects they found that fluoridation of water supplies reduced dental decay by an average of about 15%. There was a useful effect, even for those people already using fluoride toothpaste.

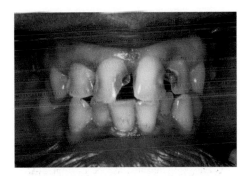

Fluorides in toothpaste or drinking water help to prevent tooth decay.

However, there was a marked increase in dental fluorosis. This is a mottling of the teeth caused by absorption of fluorine. In about 1 in 8 people this was enough to be thought visually unattractive.

There was no evidence for any effects on bone structure or development, and no evidence that fluoridation causes cancer. On the other hand, the studies were largely short term, and there could be hidden long term effects.

Many people object strongly to having medication forced on them via the water supply, whether they want it or not.

Chlorine

Chlorine is added to the water supply in very small quantities to kill bacteria.

Chlorine is also a powerful bleach. A solution of chlorine in water ("chlorine water") could be used as a bleach, but would be dangerous because of the free chlorine. Household bleach is a chlorine compound, sodium chlorate(I), NaClO.

The manufacture of chlorine and some other uses are discussed in Chapter 15.

Chlorine is used to sterilise water.

Iodine

A dilute solution of iodine in potassium iodide solution is used as an antiseptic.

Silver halides

If you are interested, you might like to do some research on what happens during the formation of the negative and the final print. You might also like to find out how colour images are formed.

Three of the silver halides (silver chloride, silver bromide and silver iodide) are important in photography.

For example, when silver bromide is exposed to light, a redox reaction happens in which silver ions gain electrons from the bromide ions. This is called a **photochemical reaction** because it involves light.

Small amounts of metallic silver are formed.

$$2Ag^+Br^-_{(s)} \rightarrow 2Ag_{(s)} + Br_{2(g)}$$

The presence of the silver leaves a faint image on the photographic film or paper. This image is strengthened during the process of developing the film or paper.

X-rays and radiation from radioactive substances affect photographic film or paper in the same way.

Transition Metals

The bold type simply picks out some of the more familiar transition elements. You may not have heard of many of the inner transition elements except uranium and plutonium.

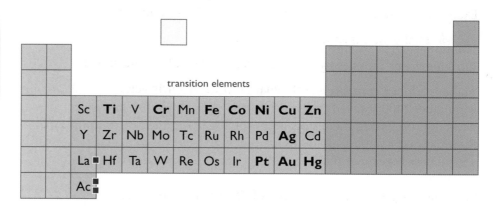

transition elements

Sc	**Ti**	V	**Cr**	Mn	**Fe**	**Co**	**Ni**	**Cu**	**Zn**
Y	Zr	Nb	Mo	Tc	Ru	Rh	Pd	**Ag**	Cd
La ▪	Hf	Ta	W	Re	Os	Ir	**Pt**	**Au**	**Hg**
Ac ▪									

inner transition elements

▪	Ce	Pr	Nd	Pm	Sm	Eu	Gd	Tb	Dy	Ho	Er	Tm	Yb	Lu
▪	Th	Pa	**U**	Np	**Pu**	Am	Cm	Bk	Cf	Es	Fm	Md	No	Lr

If you aren't sure about typical metal properties, read pages 17–18 and pages 22–23.

These are all typically metallic elements. They are good conductors of heat and electricity, workable, strong, and mostly with high densities. With the exception of liquid mercury, Hg, they have melting points which range from fairly high to very high.

They are much less reactive than the metals in Groups 1 and 2, and so they don't react as rapidly with air or water.

Because of their useful physical properties and relative lack of reactivity, several of the transition elements are important in everyday life.

You will find examples of uses of titanium, iron, nickel and copper in Chapter 14.

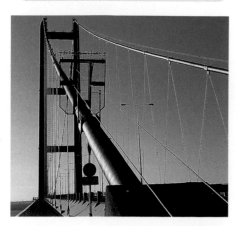

Iron (as steel) is cheap, strong and easily worked.

Transition metals form coloured compounds

The majority of transition metal compounds are coloured. The photographs show some examples, both in and out of the lab.

The blue on this Yuan Dynasty vase was created using cobalt compounds in the glaze.

Solutions of some common transition metal compounds.

Transition metals and their compounds are often useful catalysts

Examples you will come across elsewhere in this book include:

- Iron in the manufacture of ammonia.
- Vanadium(V) oxide, V_2O_5, in the manufacture of sulphuric acid.
- Platinum and rhodium in the manufacture of nitric acid.
- Platinum and palladium in catalytic converters for cars.
- Nickel in the manufacture of margarine.
- Manganese(IV) oxide, MnO_2, in the decomposition of hydrogen peroxide.

Weathering on the copper covered spire of Truro Cathedral.

Summarising the main features of the transition elements

Transition elements:

- are the group of elements found in the centre of the Periodic Table
- are metals with a high melting point (except mercury)
- are hard, tough and strong
- are much less reactive than Group 1 metals
- have many uses in everyday life
- form coloured compounds
- are often useful catalysts.

End of Chapter Checklist

If you haven't got a copy of your specification, read the introduction on page vi.

You will need to be able to do some or all of the following. Check your Awarding Body's specification (syllabus) to find out exactly what you need to know.

- Understand the contributions made by Döbereiner, Newlands and Mendeleev to the development of the Periodic Table.

- Know that in the modern Periodic Table the 100 or so elements are arranged in order of their atomic (proton) numbers.

- Understand the use of the terms *group*, *period* and *transition element* (*metal*).

- Determine whether an element is a metal or a non-metal from its position in the Periodic Table.

- Know that the Group 0 elements are generally unreactive gases, and be able to explain that lack of reactivity.

- Know the trends in melting and boiling points and in density for the Group 0 gases.

- State some uses of the Group 0 gases.

- Know the trends in physical properties for the Group 1 metals, and the precautions that have to be taken when using them.

- Describe and explain the reactions of the Group 1 metals with water.

- Know that reactivity increases as you go down Group 1, and be able to explain that increase in reactivity.

- Know that compounds of the Group 1 metals are normally white (or colourless) soluble solids.

- Know the trends in physical properties for the Group 7 elements (the halogens), and the precautions that have to be taken when using them.

- Describe and explain the reactions of the halogens with hydrogen and metals, and in displacement reactions involving other members of the Group.

- Know that the halogens become less reactive as you go down the Group, and be able to explain that fall in reactivity for chlorine, bromine and iodine.

- Describe some uses of the halogens and their compounds.

- Know that the transition elements are typical metals and are less reactive than Group 1 metals.

- Know that the transition elements form coloured compounds and that the metals or their compounds may act as catalysts.

Questions

You will need to use the Periodic Table on page 306. More questions on the Periodic Table can be found at the end of Section B on page 142.

I Answer the questions which follow using *only* the elements in this list:

caesium, chlorine, molybdenum, neon, nickel, nitrogen, strontium, tin

a) Name an element which is in

   i) Group 2

   ii) the same period as silicon

   iii) the same group as phosphorus.

b) How many electrons are there in the outer levels of atoms of: *i)* strontium, *ii)* chlorine, *iii)* nitrogen?

c) Divide the list of elements at the beginning of the question into metals and non-metals.

d) Name the two elements which are likely to have the greatest number of coloured compounds.

e) Name *i)* the most reactive metal, *ii)* the least reactive element.

2 This question concerns the chemistry of the elements Li, Na, K, Rb and Cs on the extreme left-hand side of the Periodic Table. In each case, you should identify the substances represented by letters.

a) **A** is the lightest of all metals.

b) When metal **B** is dropped onto water it melts into a small ball and rushes around the surface. A gas **C** is given off and this burns with a lilac flame. A white trail dissolves into the water to make a solution of **D**.

c) When metal **E** is heated in a green gas **F**, it burns with an orange flame and leaves a white solid product **G**.

d) Write equations for:

   i) the reaction of **B** with water

   ii) the reaction between **E** and **F**.

e) What would you expect to see if solution **D** was tested with red and blue litmus paper?

f) Explain why **B** melts into a small ball when it is dropped onto water.

3 This question is about astatine, At, at the bottom of Group 7 of the Periodic Table. Astatine is radioactive, and extremely rare. You are asked to make some predictions about astatine and its chemistry.

a) Showing only the outer electrons, draw dot-and-cross diagrams to show the arrangement of electrons in an astatine *atom*, an astatide *ion* and an astatine *molecule*.

b) What physical state would you expect astatine to be in at room temperature?

c) Would you expect astatine to be more or less reactive than iodine?

d) Describe hydrogen astatide, and suggest a likely pH for a reasonably concentrated solution of it in water.

e) What would you expect caesium astatide to look like? Will it be soluble in water? Explain your reasoning.

f) Write an ionic equation for the reaction which will occur if you add chlorine water to a solution of sodium astatide. Assume that astatine is insoluble in water. Explain clearly why this reaction would be counted as a redox reaction.

4 Predict any *five* properties of the element palladium, Pd (atomic number 46) or its compounds. The properties can either be physical or chemical.

5 Explain as fully as you can why:

a) neon is an unreactive monatomic gas

b) potassium is more reactive than sodium

c) chlorine displaces bromine from potassium bromide solution.

6 In this question you will be given some information about three elements. You should say, with reasons, where you might expect to find them in the Periodic Table. You can choose between:

   Group 1 or 2 element

   Transition element

   Group 7 element

   Noble gas

a) Element **A**

Melting point (°C)	−112
Boiling point (°C)	−108
Density at 0°C (g/cm³)	0.0059
Reaction with chlorine	none
Reaction with oxygen	none
Reaction with sodium	none

b) Element **B**

   B melts at 1890°C and is a good conductor of electricity. It has no reaction with cold water, but will react with chlorine on heating. It reacts very slowly with dilute hydrochloric acid and dilute sulphuric acid. Compounds of **B** are highly coloured - including blue, green, purple, orange and yellow.

c) Element **C**

Melting point (°C)	850
Boiling point (°C)	1487
Reaction with cold water	steady production of hydrogen
Reaction with dilute hydrochloric acid	very vigorous reaction producing hydrogen and a colourless solution
Reaction with oxygen	burns to give a white solid

7 Within the same decade (the 1860s) Newlands and Mendeleev both tried to find patterns in the chemistry of the elements by arranging them in order of increasing atomic weight (relative atomic mass). Newlands met with ridicule, but Mendeleev created the basis for the modern Periodic Table. Explain briefly the differences between the two attempts which explain why Mendeleev's Table was accepted whereas Newlands' wasn't.

8 By doing an internet (or other) search, find out why mixtures of helium and oxygen are used for deep diving rather than using compressed air or pure oxygen. Your account should not exceed 250 words, and should be written using a word processor, including at least one diagram or picture.

9 Suppose a public enquiry is to be held in your local area to decide whether fluorides should be added to the water supply to help prevent tooth decay. Write two short speeches (each to last as close as possible to 2 minutes), one in favour of fluoridation and one opposed to it. (Hint: The only way of judging the length is to write a bit and then say it out loud as you would if you were delivering the speech. Time yourself and then adjust the length accordingly.)

Chapter 11: Electrolysis

The photograph shows what happens if you connect a solution of potassium iodide into a simple electrical circuit. If you look at what is happening in the solution, you can see obvious signs of chemical change. Some coloured substance is being produced at the positive carbon electrode, and a gas is being given off at the negative electrode. This chapter explores the effect of electricity on chemical compounds. Examples of applications of this in industry will be found in later chapters.

For the reason that metals conduct electricity, see page 22. For the reason that graphite conducts electricity, see pages 26–27. The carbon rods used are essentially made up of very, very tiny graphite crystals.

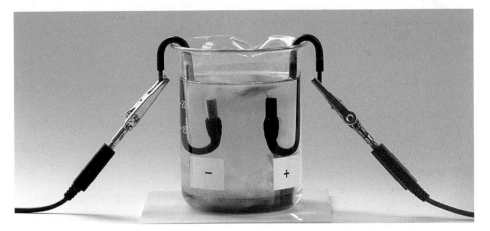

Part of an electric circuit involving conducting metals, carbon and a solution of potassium iodide.

Some Important Background

The conduction of electricity by metals and carbon

In a metal or carbon, electricity is simply a flow of electrons. The movement of the electrons doesn't produce any chemical change in the metal or the carbon.

Metals and carbon contain mobile electrons, and it is these that move. That's equally true even for a liquid metal like mercury. In an electrical circuit, you can think of a battery or a power pack as an "electron pump", pushing the electrons through the various bits of metal or carbon.

Passing electricity through compounds – electrolysis

Hardly any solid compounds conduct electricity. On the other hand, lots of compounds will conduct electricity when they are molten or when they are dissolved in water. All of these show signs of a chemical reaction while they are conducting.

Defining electrolysis

Electrolysis is a chemical change caused by passing an electric current through a compound which is either molten or in solution.

The reason that some compounds undergo electrolysis will be explored in the rest of this chapter.

Some other important words

An **electrolyte** is a substance which undergoes electrolysis. Electrolytes all contain ions. As you will see shortly, the movement of the ions is responsible for both the conduction of electricity and the chemical changes that take place.

In a solid electrolyte, the ions aren't free to move, and so nothing happens.

The electricity is passed into and out of the electrolyte via two **electrodes**. Carbon is frequently used for electrodes because it conducts electricity and is chemically fairly inert. Platinum is also fairly inert and can be used instead of carbon. Various other metals are sometimes used as well.

The positive electrode is called the **anode**. The negative electrode is called the **cathode**.

Remember **PANC** – positive anode, negative cathode.

The Electrolysis of Molten Compounds

Electrolysing molten anhydrous zinc chloride, $ZnCl_2$

The power supply can be a 6 volt battery or a power pack. It doesn't matter which. The voltage isn't very critical either.

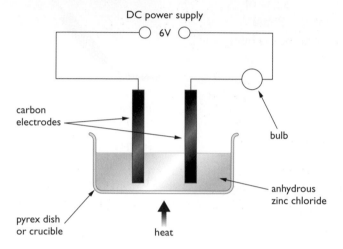

Nothing at all happens while the zinc chloride is solid, but as soon as it melts:

• The bulb lights up, showing that electrons are flowing through it.

• There is bubbling around the electrode (the anode) connected to the positive terminal of the power source as chlorine is given off.

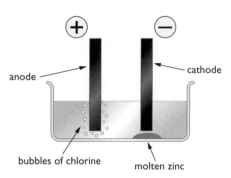

Showing what happens when the zinc chloride melts. The bulb also lights up.

• Nothing seems to be happening at the electrode (the cathode) connected to the negative terminal of the power source, but afterwards metallic zinc is found underneath it.

• When you stop heating and the zinc chloride solidifies again, everything stops – there is no more bubbling and the bulb goes out.

Explaining what is happening

Zinc chloride is an ionic compound. The solid consists of a giant structure of zinc ions and chloride ions packed regularly in a crystal lattice. It doesn't have any mobile electrons, and the ions are locked tightly in the lattice and aren't free to move. The solid zinc chloride doesn't conduct electricity.

If you aren't happy about giant ionic structures, read pages 23–25. Because zinc chloride is $ZnCl_2$, its structure will be more complicated than that of sodium chloride, $NaCl$.

As soon as the solid melts, the ions *do* become free to move around, and it is this movement which enables the electrons to flow in the external circuit. This is how it works...

As soon as you connect the power source, it pumps any mobile electrons away from the left-hand electrode towards the right-hand one as we've drawn it in the diagram.

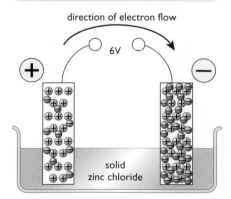

The excess of electrons on the right-hand electrode makes it negatively charged. The left-hand electrode is positively charged because it is short of electrons. There is a limit to how many extra electrons the "pump" can squeeze into the negative electrode because of the repulsion by the electrons already there.

Things change when the zinc chloride melts, and the ions become free to move.

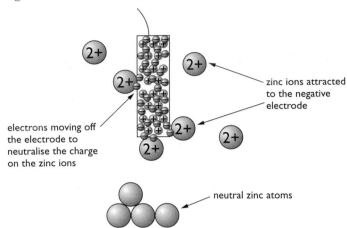

electrons moving off the electrode to neutralise the charge on the zinc ions

zinc ions attracted to the negative electrode

neutral zinc atoms

> In this (and the next) diagram, the zinc ions and chloride ions are drawn much bigger than the remainder of the atoms in the electrode (the green positively charged circles which represent the nuclei and the non-mobile electrons). This is so it is easier to see what is happening.

Showing what happens at the cathode.

The positive zinc ions are attracted to the cathode. When they get there, each zinc ion picks up two electrons from the electrode and forms neutral zinc atoms. These fall to the bottom of the container as liquid zinc.

$$Zn^{2+}_{(l)} + 2e^- \rightarrow Zn_{(l)}$$

This leaves spaces in the electrode that more electrons can move into. The power source pumps new electrons along the wire to fill those spaces.

Chloride ions are attracted to the positive anode. When they get there, the extra electron which makes the chloride ion negatively charged moves onto the electrode because this electrode is short of electrons.

The loss of the extra electron turns each chloride ion into a chlorine atom. These join in pairs to make chlorine molecules. Overall:

$$2Cl^-_{(l)} \rightarrow Cl_{2(g)} + 2e^-$$

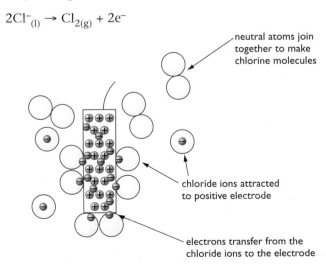

neutral atoms join together to make chlorine molecules

chloride ions attracted to positive electrode

electrons transfer from the chloride ions to the electrode

Showing what happens at the anode.

The new electrons on the electrode are pumped away by the power source to help fill the spaces being created at the cathode. Because electrons are flowing in the external circuit, the bulb lights up.

Electrons can flow in the external circuit because of the chemical changes to the ions arriving at the electrodes. We say that the ions are **discharged** at the electrodes. Discharging an ion simply means that it loses its charge – either giving up electron(s) to the electrode or receiving electron(s) from it.

Electrolysis and redox

The zinc ions gain electrons at the cathode:

$$Zn^{2+}_{(l)} + 2e^- \rightarrow Zn_{(l)}$$

Gain of electrons is reduction. The zinc ions are reduced to zinc atoms.

The chloride ions lose electrons at the anode.

$$2Cl^-_{(l)} \rightarrow Cl_{2(g)} + 2e^-$$

Loss of electrons is oxidation. Chloride ions are oxidised to chlorine molecules.

In any example of electrolysis (whether molten or in solution), reduction happens at the cathode and oxidation happens at the anode.

The electrolysis of other molten substances

In each case, the positive ions are attracted to the negative cathode, where they are discharged by gaining electrons. Positive ions are known as **cations** because they are attracted to the cathode. The negative ions move to the anode, where they are discharged by giving electrons to the electrode. Negative ions are known as **anions**.

For example, with molten lead(II) bromide:

| At the cathode: | $Pb^{2+}_{(l)} + 2e^- \rightarrow Pb_{(l)}$ | Molten lead is produced. |
| At the anode: | $2Br^-_{(l)} \rightarrow Br_{2(g)} + 2e^-$ | Bromine gas is produced. |

Not all ionic compounds can be electrolysed molten. Some break up into simpler substances before their melting point. For example, copper(II) carbonate breaks into copper(II) oxide and carbon dioxide even on gentle heating. It is impossible to melt it.

Other ionic compounds have such high melting points that it isn't possible to melt them in the lab, although it can be done industrially. For example, it is difficult to keep sodium chloride molten in the lab because its melting point is 801°C. If you could keep it molten you would get sodium at the cathode and chlorine at the anode. Sodium is manufactured by electrolysing molten sodium chloride.

| At the cathode: | $Na^+_{(l)} + e^- \rightarrow Na_{(l)}$ | Molten sodium is produced. |
| At the anode: | $2Cl^-_{(l)} \rightarrow Cl_{2(g)} + 2e^-$ | Chlorine gas is produced. |

The Electrolysis of Aqueous Solutions

The electrolysis of sodium chloride solution

You might have thought that you would get the same products if you electrolysed sodium chloride solution as if you electrolysed it molten. You would be wrong!

Remember OIL RIG.

Remember that the cathode is negative and that a cation is attracted to it – so must be positive.

Remember that all these reactions are being done very hot. You won't get solid lead or liquid bromine at these temperatures.

Although chlorine is formed at the anode as you might expect, hydrogen is produced at the cathode rather than sodium. The hydrogen at the cathode is coming from the water.

Water is a very **weak electrolyte**. It ionises very slightly to give hydrogen ions and hydroxide ions.

$$H_2O_{(l)} \rightleftharpoons H^+_{(aq)} + OH^-_{(aq)}$$

Whenever you have water present, you have to consider these ions *as well* as the ions in the compound you are electrolysing.

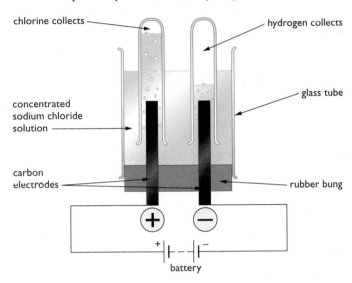

Electrolysis of sodium chloride solution.

> The reason that you don't appear to get as much chlorine as you do hydrogen is that chlorine is slightly soluble and some of it dissolves in the water present.

At the cathode

The solution contains $Na^+_{(aq)}$ and $H^+_{(aq)}$, and these are both attracted to the negative cathode. The $H^+_{(aq)}$ gets discharged because it is much easier to persuade a hydrogen ion to accept an electron than it is a sodium ion. Each hydrogen atom formed combines with another one to make a hydrogen molecule.

$$2H^+_{(aq)} + 2e^- \rightarrow H_{2(g)}$$

Remember that the hydrogen ions come from water molecules splitting up. Each time a water molecule ionises, it also produces a hydroxide ion. There is a build up of these in the solution around the cathode.

> For positive ions the lower an element is in the Reactivity Series, the more easily it will accept an electron.

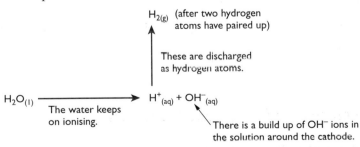

> Because the hydrogen ions are discharged, they can no longer react with the hydroxide ions and re-form water. The ionisation of the water becomes a one-way process.

These hydroxide ions make the solution strongly alkaline in the region around the cathode. Because of the presence of the sodium ions attracted to the cathode, you can think of the electrolysis as also forming sodium hydroxide solution.

Chapter 11: Electrolysis

At the anode

$Cl^-_{(aq)}$ and $OH^-_{(aq)}$ are both attracted by the positive anode. The hydroxide ion is *slightly* easier to discharge than the chloride ion is, but there isn't that much difference. There are far, far more chloride ions present in the solution, and so it is mainly these which get discharged.

$$2Cl^-_{(aq)} \rightarrow Cl_{2(g)} + 2e^-$$

If the sodium chloride solution is dilute, you get noticeable amounts of oxygen produced as well as chlorine. This comes from the discharge of the hydroxide ions.

$$4OH^-_{(aq)} \rightarrow 2H_2O_{(l)} + O_{2(g)} + 4e^-$$

The electrolysis of some other solutions using carbon electrodes

	Cathode		Anode	
	Product	Equation	Product	Equation
$KI_{(aq)}$	hydrogen	$2H^+_{(aq)} + 2e^- \rightarrow H_{2(g)}$	iodine	$2I^-_{(aq)} \rightarrow I_{2(aq)} + 2e^-$
$MgBr_{2(aq)}$	hydrogen	$2H^+_{(aq)} + 2e^- \rightarrow H_{2(g)}$	bromine	$2Br^-_{(aq)} \rightarrow Br_{2(aq)} + 2e^-$
$H_2SO_{4(aq)}$	hydrogen	$2H^+_{(aq)} + 2e^- \rightarrow H_{2(g)}$	oxygen	$4OH^-_{(aq)} \rightarrow 2H_2O_{(l)} + O_{2(g)} + 4e^-$
$CuSO_{4(aq)}$	copper	$Cu^{2+}_{(aq)} + 2e^- \rightarrow Cu_{(s)}$	oxygen	$4OH^-_{(aq)} \rightarrow 2H_2O_{(l)} + O_{2(g)} + 4e^-$

In the electrolysis of dilute sulphuric acid, the sulphate ions are too stable to be discharged. Instead, you get oxygen from discharge of hydroxide ions from the water. Twice as much hydrogen is produced as oxygen.

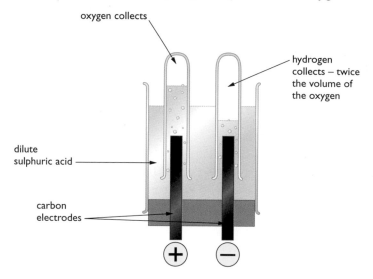

oxygen collects

hydrogen collects – twice the volume of the oxygen

dilute sulphuric acid

carbon electrodes

Look at the sulphuric acid equations in the table. For every 4 electrons which flow around the circuit, you would get 1 molecule of oxygen. But 4 electrons would produce 2 molecules of hydrogen.

This is the result of Avogadro's Law. See page 275 if you are interested.

You get twice the number of molecules of hydrogen as of oxygen. Twice the number of molecules occupy twice the volume.

Summary of the electrolysis of solutions using carbon electrodes

This also applies to using any other *inert* electrodes – like platinum, for example. As you will see below, some metal electrodes are changed during electrolysis.

- If the metal is high in the Reactivity Series, hydrogen is produced instead of the metal.

- If the metal is below hydrogen in the Reactivity Series, the metal is produced.

- If you have reasonably concentrated solutions of halides (chlorides, bromides or iodides), the halogen (chlorine, bromine or iodine) is produced. With other common negative ions, oxygen is produced.

This leaves the problem of what you get if you have a moderately reactive metal like zinc, for example. Reasonably concentrated solutions will give you the metal. Very dilute solutions will give you mainly hydrogen. In between, you will get both.

At GCSE you probably won't have to worry about this. The examples you will meet in exams are always clear-cut.

The electrolysis of copper(II) sulphate solution using copper electrodes

If you use metal electrodes rather than carbon, different things can happen at the anode, unless the metal is extremely unreactive – like platinum.

Remember that positive ions are turned into atoms by taking electrons from the cathode. Electrons are pumped around the circuit from the anode to replace them. The reactions at the anode act as a source of these electrons.

So far, these electrons have come from negative ions giving up electrons to the anode, but there is another possibility. They can come from atoms in the electrode itself if that is an easier process.

A copper atom breaks away from the electrode forming a copper(II) ion, leaving its electrons behind on the electrode. Those electrons can then be pumped away by the power source around the circuit to the cathode.

$$Cu_{(s)} \rightarrow Cu^{2+}_{(aq)} + 2e^-$$

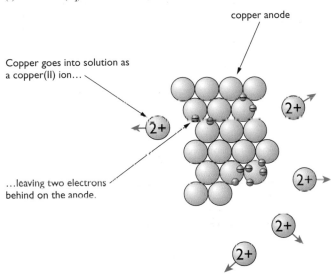

copper anode

Copper goes into solution as a copper(II) ion...

...leaving two electrons behind on the anode.

For every copper(II) ion that breaks away from the anode, 2 electrons are made available to pump around the circuit. That's exactly what you need to discharge a copper(II) ion arriving at the cathode.

$$Cu^{2+}_{(aq)} + 2e^- \rightarrow Cu_{(s)}$$

The overall effect is that:

- the anode loses mass

- the cathode gains exactly the same mass

- the number of copper(II) ions in the solution doesn't change at all. For every one that is discharged at the cathode, another one goes into solution at the anode.

This process is used in purifying copper. You will find that described on pages 154–155.

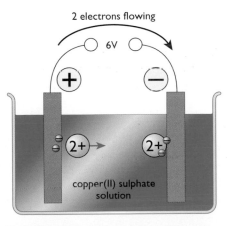

2 electrons flowing

6V

copper(II) sulphate solution

Electrolysing copper(II) sulphate solution using copper electrodes.

Showing that ions move during electrolysis

So far, we have explained electrolysis in terms of the movement and discharge of ions. Is there any proof that they actually do move? You can show it simply, using substances with coloured ions.

Potassium manganate(VII) contains colourless K^+ ions and dark purple MnO_4^- ions.

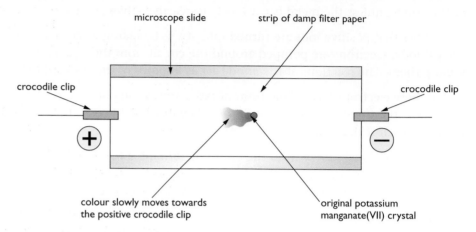

If a crystal of potassium manganate(VII) is put on a piece of damp filter paper and connected into a circuit, the purple colour spreads noticeably towards the positive end over the course of half an hour or so. This is because the purple MnO_4^- ions are attracted to the positive electrode.

If you had a strongly coloured positive ion, it would obviously move the other way.

End of Chapter Checklist

If you haven't got a copy of your specification, read the introduction on page vi.

You will need to be able to do some or all of the following. Check your Awarding Body's specification (syllabus) to find out exactly what you need to know.

- Understand that the flow of electricity in metals and carbon is due to movement of delocalised (mobile) electrons, and that there is no chemical change involved.

- Understand what is meant by *electrolysis*, *electrolyte*, *electrode*, *anode*, *cathode*.

- Know that electricity is conducted through electrolytes (molten or in solution) by the movement and discharge of ions.

- Know the products of electrolysis of simple molten electrolytes like zinc chloride, explaining the electrolysis with the help of electrode equations.

- Describe and explain the electrolysis of sodium chloride solution using carbon electrodes, and copper(II) sulphate solution using copper electrodes, including electrode equations.

- Deduce the products of electrolysis of other aqueous solutions in simple cases.

- Describe a simple experiment to show the movement of coloured ions.

You will find calculations involving electrolysis on pages 281–286.

Questions

More questions on electrolysis can be found at the end of Section B on page 142.

1 Say what is formed at the cathode and at the anode during the electrolysis of the following substances. Assume that carbon electrodes were used each time. You don't need to write electrode equations.

   a) molten zinc chloride

   b) molten lead(II) bromide

   c) sodium iodide solution

   d) molten sodium iodide

   e) copper(II) chloride solution

   f) dilute hydrochloric acid

   g) magnesium sulphate solution

   h) sodium hydroxide solution.

2 Some solid potassium iodide was placed in a pyrex evaporating basin. Two carbon electrodes were inserted and connected to a 12 volt DC power source and a light bulb.

The potassium iodide was heated. As soon as the potassium iodide was molten, the bulb came on. Purple fumes were seen coming from the positive electrode, and lilac flashes were seen around the negative one.

   a) Explain why the bulb didn't come on until the potassium iodide melted.

   b) What name is given to the positive electrode?

   c) Name the purple fumes seen at the positive electrode, and write the electrode equation for their formation.

   d) The lilac flashes seen around the negative electrode are caused by the potassium which is formed. The potassium burns with a lilac flame. Write the electrode equation for the formation of the potassium.

   e) What differences would you expect to observe if you used molten sodium bromide instead of potassium iodide?

   f) Write the electrode equations for the reactions occurring during the electrolysis of molten sodium bromide.

3 For each of the following electrolytes: *i)* write the cathode equation; *ii)* write the anode equation; *iii)* say what has been oxidised and what has been reduced.

a) molten lead(II) bromide using carbon electrodes

b) sodium chloride solution using carbon electrodes

c) calcium bromide solution using carbon electrodes

d) copper(II) sulphate solution using carbon electrodes

e) copper(II) sulphate solution using copper electrodes

f) molten magnesium iodide using carbon electrodes

g) dilute hydrochloric acid using carbon electrodes

h) silver nitrate solution using silver electrodes.

4 Suppose you were asked to investigate the transfer of copper from the anode to the cathode during the electrolysis of copper(II) sulphate solution using copper electrodes. As a part of your investigation you will need to measure the mass of copper which is deposited on the cathode. The copper doesn't stick properly to the cathode unless the cathode is very clean. If you use too high a current, the copper deposits as a spongy mass which again won't stick properly to the electrode. If you use too small a current, it takes a very long time to produce a mass of copper which you could weigh with reasonable accuracy.

Describe in detail the preliminary experiments you would do to find the best conditions for plating a sensible mass of copper securely on to the cathode. You should consider the various ways in which you might control the current, and how you would treat the cathode before and after the electrolysis.

5 A deep blue solution was known to contain one of the following combinations of ions:

either **A** $[Cu(NH_3)_4(H_2O)_2]^{2+}_{(aq)}$ and $SO_4^{2-}_{(aq)}$

or **B** $2H^+_{(aq)}$ and $[CoCl_4]^{2-}_{(aq)}$

Describe how you would find out which combination of ions you had using a battery, some connecting wire and a piece of filter paper. (Hints: Dilute sulphuric acid contains hydrogen ions and sulphate ions and is colourless. Don't worry that the metal-containing ions look complicated and unfamiliar. That doesn't matter in the least.)

Chapter 12: Electrochemical Cells

This chapter looks at how you can use chemical reactions to produce electricity. You only need to know about this for exam purposes if you are doing the OCR specification using Extension Block B.

The porous container (made of unglazed ceramic) allows ions to move between the two parts of the cell without getting too much mixing of the solutions. You have to allow ion movement to complete the electrical circuit.

Important! You don't need to remember details of the Daniel Cell for exam purposes. It is included because it is much easier to explain what is happening than in the "simple" cases the syllabus demands!

Getting Electricity from Chemical Reactions

Any battery – of whatever type – uses chemical reactions to produce lots of electrons at one terminal (the negative one), and a relative lack of electrons at the positive one.

When you connect a bulb or anything else to the battery, electrons flow from where there is a lot of them to where they are more scarce.

The Daniel Cell

The Daniel Cell uses a simple Reactivity Series reaction to generate electricity.

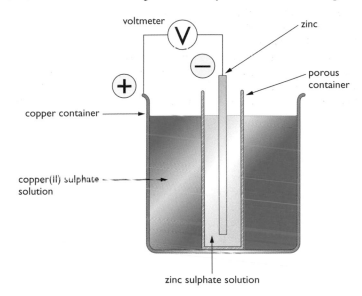

The zinc is the negative terminal of the cell, and the copper is the positive one. If you measured the voltage under standard conditions (25°C and a particular concentration of solutions), you would find the voltmeter would read 1.10 volts.

If you now replaced the voltmeter by a simple piece of wire connecting negative to positive, electrons will flow along it. You could use this flow of electrons to light a (very small!) bulb.

What's happening?

When metals react they give away electrons and form positive ions. The more reactive a metal is, the more easily it releases electrons and forms its ions. Zinc is more reactive than copper. Therefore zinc from the electrode goes into solution as zinc ions, releasing electrons as it does so.

$$Zn_{(s)} \rightarrow Zn^{2+}_{(aq)} + 2e^-$$

These electrons are left behind on the electrode and make it negative. If the electrons are allowed to flow around the external circuit, they flow to the copper and are picked up by the copper(II) ions in the copper(II) sulphate solution to produce extra copper metal.

$$Cu^{2+}_{(aq)} + 2e^- \rightarrow Cu_{(s)}$$

Compare this with the reaction between zinc and copper(II) sulphate solution on page 56.

Although the reaction is split into two separate bits, this is essentially just a Reactivity Series displacement reaction. The more reactive zinc goes into solution as zinc ions, and the copper is displaced from solution to form copper metal.

Changing the metals

If you use different metals and solutions, you will change the voltage. Suppose you replaced the zinc and zinc sulphate solution in the porous container by magnesium in magnesium sulphate solution, or by iron in iron(II) sulphate solution.

In each case, the copper container, the copper(II) sulphate solution, the temperature and all the concentrations are kept constant. If any of these varied, you wouldn't be comparing like with like.

Changing the concentrations or the temperature makes small differences to the voltages.

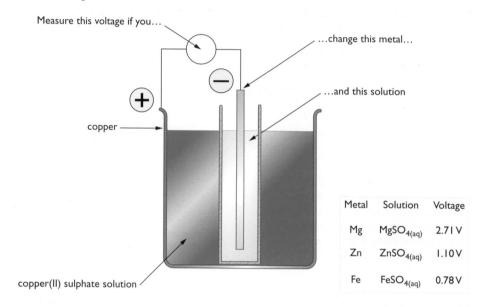

Measure this voltage if you...

...change this metal...

...and this solution

copper

copper(II) sulphate solution

Metal	Solution	Voltage
Mg	$MgSO_{4(aq)}$	2.71 V
Zn	$ZnSO_{4(aq)}$	1.10 V
Fe	$FeSO_{4(aq)}$	0.78 V

In each case, the voltage compares the reactivity of the other metal with the reactivity of copper.

Magnesium is very reactive, and so forms its ions even more readily than zinc does. Many more electrons would be left behind on the magnesium electrode and so it would be even more negative than the zinc one. The greater voltage simply measures this greater difference between the magnesium electrode and the copper electrode.

Iron is less reactive than zinc – although it is still more reactive than copper. The iron doesn't ionise as readily as the zinc does, so doesn't leave so many electrons behind on the electrode. The voltage measured is less.

What if you compared copper with a less reactive metal like silver?

You might need to look carefully at the diagram to spot the difference. This time, the copper is the *negative* electrode.

Of the two metals, copper is the more reactive and so forms its ions more readily.

$$Cu_{(s)} \rightarrow Cu^{2+}{}_{(aq)} + 2e^-$$

Extra electrons build up on the copper this time, and so this is now the negative electrode. If those electrons are allowed to flow to the silver electrode, they react with silver ions from the solution to produce extra silver metal.

$$Ag^+{}_{(aq)} + e^- \rightarrow Ag_{(s)}$$

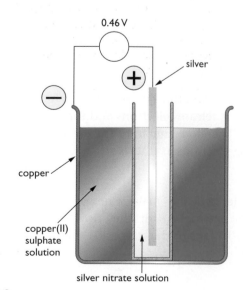

0.46 V

silver

copper

copper(II) sulphate solution

silver nitrate solution

Summary

The size of the voltage is a measure of the gap between the two metals in the Reactivity Series. The larger the gap, the larger the voltage. The more reactive metal will always form the negative terminal of the cell.

Calculating the voltage for a given combination of metals

In the examples, we have looked at the voltages produced by combining various other metal electrodes with a copper electrode. Copper in the copper(II) sulphate solution would be described as a **reference electrode**.

Books of chemical data list voltage values where the reference electrode is a **standard hydrogen electrode**. You don't need to know anything about this for GCSE, although you might be expected to use the data to calculate a voltage for any combination of metal electrodes.

The values listed are known as **standard electrode potentials**. Here are some of them:

Metal	In a solution of	Standard electrode potential (volts)
Mg	magnesium sulphate	–2.37
Al	aluminium sulphate	–1.66
Zn	zinc sulphate	–0.76
Fe	iron(II) sulphate	–0.44
Cu	copper(II) sulphate	+0.34
Ag	silver nitrate	+0.80

The more reactive the metal, the more negative the value. Those metals with positive values are the ones which are less reactive than hydrogen.

Finding the voltage for a magnesium and zinc combination

To find the voltage for a cell using a combination of, say, a magnesium electrode and a zinc electrode, you simply find the difference between the two values. To find the difference between –2.37 and –0.76, take –0.76 away from –2.37.

voltage = –2.37 – (–0.76) = –1.61 volts

The voltage would be 1.61 volts, and the more reactive metal (the magnesium) would be the negative electrode.

Finding the voltage for an aluminium and silver combination

voltage = –1.66 – (+0.80) = –2.46 volts

The voltage would be 2.46 volts, and the more reactive aluminium would be the negative electrode.

Very simple cells

You can detect the differences between the reactivity of the metals using the very simple cell in the diagram overleaf. The greater the difference in reactivity of the metals, the higher the voltage. As before, the more reactive metal will form the negative electrode. It is, however, more difficult to explain what is going on.

What matters here is the **difference** between the two values. Whether you get a positive or negative answer depends on which you take away from which. At GCSE you can ignore the sign of the answer. The voltage produced by the cell in the example would be 1.61 volts.

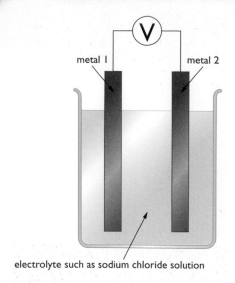

metal 1   metal 2

electrolyte such as sodium chloride solution

Suppose you use magnesium and copper. The magnesium is the more reactive metal and it readily forms ions, leaving its electrons behind on the electrode.

$$Mg_{(s)} \rightarrow Mg^{2+}_{(aq)} + 2e^-$$

The magnesium will be the negative electrode. If the voltmeter is replaced by a bit of wire, electrons can flow along it towards the copper.

It is more difficult to explain what happens at the copper electrode. In the proper case we described before (on page 129), copper(II) ions picked up the electrons which were arriving and formed extra copper. This time there aren't any copper(II) ions in the solution!

If you think of it as a simple case of electrolysis happening at this electrode, hydrogen ions from the solution can pick up the electrons to form hydrogen gas.

$$2H^+_{(aq)} + 2e^- \rightarrow H_{2(g)}$$

The voltages which you measure in simple cells like this are not the same as the voltages which you can calculate, because the cells aren't operating under the same conditions as before. The electric current that you can get from the cell is usually extremely small.

Fuel Cells

Most electricity is currently generated by burning fuels – an extremely wasteful and usually polluting process. The heat produced when the fuel burns is used to generate steam, which is used to drive turbines. The motion of the turbines is used to produce electricity.

Every stage in this sequence is inefficient, so that the amount of electrical energy finally generated is only a fraction of the energy available from burning the fuel.

Fuel cells overcome this problem by using the fuel and oxygen to produce electricity directly.

A hydrogen fuel cell

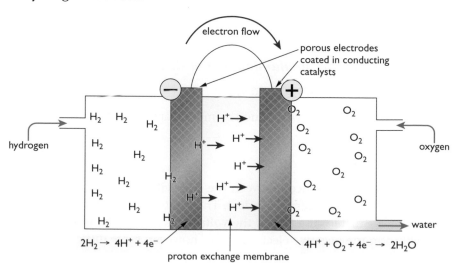

In the left-hand compartment, hydrogen gas diffuses into the porous electrode where it releases electrons to make hydrogen ions (protons).

$$2H_2 \rightarrow 4H^+ + 4e^-$$

The electrons move away around the external circuit as a flow of electricity. The protons (hydrogen ions) diffuse through the membrane (a complex polymer-based material) towards the right-hand electrode, where they combine with oxygen gas and the electrons arriving from the circuit to make water.

$$4H^+ + O_2 + 4e^- \rightarrow 2H_2O$$

The overall effect of all this is that hydrogen and oxygen have combined to make water, and generated a flow of electricity in the process.

Some practical considerations

Fuel cells convert a fuel into electricity extremely efficiently, and at ordinary temperatures. Using hydrogen has two important advantages. Hydrogen produces more energy per gram when it reacts with oxygen than any other chemical fuel, and the only product is water – which is completely non-polluting.

Hydrogen fuel cells are used to provide electricity for spacecraft, and there are already electric cars being fuelled in this way.

One of the problems that has to be resolved is storage of the hydrogen. It could be carried as a compressed gas, but there are other possibilities. For example, research is being done on metals which can absorb hydrogen.

Magnesium alloy can bond chemically to hydrogen to form magnesium hydride. When this is heated to 300°C, the hydrogen is released again. A unit the size of a normal petrol tank could provide enough hydrogen to power the car for about 300 miles.

We are starting with $2H_2$ because we need to generate 4 electrons to flow around the external circuit for the reaction in the right-hand compartment.

The Ford Focus FCV (fuel cell vehicle).

During the lifetime of this book there are likely to be major advances in fuel cell technology and in the development of hydrogen storage for use in cars. If you are interested, try an internet search on **fuel cells**.

End of Chapter Checklist

Remember: You only have to do this if you are doing OCR Extension Block B. If you haven't got a copy of your specification, read the introduction on page vi.

You will need to be able to do some or all of the following.

● Describe the structure of a simple electrochemical cell.

● Explain how a cell produces electricity by considering the processes happening at the electrodes, including writing electrode equations.

● Understand the relationship between the reactivities of the metals used and the voltage produced by the cell.

● Calculate the voltage of a simple cell from data from reference cells.

● Describe how a hydrogen–oxygen fuel cell works, and discuss its advantages and possible uses.

Questions

1   Some students are given pieces of different metals. They are asked to investigate which two metals would give the highest voltage when used in the cell. They use copper, zinc and magnesium.

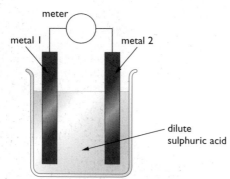

a)   i)   Which pair of metals gives the highest voltage?

ii) Give reasons for your choice.

b)   What other change would affect the voltage of the cell?

c)   In one cell zinc is used as metal 1 and copper is used as metal 2.

i)   State and explain which way the electrons flow through the meter in this cell.

ii) Write an ionic equation for the reaction involving zinc in this cell.

(OCR Specimen Paper 2001, Paper 6, Question 3)

2   Use the table of standard electrode potentials on page 131 to work out the voltages produced by the following electrode combinations. In each case say which piece of metal would be the negative terminal of the cell.

a)   iron in iron(II) sulphate solution in contact with silver in silver nitrate solution

b)   zinc in zinc sulphate solution in contact with aluminium in aluminium sulphate solution

c)   iron in iron(II) sulphate solution in contact with magnesium in magnesium sulphate solution

d)   aluminium in aluminium sulphate solution in contact with copper in copper(II) sulphate solution.

3   You are given 5 unlabelled metals in the form of rods of about 3 mm diameter, a voltmeter, and any equipment or chemicals that you might reasonably expect to find in an ordinary household. Describe how you would put the metals into Reactivity Series order. You should explain how you would control any variables in order to do a "fair test".

4   a)   With the help of a simple diagram or diagrams, explain how a hydrogen–oxygen fuel cell produces electricity. Write equations for any chemical changes you describe.

b)   Give two advantages of using a hydrogen–oxygen fuel cell to power a car compared with conventional fuels like petrol or diesel.

Chapter 13: Energy Changes in Reactions

Some chemical reactions produce heat. Others need to be heated constantly to make them occur at all. This chapter explores some of the ideas involved in energy transfers during reactions, and looks at some simple calculations.

Exothermic and Endothermic Changes

Exothermic reactions

It is common experience that lots of chemical reactions give out energy – often in the form of heat. A reaction which gives out energy is said to be **exothermic**.

Combustion reactions

Any reaction which produces a flame must be exothermic.

Burning fuel produces enough energy to launch a rocket.

The flare stack at an oil refinery is a safety device. If a process goes wrong (for example, if pressures get too high), reactants and products can be vented to the flare stack where they burn off safely.

You will be familiar with the test for hydrogen by lighting it and getting a squeaky pop. That is an obvious sign of energy being released – an exothermic change. This can be harnessed in oxy-hydrogen cutting equipment which can be used underwater.

$$2H_{2(g)} + O_{2(g)} \rightarrow 2H_2O_{(l)}$$

Apart from burning, other simple exothermic changes include:

- The reactions of metals with acids.

- Neutralisation reactions.

- Adding water to calcium oxide.

The burning of hydrogen is used in oxy-hydrogen cutting equipment underwater.

This reaction is described in detail on page 65.

The reactions of metals with acids

For example, when magnesium reacts with dilute sulphuric acid, the mixture gets very warm.

$$Mg_{(s)} + H_2SO_{4(aq)} \rightarrow MgSO_{4(aq)} + H_{2(g)}$$

Neutralisation reactions

You can read about this reaction on pages 67–69.

About the only interesting thing that you can observe happening when sodium hydroxide solution reacts with dilute hydrochloric acid is that the temperature rises!

$$NaOH_{(aq)} + HCl_{(aq)} \rightarrow NaCl_{(aq)} + H_2O_{(l)}$$

Adding water to calcium oxide

This is described on page 163.

If you add water to solid calcium oxide, the heat produced is enough to boil the water and produce steam.

$$CaO_{(s)} + H_2O_{(l)} \rightarrow Ca(OH)_{2(s)}$$

Showing an exothermic change on an energy diagram

In an exothermic reaction the reactants have more energy than the products. As the reaction happens, energy is given out in the form of heat. That energy warms up both the reaction itself and the surroundings.

Remember that in a chemical reaction the reactants are the chemicals you start with.

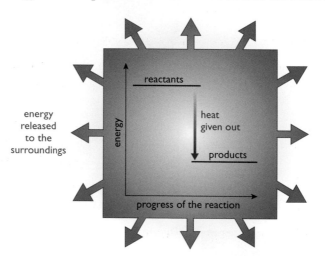

You can measure the amount of heat energy released. It is given the symbol ΔH.

ΔH is read as "delta H". The Greek letter Δ is used to mean "change in". ΔH means "change in heat".

The mole is a particular quantity of a substance. You can read about it on pages 264–267, but you probably don't need to worry about it at the moment. In the magnesium case, 1 mole weighs 24.3 g.

ΔH is given a minus or a plus sign to show whether heat is being given out or absorbed by the reaction. You always look at it from the point of view of the substances taking part. For an exothermic reaction, ΔH is given a *negative* number, because the reactants are *losing* energy as heat. That heat is transferred to the surroundings which then get warmer. ΔH is measured in units of $kJ\,mol^{-1}$ ("kilojoules per mole").

In an equation, this would be shown as, for example:

$$Mg_{(s)} + H_2SO_{4(aq)} \rightarrow MgSO_{4(aq)} + H_{2(g)} \qquad \Delta H = -466.9 \text{ kJ}\,mol^{-1}$$

This means that 466.9 kJ of heat is given out when 1 mole of magnesium reacts in this way. You know it has been given out because ΔH has a negative sign.

Endothermic reactions

A reaction which absorbs energy is said to be **endothermic**. The energy absorbed may be in the form of light, heat taken from the surroundings, or electrical energy.

Photosynthesis

Photosynthesis uses energy in the form of light to convert carbon dioxide and water into carbohydrates (such as glucose) and oxygen.

$$6CO_{2(g)} + 6H_2O_{(l)} \rightarrow C_6H_{12}O_{6(aq)} + 6O_{2(g)}$$

This reaction requires the green plant pigment, **chlorophyll**.

Life on Earth depends on photosynthesis.

The effect of heat on calcium carbonate

Calcium carbonate doesn't decompose (split up) unless it is heated at very high temperatures. The reaction is endothermic and needs a constant input of heat.

$$CaCO_{3(s)} \rightarrow CaO_{(s)} + CO_{2(g)}$$

This reaction is explored further on pages 162–163.

Electrolysis

The chemical changes during electrolysis need a constant input of electrical energy. If you stop the flow of electricity, the reaction stops at once. All electrolysis reactions must be endothermic.

The reaction between sodium carbonate and ethanoic acid

All acids react with carbonates to give off carbon dioxide, and ethanoic acid is no exception. Despite the vigorous reaction, the temperature of the mixture falls.

The reaction is endothermic, and takes heat from the water, the beaker and the surroundings. You can tell that the reaction is endothermic by putting a thermometer in the mixture, or because the beaker will feel cold as heat is taken from your skin.

$$Na_2CO_{3(s)} + 2CH_3COOH_{(aq)} \rightarrow 2CH_3COONa_{(aq)} + CO_{2(g)} + H_2O_{(l)}$$

Showing an endothermic change on an energy diagram

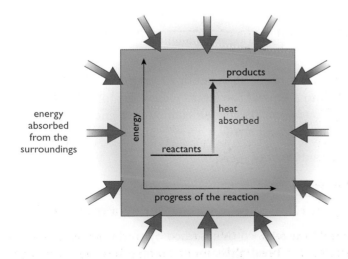

Chapter 13: Energy Changes in Reactions

In an endothermic change, the products have more energy than the reactants. That extra energy has to come from somewhere, and it is absorbed from the surroundings – making everything cooler.

Because the reactants are *gaining* energy, ΔH is given a *positive* sign.

For example:

$$CaCO_{3(s)} \rightarrow CaO_{(s)} + CO_{2(g)} \qquad \Delta H = +178 \text{ kJ mol}^{-1}$$

This means that it needs 178 kJ of heat energy to convert 1 mole of calcium carbonate (in this case 100 g) into calcium oxide and carbon dioxide.

Again, don't worry if you don't understand the units of ΔH. It isn't important for now.

Calculations Involving Heat Changes During Reactions

Bond energies (bond strengths)

It needs energy to break chemical bonds. The stronger the bond is, the more energy is needed to break it. Bond energy measures the amount of energy needed to break a particular bond.

Bond energies are measured in kilojoules per mole. For now, all you need to do is to realise that the "mole" is related to the formula of a substance. For example, if you see "Cl_2" written in an equation, it will take 346 kJ to break all the bonds in it. If you see "$2HCl$", it will take 2×432 kJ to break all the bonds in that.

Bond	C–H	C–Cl	C–I	Cl–Cl	I–I	H–Cl	H–I
Bond energy (kJ mol^{-1})	+413	+346	+234	+243	+151	+432	+298

You can see that some bonds are much stronger than others – for example, the bond between iodine and hydrogen is about twice as strong as the bond between two iodine atoms.

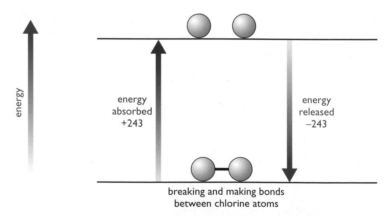

breaking and making bonds between chlorine atoms

The diagram shows that it needs an input of 243 kJ per mole to break chlorine molecules into atoms. If the atoms recombine into their original molecules, then obviously exactly the same amount of energy will be released again. When bonds are made, energy is given out.

- Breaking bonds needs energy.
- Making bonds releases energy.

Calculating the heat released or absorbed during a reaction

You can estimate the heat released or absorbed by working out how much energy would be needed to break the substances up into individual atoms, and then how much would be given out when those atoms recombine into new arrangements. For example:

If   heat needed to break all the bonds   =   +1000 kJ
        heat released when new bonds are made   =   −1200 kJ
then   overall change   =   −200 kJ

The reaction between methane and chlorine

$$CH_{4(g)} + Cl_{2(g)} \rightarrow CH_3Cl_{(g)} + HCl_{(g)}$$

Methane reacts with chlorine in the presence of ultra-violet light to produce chloromethane and hydrogen chloride. You can picture all the bonds being broken in the methane and chlorine and then being reformed in new ways in the products.

You can work out the heat needed to break all the bonds, and the heat given out as new ones are made.

Bonds that need to be broken:

4 C–H bonds   =   4 × (+413)   =   +1652 kJ
1 Cl–Cl bond =   1 × (+243)   =   + 243 kJ
Total   =   +1895 kJ

New bonds made:

3 C–H bonds   =   3 × (−413)   =   −1239 kJ
1 C–Cl bond   =   1 × (−346)   =   − 346 kJ
1 H–Cl bond   =   1 × (−432)   =   − 432 kJ
Total   =   −2017 kJ

The overall energy change is +1895 + (−2017) kJ = −122 kJ

The negative sign of the answer shows that, overall, heat is given out as the bonds rearrange. More energy is released when the new bonds were made than is used to break the old ones.

The excess heat given out means that the reaction is exothermic.

You can show all this happening on an energy diagram:

> **Important!** The real reaction doesn't happen by all the bonds being broken in this way. That doesn't matter.
>
> The *overall* amount of heat released or absorbed is the same however you do the reaction – even if one of the ways you use to work it out is entirely imaginary! This is summarised in an important law called Hess's Law which you will meet if you do A-level Chemistry.

> Remember that when heat is given out, you show this by putting a negative sign in front of the value.

bonds being broken

4 (C–H) = 4 × (+413) kJ
1 (Cl–Cl) = +243 kJ
total   = +1895 kJ

bonds being made

3 (C–H) = 3 × (−413) kJ
1 (C–Cl) = −346 kJ
1 (H–Cl) = − 432 kJ
total   = −2017 kJ

energy

reactants

products

Overall change: ΔH= +1895 + (−2017) = −122 kJ

The reaction between hydrogen and fluorine

Calculate the heat released or absorbed when this reaction occurs:

$$H_{2(g)} + F_{2(g)} \rightarrow 2HF_{(g)}$$

The bond energies (in $kJ\,mol^{-1}$) are: H–H 436, F–F 158, H–F 568.

Bonds that need to be broken:

1 H–H bond	= 1 × (+436)	=	+436 kJ
1 F–F bond	= 1 × (+158)	=	+158 kJ
Total		=	+594 kJ

New bonds made:

2 H–F bonds	= 2 × (–568)	= –1136 kJ

The overall energy change is +594 +(–1136) = –542 kJ

Because the answer is negative, the reaction is exothermic.

Calculating heat given out during experimental work

If you do a reaction using a known mass of solution and measure the temperature rise, the amount of heat given out during the reaction is given by

heat given out = mass × specific heat × temperature rise

The specific heat is the amount of heat which is needed to raise the temperature of 1 gram of a substance by 1°C. For water, the value is $4.18\,J\,g^{-1}\,°C^{-1}$ (joules per gram per degree Celsius).

You can normally assume that dilute solutions have the same specific heat ($4.18\,J\,g^{-1}\,°C^{-1}$) and density ($1\,g\,cm^{-3}$ – 1 gram per cubic centimetre) as water. You also assume that negligibly small amounts of heat are used to warm up the cup and the thermometer.

If you had, say, $50.0\,cm^3$ of solution, you could assume that it weighed 50.0 g. If the temperature rose by 11.2 °C, the amount of heat given out would be as follows:

heat given out = 50.0 × 4.18 × 11.2 joules
            = 2340 joules

You will find much more about measuring the heat given out in an experimental situation in the appendix on coursework starting on page 297.

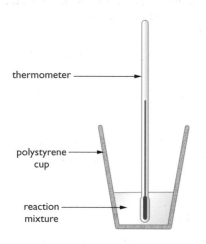

thermometer

polystyrene cup

reaction mixture

Important! This isn't something you are likely to need in a Chemistry exam, but you might find it useful if you are asked to do an investigation involving energy changes.

End of Chapter Checklist

If you haven't got a copy of your specification, read the introduction on page vi.

You will need to be able to do some or all of the following. Check your Awarding Body's specification (syllabus) to find out exactly what you need to know.

- Understand the meaning of the terms *exothermic* and *endothermic* and give examples of reactions of each type.

- Understand simple energy diagrams showing the relative energies of reactants and products.

- Understand what is meant by bond energies (bond strengths), and know that energy is needed to break bonds and is released when bonds are made.

- Use bond energies to estimate the heat released or absorbed in simple reactions.

Questions

More questions on energy changes can be found at the end of Section B on page 142.

1 a) Explain what is meant by an exothermic reaction, and write balanced equations for any *two* exothermic changes (apart from the combustion of heptane given in part **b**)).

   b) Heptane, C_7H_{16}, is a hydrocarbon found in petrol. The equation for the combustion of heptane is:

   $$C_7H_{16(l)} + 11O_{2(g)} \rightarrow 7CO_{2(g)} + 8H_2O_{(l)}$$

   $$\Delta H = -4817 \text{ kJ mol}^{-1}$$

   Draw a simple energy diagram to show the combustion of heptane. Show clearly how the figure of -4817 fits onto your diagram.

   c) Explain what is meant by an endothermic reaction, and write balanced equations for any *two* endothermic changes (apart from the photosynthesis reaction given in part **d**)).

   d) Photosynthesis involves the conversion of carbon dioxide and water into carbohydrates such as glucose, $C_6H_{12}O_6$, and oxygen.

   $$6CO_{2(g)} + 6H_2O_{(l)} \rightarrow C_6H_{12}O_{6(aq)} + 6O_{2(g)}$$

   $$\Delta H = +2820 \text{ kJ mol}^{-1}$$

   Draw a simple energy diagram to show the process of photosynthesis. Show clearly how the figure of +2820 fits onto your diagram.

2 Use the bond energies in the table to estimate the amount of heat released or absorbed when the following reactions take place. In each case, say whether the change is exothermic or endothermic.

Bond	C–H	C–Br	Br–Br	H–Br	H–H	Cl–Cl	H–Cl	O=O	O–H
Bond energy (kJ mol^{-1})	+413	+290	+193	+366	+436	+243	+432	+498	+464

   a) $CH_{4(g)} + Br_{2(g)} \rightarrow CH_3Br_{(g)} + HBr_{(g)}$. (The structure of CH_3Br is the same as that of CH_3Cl. See the diagram on page 139.)

   b) $H_{2(g)} + Cl_{2(g)} \rightarrow 2HCl_{(g)}$

   c) $2H_{2(g)} + O_{2(g)} \rightarrow 2H_2O_{(g)}$

3 Self heating cans are used to provide warm food in situations where it is inconvenient to use a more conventional form of heat. By doing an internet (or other) search, find out how self heating cans work. Write a short explanation of your findings (not exceeding 200 words) using a word processor. You should include equation(s) for any reaction(s) involved, and a diagram or picture if it is useful.

End of Section Questions

You may need to refer to the Periodic Table on page 306.

1   This question is about copper(II) sulphate solution which contains $Cu^{2+}_{(aq)}$ and $SO_4^{2-}_{(aq)}$ ions.

a)   The presence of the sulphate ions can be shown by adding dilute hydrochloric acid and barium chloride solution to produce a white precipitate.

i)   Name the white precipitate.

ii)   Write the ionic equation for its formation.
(2 marks)

b)   Describe a test which would show the presence of the copper(II) ions in the solution.   *(2 marks)*

c)   An excess of zinc powder was added to some copper(II) sulphate solution and the mixture was shaken thoroughly. A displacement reaction occurred. The ionic equation for the reaction is:

$$Zn_{(s)} + Cu^{2+}_{(aq)} \rightarrow Zn^{2+}_{(aq)} + Cu_{(s)}$$

i)   Describe two changes which you would expect to see during the reaction.

ii)   Which substance in the reaction has been oxidised? Explain your answer.   *(4 marks)*

d)   Copper can also be produced from copper(II) sulphate by electrolysis.

i)   At which electrode is the copper produced?

ii)   Write the equation for the reaction taking place at the electrode.

iii)   What type of reaction is occurring at this electrode? Explain your answer.   *(4 marks)*

e)   A drop of concentrated copper(II) sulphate solution was placed in the middle of a strip of damp filter paper. The ends of the paper were connected to the terminals of a battery.

i)   Describe what you would expect to see happen.

ii)   Explain your answer.   *(3 marks)*

Total 15 marks

2   This question is about salts.

a)   What would you add to dilute hydrochloric acid to make each of the following salts? In each case, say whether you would add it as a solid or in solution.

i)   copper(II) chloride

ii)   sodium chloride

iii)   silver chloride.   *(3 marks)*

b)   Silver chloride is a white solid which turns greyish on exposure to light. State one use for silver chloride based on this property.   *(1 mark)*

c)   Potassium sulphate is produced when dilute sulphuric acid reacts with potassium carbonate solution.

$$K_2CO_{3(aq)} + H_2SO_{4(aq)} \rightarrow K_2SO_{4(aq)} + CO_{2(g)} + H_2O_{(l)}$$

Given solutions of potassium carbonate and dilute sulphuric acid, an indicator, and suitable titration apparatus, describe how you would make a pure, neutral solution of potassium sulphate by a titration method. You should name the indicator you would choose to use, and state any important colour change(s).   *(6 marks)*

Total 10 marks

3   Sodium chloride solution was electrolysed using the apparatus in the diagram.

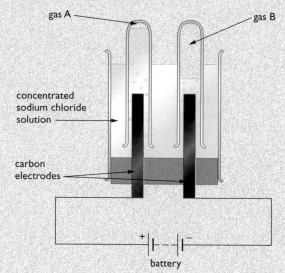

a)   Name i) gas A, ii) gas B.   *(2 marks)*

b)   Describe how you would test for i) gas A, ii) gas B.   *(4 marks)*

c)   Suppose the sodium chloride solution was replaced by potassium iodide solution. What differences would you observe (if any) at i) the positive electrode, ii) the negative electrode?   *(3 marks)*

d)   The same apparatus was used to electrolyse another solution, C. A brown solid was formed on the negative electrode, and an orange solution around the positive one. Suggest a possible identity for solution C.   *(2 marks)*

Total 11 marks

4 a) Chlorine was bubbled through a solution of potassium iodide.

  i) Describe, with a reason, any precaution that you would have to take in using the chlorine.

  ii) Balance the ionic equation for the reaction involved.

  $$Cl_{2(g)} + I^-_{(aq)} \rightarrow Cl^-_{(aq)} + I_{2(s)}$$

  iii) Describe what you would expect to see happen in the solution.

  iv) Describe the function of the chlorine in the reaction with the iodide ions.

  (7 marks)

b) Samples of a very pale green solution, **G**, were tested as follows:

Test	Observation
A sample of solution was acidified with dilute nitric acid and silver nitrate solution was added	A white precipitate (**H**) was formed
A small amount of sodium hydroxide solution was added to a sample of **G**	A dark green precipitate (**I**) was formed
Chlorine was bubbled through a sample of **G**	The pale green solution turned yellow (solution **J**)
A small amount of sodium hydroxide solution was added to solution **J**	An orange-brown precipitate (**K**) was formed

  i) Use the results from the first two tests to identify solution **G**.

  ii) Identify precipitates **I** and **K**.

  iii) Suggest the identity of solution **J**.

  iv) Write an ionic equation for the formation of **I**. *(6 marks)*

  Total 13 marks

5 Mendeleev produced his Periodic Table by arranging the elements in order of their atomic masses. When argon was discovered, its atomic mass turned out to be slightly higher than potassium's. In this instance, Mendeleev reversed the usual order in the Periodic Table.

Mendeleev's order			Atomic mass order		
Group 0	Group 1	Group 2	Group 0	Group 1	Group 2
Ne	Na	Mg	Ne	Na	Mg
Ar	K	Ca	K	Ar	Ca

a) State one physical property of potassium which suggests that it should be in the same Group as sodium rather than with neon. *(1 mark)*

b) Give any one chemical property of potassium that is similar to that of sodium. Say what the potassium reacts with and what is formed. Write the balanced equation for the reaction. *(4 marks)*

c) i) Draw dot-and-cross diagrams to show the electronic structures of sodium and potassium atoms.

  ii) What happens to these structures when sodium or potassium react to form compounds?

  iii) Explain why potassium is more reactive than sodium. *(6 marks)*

d) Argon is chemically unreactive and its molecules are monatomic. What is a monatomic molecule? Explain why argon's molecules are monatomic. *(3 marks)*

e) Give a use for argon. *(1 mark)*

  Total 15 marks

6 When nitrogen and hydrogen react together under suitable conditions, ammonia is formed.

$$N_{2(g)} + 3H_{2(g)} \rightarrow 2NH_{3(g)}$$

a) Use the following bond energies to calculate the energy change during the reaction.

bond	N≡N	H–H	N–H
bond energy (kJ mol^{-1})	945	436	391

  Hint: Be careful in adding up the number of N–H bonds in 2NH$_3$.

  (3 marks)

b) Is the reaction exothermic or endothermic? Explain your answer. *(2 marks)*

  Total 5 marks

7 Reactions can be described as (amongst other things): neutralisation, precipitation, redox, thermal decomposition. Decide which of these types of reaction each of the following equations represents.

a) $Zn_{(s)} + CuO_{(s)} \rightarrow ZnO_{(s)} + Cu_{(s)}$ *(1 mark)*

b) $ZnCO_{3(s)} \rightarrow ZnO_{(s)} + CO_{2(g)}$ *(1 mark)*

c) $ZnCO_{3(s)} + H_2SO_{4(aq)} \rightarrow ZnSO_{4(aq)} + CO_{2(g)} + H_2O_{(l)}$

                                       (1 mark)

d) $Zn^{2+}_{(aq)} + CO_3^{2-}_{(aq)} \rightarrow ZnCO_{3(s)}$ *(1 mark)*

e) $Zn_{(s)} + Pb^{2+}_{(aq)} \rightarrow Zn^{2+}_{(aq)} + Pb_{(s)}$ *(1 mark)*

f) $Zn_{(s)} + 2HCl_{(aq)} \rightarrow ZnCl_{2(aq)} + H_{2(g)}$ *(1 mark)*

Total 6 marks

Chapter 14: Metals

This chapter explores the extraction and uses of some important metals.

Extracting Metals from their Ores

Minerals and ores

Most metals are found in the Earth's crust combined with other elements. The individual compounds are called **minerals**.

Magnetite, Fe_3O_4

Haematite, Fe_2O_3

Pyrite (iron pyrites), FeS_2

The photographs show samples of some iron-containing minerals, but they are normally found mixed with other unwanted minerals in rocks. An **ore** contains enough of the mineral for it to be worthwhile to extract the metal.

The price of a metal is affected by how common the ore is and how difficult it is to extract the metal from the ore.

A few very unreactive metals like gold are found **native**. That means that they exist naturally as the uncombined element. The photograph shows some native gold. Silver and copper are also sometimes found native – although much more rarely.

Native gold on quartz.

If you have forgotten about oxidation and reduction, you would find it useful to re-read Chapter 6.

Extracting the metal

Many ores are either oxides or compounds which are easily converted to oxides. Sulphides like sphalerite (zinc blende), ZnS, can be easily converted into an oxide by heating in air.

$$2ZnS_{(s)} + 3O_{2(g)} \rightarrow 2ZnO_{(s)} + 2SO_{2(g)}$$

To obtain the metal, you have to remove the oxygen – removal of oxygen is called reduction. Metals exist as positive ions in their ionic compounds, and to produce the metal, you would have to add electrons to the ion. Addition of electrons is also called reduction.

Methods of extraction and the Reactivity Series

How a metal is extracted depends to a large extent on its position in the Reactivity Series. A manufacturer obviously wants to use the cheapest possible method of reducing an ore to the metal. There are two main economic factors to take into account:

potassium

sodium

calcium

magnesium

aluminium

(carbon)

zinc

iron

copper

A part of the Reactivity Series.

- the cost of energy
- the cost of the reducing agent.

145

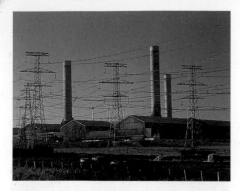

Aluminium production needs huge amounts of expensive electricity.

ideas
evidence

Bronze is an alloy of copper and tin.

Bauxite – essentially impure aluminium oxide.

For metals up to zinc in the Reactivity Series, the cheapest method of reducing the ore is often to heat it with carbon or carbon monoxide. Carbon is cheap and can also be used as the source of heat. The extraction of iron is a good example of this.

Ores of metals higher than zinc in the Reactivity Series can't be reduced using carbon at reasonable temperatures.

Metals above zinc are usually produced by electrolysis. The metal ions are given electrons directly by the cathode. Unfortunately, the large amounts of electricity involved make this an expensive process – and so a metal like aluminium is much more expensive than one like iron.

Some metals, like titanium, are extracted by heating the compound with a more reactive metal. This is also bound to be an expensive method, because the more reactive metal itself will have had to be extracted by an expensive process first.

The Reactivity Series and the history of metal use

Why was the Bronze Age before the Iron Age? Why wasn't aluminium discovered until 1827?

Bronze is an alloy of copper and tin, both of which are low in the Reactivity Series. Both can be made easily from their ores by heating them with carbon. You can imagine the metals being found accidentally when charcoal (a form of impure carbon) in a fire came into contact with stones containing copper or tin ores.

Iron can also be made from its ores by heating them with carbon, but higher temperatures are needed. The iron produced is also more difficult to purify into a useful form than copper is. Iron therefore wasn't in common use until much later than bronze.

Because aluminium is above carbon in the Reactivity Series, it can't be made accidentally by heating aluminium oxide with carbon. It has to be extracted using electrolysis, and so it was impossible to get metallic aluminium before the discovery of electricity. This is true of all the metals from aluminium upwards in the Reactivity Series.

Aluminium

Extraction

Aluminium is the most common metal in the Earth's crust, making up 7.5% by mass. Its main ore is **bauxite** – a clay mineral which you can think of as impure aluminium oxide.

The bauxite is first treated to produce pure aluminium oxide. You don't need to know how this is done for GCSE purposes.

Because aluminium is a fairly reactive metal it has to be extracted using electrolysis. Aluminium oxide, however, has a very high melting point and it isn't practical to electrolyse molten aluminium oxide.

Instead the aluminium oxide is dissolved in molten **cryolite**. This is another aluminium compound which melts at a more reasonable temperature. The electrolyte is a solution of aluminium oxide in molten cryolite at a temperature of about 1000°C.

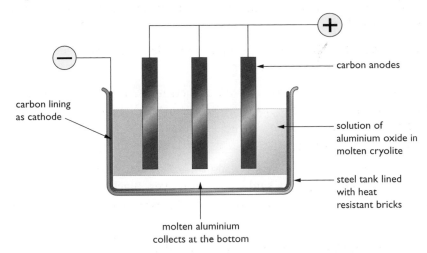

The cell room in an aluminium smelter.

The diagram shows a very simplified view of the electrolysis cell. The molten aluminium is siphoned off from time to time, and fresh aluminium oxide is added to the cell. The cell operates at about 5–6 volts, but with currents of up to about 100,000 amps. The heat generated by the huge current keeps the electrolyte molten.

The chemistry of the process

Aluminium ions are attracted to the cathode and are reduced to aluminium by gaining electrons.

$$Al^{3+}_{(l)} + 3e^- \rightarrow Al_{(l)}$$

> If you aren't sure about electrolysis, you ought to read Chapter 11 before you go on.

The molten aluminium produced sinks to the bottom of the cell.

The oxide ions are attracted to the anode and lose electrons to form oxygen gas.

$$2O^{2-}_{(l)} \rightarrow O_{2(g)} + 4e^-$$

This creates a problem. Because of the high temperatures, the carbon anodes burn in the oxygen to form carbon dioxide. The anodes have to be replaced regularly and this adds to the expense of the process.

Anodising aluminium

Given its position in the Reactivity Series, aluminium is a surprisingly unreactive metal under normal circumstances. Even apparently shiny aluminium is covered in a microscopically thin, but very strong, layer of aluminium oxide which prevents air and water getting at the aluminium and corroding it. This layer can be made even stronger by **anodising** the aluminium.

The aluminium is etched by treating it with sodium hydroxide solution. This removes existing aluminium oxide and produces the satin or matt finishes which are commonly seen in anodised aluminium.

The aluminium article is then made the anode in an electrolysis of dilute sulphuric acid. Oxygen is given off at the anode and this reacts with the aluminium to build up a thin film (about 0.02 mm thick) of aluminium oxide.

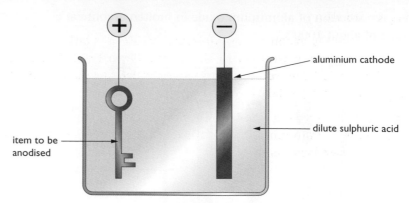

aluminium cathode

dilute sulphuric acid

item to be anodised

Dyed anodised aluminium.

Apart from a very thin barrier layer of aluminium oxide next to the aluminium, the film is porous, and will absorb dyes. Black, gold and bronze coloured dyes are typical of anodised aluminium. After the aluminium has been dyed, further treatment makes the surface completely non-porous.

Summary

Anodising aluminium:

- makes it more corrosion resistant by strengthening the oxide film
- enables its surface to be dyed.

Uses of aluminium

Pure aluminium isn't very strong, so aluminium alloys are normally used instead. The aluminium can be strengthened by mixing it with other elements such as silicon, copper or magnesium.

Aluminium's uses depend on its low density and strength (when alloyed), its ability to conduct electricity and heat, its appearance, and its ability to resist corrosion.

Aluminium resists corrosion, is light and strong...

...and light vehicles need less energy to move them.

Aluminium has a shiny appearance, resists corrosion, is light and a good conductor of heat.

Aluminium resists corrosion, is light and a good conductor of electricity. The aluminium in the cables is strengthened by a core of steel.

Recycling aluminium

Although aluminium is so common in the Earth's crust, it is only concentrated enough to be worth extracting in ores like bauxite. As with all other ores, these are a **finite resource** and will eventually run out – although in this case, that is going to take a very long time.

The main reason for recycling aluminium is that it is so expensive to produce. Recycling aluminium uses only about 5% of the energy needed to extract the same mass from its ore.

Iron

Extraction of iron using a blast furnace

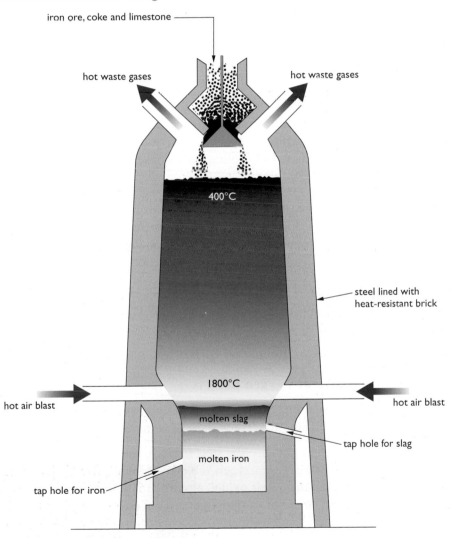

The hot waste gases at the top of the furnace are piped away and used to heat the air blast at the bottom.

Coke is impure carbon, and it burns in the hot air blast to form carbon dioxide. This is a strongly exothermic reaction.

$$C_{(s)} + O_{2(g)} \rightarrow CO_{2(g)}$$

An exothermic reaction is one which gives out heat.

Molten iron being tapped from a blast furnace.

At the high temperatures in the furnace, the carbon dioxide is reduced by more carbon to give carbon monoxide.

$$CO_{2(g)} + C_{(s)} \rightarrow 2CO_{(g)}$$

It is the carbon monoxide which is the main reducing agent in the furnace – especially in the cooler parts. Assuming that the iron ore is haematite, Fe_2O_3:

$$Fe_2O_{3(s)} + 3CO_{(g)} \rightarrow 2Fe_{(l)} + 3CO_{2(g)}$$

The iron melts and flows to the bottom of the furnace where it can be tapped off.

In the hotter parts of the furnace, some of the iron oxide is also reduced by carbon itself.

$$Fe_2O_{3(s)} + 3C_{(s)} \rightarrow 2Fe_{(l)} + 3CO_{(g)}$$

Notice that carbon monoxide, rather than carbon dioxide, is formed at these temperatures.

The limestone is added to the furnace to remove impurities in the ore which would otherwise clog the furnace with solid material.

> Thermal decomposition is splitting a compound into simpler bits using heat.

The furnace is hot enough for the limestone (calcium carbonate) to undergo thermal decomposition. It splits up into calcium oxide and carbon dioxide. This is an endothermic reaction (it absorbs heat) and it is important not to add too much limestone to avoid cooling the furnace.

$$CaCO_{3(s)} \rightarrow CaO_{(s)} + CO_{2(g)}$$

> Calcium oxide is a basic oxide because it reacts with acids to form salts. Calcium silicate is a salt formed when calcium oxide and silicon dioxide react. Silicon dioxide is therefore described as an acidic oxide.

Calcium oxide is a basic oxide, and its function is to react with acidic oxides like silicon dioxide, SiO_2. Silicon dioxide occurs naturally as quartz, and is typical of the sort of impurities that need to be removed from the furnace.

$$CaO_{(s)} + SiO_{2(s)} \rightarrow CaSiO_{3(l)}$$

The product is calcium silicate. This melts and trickles to the bottom of the furnace as a molten **slag**, which floats on top of the molten iron, and can be tapped off separately.

Slag is used in road making and in the manufacture of cement.

Steel making – the basic oxygen process

The iron from the blast furnace contains lots of impurities including carbon, sulphur, silicon, and phosphorus. These have to be removed to make steel.

Recycled scrap iron is first put into the steel furnace, followed by molten iron from the blast furnace. Pure oxygen is blown through, and the impurities are converted into their oxides. Some of these disappear as gases (CO_2 and SO_2). (In fact, most of the sulphur is removed first to avoid the production of poisonous sulphur dioxide. Knowledge of this process isn't required at GCSE.)

To remove the other inpurities, quicklime (calcium oxide) is added; this combines with them to make a slag which separates from the iron. For example, silicon dioxide would combine to make calcium silicate, exactly as in the blast furnace.

Blowing oxygen through the impure molten iron.

The slag floats on top of the iron and can be skimmed off.

Finally, small calculated amounts of other metals and non-metals are added to the pure iron to produce various different kinds of steel.

Properties and uses of the different kinds of iron

Cast iron

Molten iron straight from the furnace can be cooled rapidly and solidified by running it into sand moulds. This is known as **pig iron**. If the pig iron is re-melted and cooled under controlled conditions, **cast iron** is formed. This is very impure iron, containing about 4% carbon as its main impurity.

Cast iron is very fluid when it is molten and doesn't shrink much when it solidifies, and that makes it ideal for making castings. Unfortunately, although cast iron is very hard, it is also very brittle – tending to shatter if it is hit hard. It is used for things like manhole covers, guttering and drainpipes, and cylinder blocks in car engines.

Wrought iron

Pure iron is known as **wrought iron**. It was once used to make decorative gates and railings, but has now been largely replaced by mild steel. The purity of the iron makes it very easy to work because it is fairly soft, but the softness and lack of strength mean that it isn't useful for structural purposes.

Mild steel

Mild steel is iron containing up to about 0.25% of carbon. This small amount of carbon increases the hardness and strength of the iron. It is used for (amongst other things) wire, nails, car bodies, ship building, girders and bridges.

The first ever iron bridge (at Ironbridge in the Severn gorge) was made of cast iron.

Mild steel.

More mild steel.

High carbon steel

High carbon steel is iron containing up to 1.5% carbon. Increasing the carbon content makes the iron even harder, but at the same time it gets more brittle. High carbon steel is used for cutting tools and masonry nails. Masonry nails are designed to be hammered into concrete blocks or brickwork, where a mild steel nail would bend. If you miss-hit a masonry nail, it tends to fracture into two bits because of its increased brittleness.

Stainless steel

Stainless steel is an alloy of iron with chromium and nickel. Chromium and nickel form strong oxide layers in the same way as aluminium, and these oxide layers protect the iron as well. Stainless steel is therefore very resistant to corrosion.

Even more mild steel.

Stainless steel in a modern winery.

A mild steel spade tarnishes – a stainless steel one doesn't.

Obvious uses include kitchen sinks, saucepans, knives and forks, and gardening tools, but there are also major uses for it in the brewing, dairy and chemical industries where corrosion-resistant vessels are essential.

Other alloy steels

There are lots of other specialist steels in which the iron is mixed with other elements. For example...

Titanium steel (iron alloyed with titanium) has the ability to withstand high temperatures and so is used in gas turbines and spacecraft, for example.

Manganese steel is iron alloyed with about 13% manganese and is intensely hard. It is used in rock-breaking machinery, for railway track subjected to very hard wear (for example, points), and for military helmets.

Types of steel	Iron mixed with	Some uses
mild steel	up to 0.25% carbon	nails, car bodies, ship building, girders
high carbon steel	0.25–1.5% carbon	cutting tools, masonry nails
stainless steel	chromium and nickel	cutlery, cooking utensils, kitchen sinks
titanium steel	titanium	gas turbines, spacecraft
manganese steel	manganese	rock-breaking machinery, military helmets

Rusting

Iron rusts in the presence of oxygen and water. Rusting is accelerated in the presence of electrolytes like salt. **Warning!** Many metals corrode, but it is only the corrosion of iron that is referred to as *rusting*.

Rusting is accelerated by salty water.

152

The formula of rust is $Fe_2O_3 \cdot xH_2O$, where x is a variable number. It simply behaves as a mixture of iron(III) oxide and water.

Forming this from iron is a surprisingly complicated process. The iron loses electrons to form iron(II) ions, Fe^{2+}, which are then oxidised by the air to iron(III) ions, Fe^{3+}. Reactions involving the water produce the actual rust.

Preventing rusting by using barriers

The most obvious way of preventing rusting is to keep water and oxygen away from the iron. You can do this by painting it or coating it in oil or grease, but once the coating is broken the iron will rust.

Preventing rusting by alloying the iron

We have already seen that alloying the iron with chromium and nickel to produce stainless steel prevents the iron from rusting. Even if the surface is scratched the stainless steel still won't rust. Unfortunately, stainless steel is expensive.

Preventing rusting by using sacrificial metals

Galvanised iron is steel which is coated with a layer of zinc. As long as the zinc layer is unscratched, it serves as a barrier to air and water. However, the iron still doesn't rust even when the surface is broken.

Zinc is more reactive than iron, and so corrodes instead of the iron. During the process, the zinc loses electrons to form zinc ions.

$$Zn_{(s)} \rightarrow Zn^{2+}_{(aq)} + 2e^-$$

Galvanised steel doesn't rust, even in constant contact with air and water.

These electrons flow into the iron. Any iron atom which has lost electrons to form an ion immediately regains them. If the iron can't form ions, it can't rust.

Zinc blocks are attached to metal hulls or keels for the same reason. The corrosion of the more reactive zinc prevents the iron rusting. Such blocks are called **sacrificial anodes**.

Underground pipelines are also protected using sacrificial anodes. In this case, sacks containing lumps of magnesium are attached at intervals along the pipe. The very reactive magnesium corrodes in preference to the iron. The electrons produced as the magnesium forms its ions prevent ionisation of the iron.

A sacrificial anode attached to a boat.

Titanium

Extraction

Titanium makes up about 0.6% of the Earth's crust. It is corrosion resistant and very strong, with a high melting point. Its density is only about 60% that of iron. Despite all this, it isn't commonly used because it is so expensive. Titanium is difficult to extract from its ore, rutile – titanium(IV) oxide.

Traces of other elements like carbon, oxygen or nitrogen in titanium make the metal brittle. This means that the extraction method must prevent the titanium getting contaminated with any of these.

Rutile – titanium(IV) oxide.

For GCSE, you won't need to know how the titanium(IV) chloride is produced.

It isn't easy to explain why this is a reduction in GCSE terms because $TiCl_4$ is covalent. If it were ionic, the Ti^{4+} ion would be gaining electrons to make Ti.

What is actually happening is:

$$TiCl_4 + 4e^- \rightarrow Ti + 4Cl^-$$

The titanium(IV) oxide is first converted to titanium(IV) chloride. This is then heated with a more reactive metal like sodium or magnesium to displace the titanium. Titanium is close to aluminium in the Reactivity Series. For example:

$$TiCl_{4(g)} + 4Na_{(l)} \rightarrow Ti_{(s)} + 4NaCl_{(s)}$$

The sodium reduces the titanium(IV) chloride to titanium.

The extraction is done in an inert atmosphere of argon to prevent any contamination of the titanium with either oxygen or nitrogen from the air.

Uses of titanium

The all-titanium SR-71 is the fastest and highest-flying aeroplane in the world. It uses titanium because of its strength, lightness and high melting point.

The all-titanium SR-71 Blackbird.

In more ordinary planes, titanium is used to make jet engine components for similar reasons.

Titanium is also used in other applications as varied as replacement hip joints (lightness, strength and corrosion resistance) and tubes for use in nuclear reactors (strength and corrosion resistance).

Copper

The extraction of copper isn't required by any of the GCSE syllabuses, but its purification is.

Purifying copper

The purification of copper uses the electrolysis of copper(II) sulphate solution. The anode is impure copper, and the cathode is made of pure copper.

The copper from the anode goes into solution as Cu^{2+} ions.

$$Cu_{(s)} \rightarrow Cu^{2+}_{(aq)} + 2e^-$$

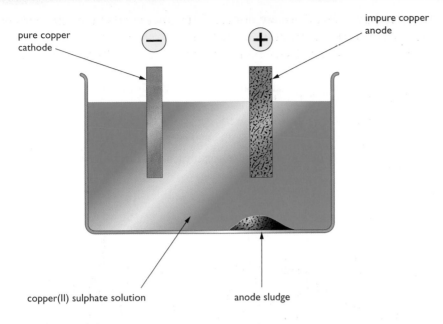

pure copper cathode

impure copper anode

copper(II) sulphate solution

anode sludge

The electrolysis of copper(II) sulphate solution using copper electrodes is explained in detail on page 125.

Copper refining.

At the cathode, Cu^{2+} ions are deposited as copper.

$$Cu^{2+}_{(aq)} + 2e^- \rightarrow Cu_{(s)}$$

The impure copper anode gradually disappears, and pure copper is plated on to the cathode.

What happens to the impurities?

There are three sorts of impurities to think about: unreactive material left from the copper ore, metals below copper in the Reactivity Series, and metals above it.

Any metal in the impure anode which is lower than copper in the Reactivity Series (like silver) doesn't go into solution as ions. It remains as a metal and falls to the bottom of the cell with other unreactive material left from the ore. This is the "anode sludge".

Metals higher than copper in the Reactivity Series *do* go into solution as their ions. They then remain in the solution. They won't be discharged when they get to the cathode as long as their concentration doesn't get too high. The copper(II) sulphate solution has to be topped up to avoid that happening.

Electroplating

Electroplating is a process very similar to the purification of copper. Suppose you wanted to plate copper onto steel. The steel object would be made the cathode in the electrolysis of copper(II) sulphate solution using a copper anode.

Copper would be transferred from the anode to the cathode, forming a layer of copper over the steel. The thickness and strength of the layer can be controlled by choice of the current and the time it is allowed to flow.

Silver, gold, nickel or chromium can all be plated on to other metals in a similar way.

To silver plate an object, it is hung in a solution of silver nitrate and is made the cathode in an electric circuit. The anode is a piece of silver.

Silver plate. Silver can easily be electroplated onto other metals. This is much cheaper than using solid silver.

Brass is a copper/zinc alloy.

"Silver" coins.

At the cathode, silver ions are discharged by picking up an electron to form metallic silver which plates on to the article.

$$Ag^+_{(aq)} + e^- \rightarrow Ag_{(s)}$$

At the anode, exactly the same amount of silver goes into solution as silver ions.

$$Ag_{(s)} \rightarrow Ag^+_{(aq)} + e^-$$

The concentration of the silver nitrate solution is unchanged.

Uses of copper

Uses of copper include:

- Electrical wiring, because it is a good conductor of electricity and is easily drawn into wires.

- Domestic plumbing, because it doesn't react with water, and is easily bent into shape.

- Making brass. Brass is an alloy of copper and zinc in various proportions. Brass is harder than either copper or zinc individually.

- Coinage. "Silver" coins are made from an alloy of copper and nickel ("cupronickel"), containing 75% copper and 25% nickel (except for 20p coins which are 84% copper and 16% nickel. If you look closely, you should be able to see the difference).

End of Chapter Checklist

If you haven't got a copy of your specification, read the introduction on page vi.

You will need to be able to do some or all of the following. Check your Awarding Body's specification (syllabus) to find out exactly what you need to know.

- Know what is meant by a *mineral* and an *ore* and understand that some metals are found *native*.

- Understand that the method of extraction of a metal, and how long it has been in use are related to its position in the Reactivity Series.

- Describe and explain the extraction of aluminium, including electrode equations. Explain why the process is expensive.

- Know why aluminium appears less reactive than its position in the Reactivity Series would suggest.

- Describe and explain the process of anodising aluminium.

- Relate the uses of aluminium to its properties.

- Describe and explain the extraction of iron using a blast furnace, including equations for the reactions involved.

- Explain the formation of slag, and give a use for it.

- Describe and explain the production of steel by the basic oxygen process.

- Give uses for cast iron, wrought iron, mild steel, high carbon steel, stainless steel, titanium steel and manganese steel, and relate these uses to the properties of the metal.

- State the conditions necessary for iron to rust and explain the various ways of preventing rusting.

- Describe and explain the extraction and uses of titanium.

- Describe the purification of copper and relate its uses to its properties.

- Describe and explain how electroplating is carried out.

- Know the composition of brass and cupronickel alloys.

Questions

More questions on metals can be found at the end of Section C on page 177.

1   Give two uses for each of the following metals. In each case explain why the metal is particularly suitable for that purpose.

   a) aluminium, *b)* iron (as mild steel), *c)* titanium, *d)* copper.

2   Sodium is the sixth most abundant element in the Earth's crust, occurring in large quantities as common salt, NaCl, and yet sodium metal wasn't produced until the early nineteenth century.

a)   From your knowledge of the position of sodium in the Reactivity Series, suggest a method for manufacturing sodium from sodium chloride. You aren't expected to give details of the manufacturing process, but should describe and explain (including equation(s) where relevant) how sodium is formed in your process.

b)   Explain why sodium wasn't first produced until the early nineteenth century.

c)   Suggest three other metals which might have been first isolated from their compounds at the same time.

d) What is bronze? Why has bronze been known for thousands of years?

3 a) Name the ore from which aluminium is extracted.

b) Aluminium is manufactured using electrolysis. Carbon electrodes are used. Describe the nature of the electrolyte.

c) At which electrode is the aluminium produced?

d) Write the electrode equation for the formation of the aluminium. Is this an example of oxidation or reduction?

e) Oxygen gas is formed at the other electrode. Explain why that causes a problem.

f) Aluminium is frequently *anodised*.

   i) Describe how you would anodise a piece of sheet aluminium.

   ii) Give two reasons why a manufacturer might choose to anodise an aluminium product.

g) Aluminium alloys are used in aircraft construction.

   i) What property of aluminium makes it particularly suitable for this purpose?

   ii) Why are aluminium alloys used in preference to pure aluminium?

4 The following reactions take place in a blast furnace:

A $C_{(s)} + O_{2(g)} \rightarrow CO_{2(g)}$ $\Delta H = -394 \text{ kJ mol}^{-1}$

B $CO_{2(g)} + C_{(s)} \rightarrow 2CO_{(g)}$ $\Delta H = +172 \text{ kJ mol}^{-1}$

C $Fe_2O_{3(s)} + 3CO_{(g)} \rightarrow 2Fe_{(l)} + 3CO_{2(g)}$

                                  $\Delta H = +4 \text{ kJ mol}^{-1}$

D $CaCO_{3(s)} \rightarrow CaO_{(s)} + CO_{2(g)}$ $\Delta H = +178 \text{ kJ mol}^{-1}$

a) Which one of these reactions provides the heat to maintain the temperature of the furnace?

b) What materials are put into the furnace to provide sources of *i)* carbon, *ii)* oxygen, *iii)* iron(III) oxide, *iv)* calcium carbonate?

c) What types of reaction are reactions B and C? Explain your answers.

d) The calcium oxide produced in reaction D takes part in the formation of slag.

   i) Give a chemical name for slag.

   ii) Write an equation to show its formation.

   iii) State one use for slag.

e) Some iron is also produced by reaction between iron(III) oxide and carbon. Balance the following equation:

    $Fe_2O_{3(s)} + C_{(s)} \rightarrow Fe_{(l)} + CO_{(g)}$

5 a) Cast iron or pig iron from the bottom of the blast furnace contains an important impurity which limits its usefulness.

   i) What is the impurity, and approximately what percentage of the cast iron does it make up?

   ii) What effect does this impurity have on the properties of cast iron which limits its usefulness?

b) Mild steel is made by removing this and other impurities. Describe briefly how this is done, explaining what happens to the impurities.

c) Describe the composition of stainless steel, and explain why it resists corrosion. State one use for stainless steel.

d) Car bodies used to be made from mild steel which was then painted. In more recent cars, the steel is galvanised before it is painted. Describe and explain the effect that this has on the life of the car.

6 Imagine that your local magazine or newspaper has started a science and technology section. Write a short, illustrated article not exceeding 200 words about any metal of your choice. You should use language which would be understood by a general reader, and the content should be of interest to people in your area. Your article should be entirely computer generated. Illustrations could be scanned in or taken from the internet.

7 Explain the reasons behind each of the following statements.

a) Although aluminium is fairly high in the Reactivity Series, it is slow to corrode.

b) Titanium is used to make artificial hip joints.

c) Overhead power cables are made of aluminium with a steel core.

d) Blocks of zinc are attached to the hulls of steel ships.

e) Titanium is extracted by heating titanium(IV) chloride with sodium in an atmosphere of argon.

f) Gold is found native (uncombined) in nature, and yet gold is a very expensive metal.

g) It is a good idea to recycle aluminium.

8 a) Copper is purified by electrolysis. Describe briefly how the process is carried out and write the equations for the reactions occurring at the electrodes.

b) Give two uses for pure copper, stating the properties on which these uses depend.

c) Copper is alloyed with zinc to make brass. Give two differences between the physical properties of pure copper and brass.

9 Given a nickel spatula, some nickel(II) sulphate solution and any other equipment that you might reasonably expect to find in your chemistry lab, describe how you would nickel plate a brass screw.

Section C: Large Scale Chemistry

Chapter 15: Salt and Limestone

Salt and limestone are two common materials which provide the source of a number of important chemicals. This chapter explores some of the large scale uses of salt and limestone.

The sun is used to evaporate shallow pans of sea water in Vietnam.

Salt

Sources of salt (sodium chloride)

All sorts of soluble ions are dissolved from rocks by rain water and end up in the sea. Others are produced by underwater volcanic activity.

The most common of these ions in the sea are sodium ions, Na^+, and chloride ions, Cl^-, but sea water contains many, many other ions as well – for example, potassium, K^+; magnesium, Mg^{2+}; calcium, Ca^{2+}; sulphate, SO_4^{2-}; bromide, Br^-; and hydrogencarbonate, HCO_3^-, to name just a few.

All the salt we now use is either obtained directly from the sea or from salt deposits which are the remains of ancient dried up seas. Untreated "sea salt" will, of course, contain sodium chloride mixed with small amounts of lots of other ionic compounds.

Inside a salt mine. The salt is the remains of an ancient sea.

Rock salt.

The mined salt is known as **rock salt**. Crushed rock salt is used to treat roads under icy conditions. It produces a thin layer of salt solution on the road. Salt solution has a lower freezing point than water. This means that the temperature will have to fall to well below 0°C before the solution will freeze.

There are large salt deposits under Cheshire. The majority of it is extracted using **solution mining**. Very hot water under pressure is pumped down into the salt deposits. The salt dissolves and is forced back to the surface as concentrated salt solution – **brine**.

This solution can either be evaporated to produce table salt, or used directly by the chemical industry.

The chlor-alkali industry

The salt solution is purified to remove ions other than sodium and chloride ions and is then electrolysed to produce three useful chemicals – sodium hydroxide, chlorine and hydrogen. The electrolysis can be carried out in a **membrane cell**.

The cell is designed to keep the products apart. If chlorine comes into contact with sodium hydroxide solution, it reacts to make bleach – a mixture of sodium chloride and sodium chlorate(I) solution. If chlorine comes into contact with hydrogen, it produces a mixture which would explode violently on exposure to sunlight or heat.

If you haven't done any electrolysis recently, you may need to read pages 122–124 which describe and explain the electrolysis of sodium chloride solution in the lab. The industrial process is a straightforward modification of this.

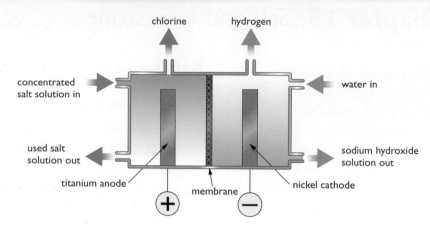

Explaining what's happening

At the anode, chloride ions are discharged to produce **chlorine** gas.

$$2Cl^-_{(aq)} \rightarrow Cl_{2(g)} + 2e^-$$

The membrane is designed so that only sodium ions can pass through it. The positive sodium ions are attracted into the right-hand compartment by the negatively charged electrode.

It is too difficult to discharge sodium ions, so hydrogen ions from the water are discharged instead to produce **hydrogen** gas.

$$2H^+_{(aq)} + 2e^- \rightarrow H_{2(g)}$$

This is the most difficult part of this explanation. If you aren't happy about it, read page 123 where it is explained in more detail.

More and more water keeps splitting up to replace the hydrogen ions as soon as they are discharged. Each time a water molecule splits up it produces a hydroxide ion as well. This means that there will be a build up of sodium ions and hydroxide ions in the right-hand compartment – **sodium hydroxide solution** is formed.

Uses of hydrogen, chlorine and sodium hydroxide

Hydrogen is used:

You might feel that this is a slightly daunting list. Don't panic! You will come across the great majority of it elsewhere in the course. Otherwise, questions are quite likely to be of the form "State one use for sodium hydroxide". The worst case is that you could lose 1 mark if you've completely forgotten this list.

* in the manufacture of ammonia (see page 169)
* in the manufacture of margarine (see page 247)
* as a fuel for the future (see pages 132–133).

Chlorine is used:

* to sterilise water (see page 195)
* to make bleach (see pages 113 and 159)
* to make hydrochloric acid (by controlled reaction with hydrogen)
* to make antiseptics and disinfectants (for example, TCP – trichlorophenol)
* to make PVC (see pages 215–216).

Sodium hydroxide is used:

* in the purification of bauxite to make aluminium oxide ("alumina"): the aluminium oxide is used to manufacture aluminium (see pages 146–147), and in the production of ceramics; the alumina industry is the largest user of sodium hydroxide in Australia

- in paper making: the sodium hydroxide helps to break the wood down into pulp
- in soap making: sodium hydroxide reacts with animal and vegetable fats and oils to make compounds like sodium stearate, which are present in soap (see page 190).

Limestone

Chemically, limestone is calcium carbonate. It is a **sedimentary rock** formed from the shells and skeletons of marine creatures, which fell to the bottom of ancient seas and were then turned into rock by more sediments forming on top of them. Some limestones were made from small, egg-like spheres (oolites) of calcium carbonate precipitated from sea water.

Geologically there are a wide range of limestones – from soft, crumbly white rocks like chalk to some very hard, dark grey ones.

Limestone extraction

In the UK, more than 100 million tonnes of limestone are mined every year, and much of this comes from areas of outstanding natural beauty. Production of limestone in the Peak District National Park, for example, exceeded 8 million tonnes in 1991, but has decreased since. There is an obvious conflict between the demands of the local tourist industry and the mining industry.

Here are some of the things that might need to be considered. You can probably think of others as well.

In favour of mining:

- good employment for local people
- provides all-year-round income for local businesses
- Peak District limestone is very pure, which makes it useful to the chemical industry.

Against mining:

- destroys landscape
- noise and dust pollution
- transport problems
- losses to the tourist industry.

Uses of limestone

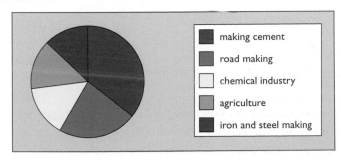

- making cement
- road making
- chemical industry
- agriculture
- iron and steel making

Limestone scenery at Malham Cove in the Yorkshire Dales.

Limestone extraction in the Peak District National Park.

The cement works at Hope in the Peak District National Park.

Cement and concrete

Most limestone is used to make **cement**. This is made by strongly heating a slurry of crushed limestone and clay in water.

Cement works are often found in areas which are on the geological boundary between clay and limestone, in order to minimise transport costs.

Don't confuse cement and **concrete**. Concrete is produced from a moistened mixture of cement, sand and aggregate – a mixture of assorted sized rock fragments. The cement hardens to bind the sand and aggregate together into a solid mass.

Glass

Amongst the "chemical industry" uses of limestone is the manufacture of **glass**. Glass is made by heating together a mixture of limestone, sand and sodium carbonate.

The thermal decomposition of limestone to give quicklime

Limestone is heated in a lime kiln, where it is thermally decomposed to give **quicklime** (calcium oxide).

$$CaCO_{3(s)} \rightarrow CaO_{(s)} + CO_{2(g)}$$

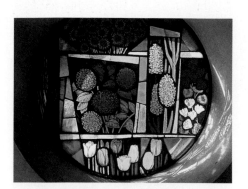

Limestone is used to make glass.

Remember that thermal decomposition is the use of heat to split a compound into simpler substances.

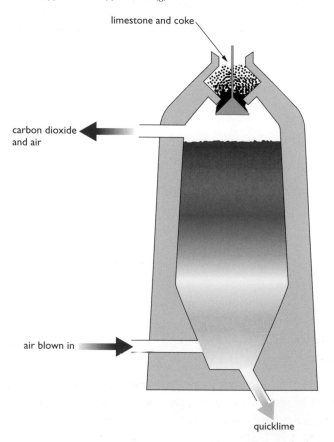

A lime kiln.

The limestone is mixed with powdered coke (an impure form of carbon); this burns in the air which is blown through the kiln to provide the heat needed for the decomposition. Coal can also be used.

The air also serves another purpose. If carbon dioxide is left in contact with the calcium oxide, it will tend to recombine with it again. The air carries away all the carbon dioxide and prevents that from happening.

Quicklime has been made by similar methods for thousands of years. Old lime kilns are quite common around the countryside.

Much quicklime is now produced in **rotary kilns**. These consist of huge rotating cylinders, with one end somewhat higher than the other. The mixture of limestone and fuel (such as coke or coal) works its way down the length of the kiln because of the rotating motion and the slope.

Apart from the different engineering of the kiln, what happens inside it is identical to what happens inside a vertical kiln.

Old lime kilns at Beadnell, Northumberland.

Converting quicklime into slaked lime (calcium hydroxide)

If water is added to quicklime, **slaked lime** (calcium hydroxide) is formed.

$$CaO_{(s)} + H_2O_{(l)} \rightarrow Ca(OH)_{2(s)}$$

The lump of calcium oxide swells and crumbles. The reaction is strongly exothermic, and the heat evolved boils the water to produce steam.

If you add a lot of water, some of the calcium hydroxide will dissolve. A solution of calcium hydroxide is known as **lime water**.

Lime water is used as a test for carbon dioxide. The white precipitate formed is calcium carbonate.

$$Ca(OH)_{2(aq)} + CO_{2(g)} \rightarrow CaCO_{3(s)} + H_2O_{(l)}$$

Adding water to calcium oxide.

Uses of quicklime and slaked lime

Quicklime (calcium oxide)

The use of quicklime in steel making was described on page 150. The calcium oxide reacts with acidic oxides produced during steel making to give an easily removable slag.

Quicklime is also used to produce slaked lime.

Slaked lime (calcium hydroxide)

Slaked lime is a base, and can be used to neutralise excess acidity either in soil or water. The calcium hydroxide will react with acids present to produce calcium salts.

Most crops grow best if the pH is just on the acid side of 7, and some, like members of the cabbage family, prefer a slightly alkaline soil. Farmers add slaked lime to raise the pH of soils which are too acidic.

Carbonates also react with acids, and so calcium carbonate could be used instead – as powdered limestone or beach sand. Sand contains shell fragments which are mainly calcium carbonate. However, you need to add a greater mass of calcium carbonate to achieve the same effect.

You can also use limestone or slaked lime to neutralise the water in lakes which have become too acidic because of acid rain, and slaked lime is added to drinking water in areas where the water is naturally too acidic.

Acid rain is dealt with on pages 183–184, and water treatment on pages 194–195.

Cabbages prefer a slightly alkaline soil.

Chapter 15: Salt and Limestone

End of Chapter Checklist

If you haven't got a copy of your specification, read the introduction on page vi.

You will need to be able to do some or all of the following. Check your Awarding Body's specification (syllabus) to find out exactly what you need to know.

- Know that salt can be extracted from sea water by evaporation, and describe some of its uses.

- Describe the industrial electrolysis of sodium chloride solution to give hydrogen, chlorine and sodium hydroxide. Give uses for each of the products.

- Understand the formation of limestone, and explain some of the environmental and economic conflicts that may arise in its extraction.

- Describe the conversion of limestone into quicklime and slaked lime, and give uses for limestone, quicklime and slaked lime.

Questions

More questions on salt and limestone can be found at the end of Section C on page 177.

1 Sodium chloride can be extracted by the evaporation of sea water or by solution mining of underground salt deposits.

   a) Why isn't salt extracted by the evaporation of sea water in Great Britain?

   b) Explain briefly how *solution mining* is carried out.

   c) Salt can also be mined conventionally from underground deposits as *rock salt*. State one use for rock salt.

   d) Electrolysis of concentrated salt solution leads to three important products, gases **A** and **B**, and solution **C**.

      i) Name **A**, **B** and **C**.

      ii) Write electrode equations to show the formation of gases **A** and **B**.

      iii) Give one use for each of **A**, **B** and **C**.

2 Imagine that you live in a village or small town near a limestone quarry. The quarry owners have applied for planning permission to double the size of the quarry. Write two brief letters to the planning enquiry, one in favour of the application, and the other against it. Each letter should fit onto one side of A4 paper, including enough space for your address at the top and signature at the bottom. Your letter should be word processed using single spacing and a 12 point Times (or similar) font. You can use your imagination to fill in details of the area and local economy, but you must use the same details for both letters. For example, if you choose to have the quarry in a National Park, it must be in the National Park for *both* letters. If the extension allows it to employ an extra 50 local people in one letter, the same must be true in the other one as well.

3 Limestone is converted into quicklime by heating it in a lime kiln with powdered coke (impure carbon). A constant stream of air is blown through the kiln.

   a) Write equations for:

      i) the reaction which produces the heat in the kiln

      ii) the reaction which produces the quicklime.

   b) Give two reasons for the stream of air blown through the kiln.

   c) Quicklime can be converted to slaked lime (calcium hydroxide) by adding water to it. Describe what you would see if you did this.

   d) Give a different use in each case for *i)* limestone, *ii)* quicklime, *iii)* slaked lime.

4 You are given samples of white chalk from the cliffs at Dover and a dark grey limestone, and you are told that they are both almost pure calcium carbonate.

   a) How would you show that they both contain *i)* calcium ions, *ii)* carbonate ions?

   b) Design a simple experiment by which you could find out how pure your samples were. The only chemicals you have available are some dilute hydrochloric acid and a sample of pure calcium carbonate from your chemistry lab stock. You can use any apparatus which you might normally have access to. Describe any preliminary experiments that you might need to do, and then describe in detail how you would carry out your main experiment, what measurements you would make, and how you would use those measurements to find out whether your samples were pure.

Chapter 16: Reversible Reactions

This chapter explores the idea of "reversibility" in a reaction, and how you can control such reactions in order to get as much as possible of what you want.

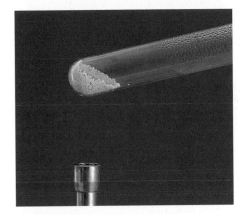

Copper(II) sulphate crystals are split into anhydrous copper(II) sulphate and water on gentle heating.

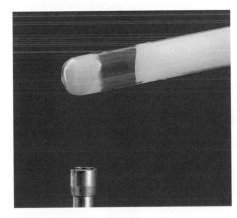

Heating ammonium chloride.

Reversibility and Dynamic Equilibria

Two simple reversible reactions

Heating copper(II) sulphate crystals

If you heat blue copper(II) sulphate crystals gently, the blue crystals turn to a white powder and water is driven off. Heating causes the crystals to lose their water of crystallisation, and white anhydrous copper(II) sulphate is formed. "Anhydrous" simply means "without water".

$$CuSO_4 \cdot 5H_2O_{(s)} \rightarrow CuSO_{4(s)} + 5H_2O_{(l)}$$

Anhydrous copper(II) sulphate is used to test for the presence of water. If you add water to the white solid, it turns blue – and also gets very warm. You will find a photograph of this on page 88.

The original change has been exactly reversed. Even the heat that you put in originally has been given out again.

$$CuSO_{4(s)} + 5H_2O_{(l)} \rightarrow CuSO_4 \cdot 5H_2O_{(s)}$$

Heating ammonium chloride

If you heat ammonium chloride, the white crystals disappear from the bottom of the tube and reappear further up. Heating ammonium chloride splits it into the colourless gases, ammonia and hydrogen chloride.

$$NH_4Cl_{(s)} \rightarrow NH_{3(g)} + HCl_{(g)}$$

These gases recombine further up the tube where it is cooler.

$$NH_{3(g)} + HCl_{(g)} \rightarrow NH_4Cl_{(s)}$$

The reaction reverses when the conditions are changed from hot to cool.

Reversible reactions under "closed" conditions

"Closed" conditions means that no substances are added to the reaction mixture and no substances escape from it. On the other hand, heat may be either given off or absorbed.

Imagine a substance which can exist in two forms – one of which we'll represent by a blue square and the other by a yellow one. Suppose you start of with a sample which is entirely blue.

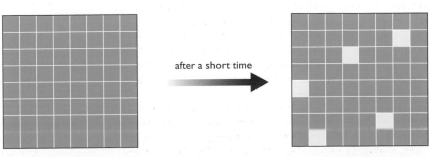

blue squares in a closed system

after a short time

some of the blue squares have turned yellow

Suppose you started with 64 blue squares, and in any second there was a 1 in 4 chance of each of them changing colour. In the first second, 16 would change colour, leaving 48 blue squares. In the next second, a quarter of these change colour – but that's only 12, leaving 36 blue ones. In the third second, 9 would change colour – and so on. The rate of change falls as the number of squares falls.

Because you are starting with a high concentration of blue squares, at the beginning of the reaction the rate at which they turn yellow will be relatively high in terms of the number of squares changing colour per second. The number changing colour per second (the rate of change) will fall as the blue gradually gets used up.

But the yellow squares can also change back to blue ones again – it is a reversible reaction. At the start, there aren't any yellow squares, so the rate of change from yellow into blue is zero. As their number increases, the rate at which yellow change to blue also increases.

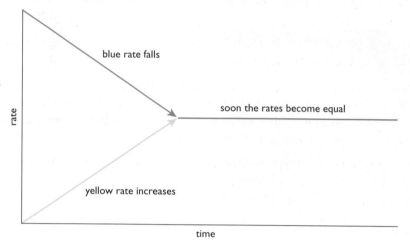

Soon the rates of both reactions become equal. At that point, blue ones are changing into yellow ones at exactly the same rate that yellow ones are turning blue.

What would you see in the reaction mixture when that happens? The total numbers of blue squares and of yellow squares would remain constant, but the reaction would still be going on. If you followed the fate of any one particular square, sometimes it would be blue and sometimes yellow.

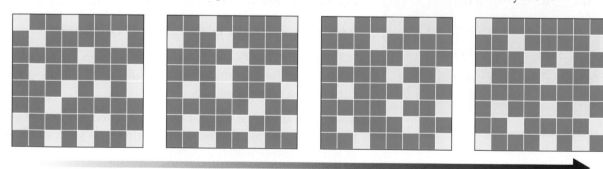

changes happening as time passes

This is an example of a **dynamic equilibrium**. It is *dynamic* in the sense that the reactions are still continuing. It is an *equilibrium* in the sense that the total amounts of the various things present are now constant.

Notice that you can only set up a dynamic equilibrium if the system is closed. If, for example, you removed the yellow ones as soon as they were formed, they would never get the chance to turn blue again. What was a reversible reaction will now go entirely in one direction as blue squares turn yellow without being replaced.

Writing equations for reactions in dynamic equilibrium

Taking a general case where A and B react reversibly to give C and D:

$$A + 2B \rightleftharpoons C + D$$

The special two-way arrows show a reversible reaction in a state of dynamic equilibrium. The reaction between A and B (the left to right reaction) is described as the **forward reaction**. The reaction between C and D (the right to left reaction) is called the **back reaction**.

Manipulating reversible reactions

If your aim in life was to produce substance C in the last equation as efficiently as possible, you might not be too pleased if it kept reacting back to produce A and B all the time.

This section looks at what can be done to alter the **position of equilibrium** so as to produce as much as possible of what you want in the equilibrium mixture. "Position of equilibrium" is just a reference to the proportions of the various things in the equilibrium mixture.

If, for example, the equilibrium mixture contains a high proportion of C and D, we would say that the "position of equilibrium lies towards C and D", or the "position of equilibrium lies to the right".

Le Chatelier's Principle

> **If a dynamic equilibrium is disturbed by changing the conditions, the reaction moves to counteract the change.**

This is a useful guide to what happens if you change the conditions in a system in dynamic equilibrium. It is essentially a "law of chemical cussedness"! The reaction sets about counteracting any changes you make.

The things that we might try to do to influence the reaction include:

- increasing or decreasing the concentrations of substances present

- changing the pressure

- changing the temperature.

Adding and removing substances

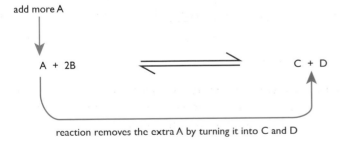

add more A

A + 2B $\rightleftharpoons$ C + D

reaction removes the extra A by turning it into C and D

If you add more A, the system responds by removing it again. That produces more C and D – which is what you probably want. You might choose to increase the amount of A if it was essential to convert as much B as possible into products – because it was expensive, for example.

The reason for using "2B" will become obvious later on when we look at the effect of pressure on the reaction.

Important! Le Chatelier's Principle is no more than a useful "rule-of-thumb" to help you to decide what happens if various conditions are changed. It is *not the reason* why the reaction responds in that way.

Alternatively, if you remove C as soon as it is formed, the reaction will respond by replacing it again by reacting more A and B. Removing a substance as soon as it is formed is a useful way of moving the position of equilibrium to generate more products.

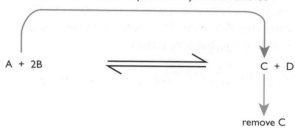

Changing the pressure

This only really applies to gas reactions and where the total number of molecules is different on each side of the equation. In our example, there are 3 molecules on the left, but only 2 on the right.

$$A_{(g)} + 2B_{(g)} \rightleftharpoons C_{(g)} + D_{(g)}$$

Pressure is caused by molecules hitting the walls of their container. If you have fewer molecules at the same temperature, you will have a lower pressure.

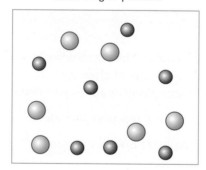

More gas molecules create a higher pressure.

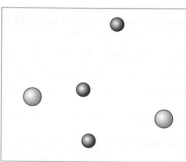

Fewer molecules create less pressure.

According to Le Chatelier's Principle, if you increase the pressure, the reaction will respond by reducing it again. It can reduce the pressure by producing fewer molecules to hit the walls of the container – in this case by creating more C and D. Increasing the pressure will always help the reaction go in the direction which produces the smaller number of molecules.

Changing the temperature

Suppose the forward reaction is *exothermic*. This is shown in an equation by writing a negative sign in front of the quantity of heat energy. For example:

$$A + 2B \rightleftharpoons C + D \qquad \Delta H = -100 \text{ kJ mol}^{-1}$$

The back reaction would be *endothermic* by exactly the same amount.

Suppose you changed the conditions by decreasing the temperature of the equilibrium – for example, if the reaction was originally in equilibrium at 500°C, you lower the temperature to 100°C. The reaction will respond in such a way as to increase the temperature again. How can it do that?

If you aren't happy about this, you ought to read Chapter 13 up to page 138 before you go on. In a reversible reaction, the value of ΔH quoted always applies to the *forward* reaction as written in the equation. The value of ΔH is given as if the reaction was a one-way process.

If more C and D is produced, more heat is given out because of the exothemic change. That extra heat which is produced will warm the reaction mixture up again – as Le Chatelier suggests. In other words, decreasing the temperature will cause more C and D to be formed.

Increasing the temperature will have exactly the opposite effect. The reaction will move to get rid of the extra heat by absorbing it in an endothermic change. This time the back reaction is favoured.

The Manufacture and Uses of Ammonia

The Haber Process

The Haber Process takes nitrogen from the air and hydrogen produced from natural gas, and combines them into ammonia, NH_3.

$$N_{2(g)} + 3H_{2(g)} \rightleftharpoons 2NH_{3(g)} \quad \Delta H = -92 \text{ kJ mol}^{-1}$$

The raw materials:	nitrogen (from the air)
	hydrogen (made from natural gas – methane)
The proportions:	1 volume of nitrogen to 3 volumes of hydrogen
The temperature:	450°C
The pressure:	200 atmospheres
The catalyst:	iron

> Remember that the negative sign for ΔH shows that the reaction is exothermic.

> The actual pressure varies in different manufacturing plants, but is always very high.

Each time the gases pass through the reaction vessel, only about 15% of the nitrogen and hydrogen combine to make ammonia. The reaction mixture is cooled and the ammonia condenses as a liquid. The unreacted nitrogen and hydrogen can simply be recycled through the reactor.

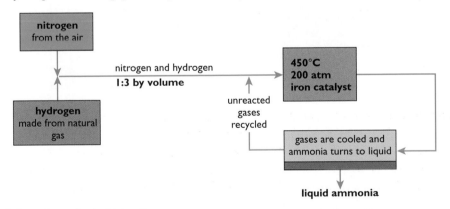

A flow scheme for the Haber Process.

The reason for the proportions of nitrogen and hydrogen

Equation proportions are used – 1 of nitrogen to 3 of hydrogen. An excess of either would clutter the reaction vessel with molecules which wouldn't have anything to react with.

The reason for the temperature

According to Le Chatelier's Principle, the forward reaction (an exothermic change) would be favoured by a low temperature, but the temperature used, 450°C, isn't a low temperature.

> If you aren't sure about this, re-read pages 167–169.

If the temperature was genuinely low, the reaction would be so slow that it would take a very long time to produce much ammonia. 450°C is a **compromise temperature**, producing a reasonable yield of ammonia reasonably quickly.

The reason for the pressure

There are 4 gas molecules on the left-hand side of the equation, but only 2 on the right-hand side. A reaction which produces fewer gaseous molecules is favoured by a high pressure. A high pressure would also produce a fast reaction rate because the molecules are brought closely together.

The 200 atmospheres actually used is high, but not *very high*. This is a compromise. Generating high pressures and building the vessels and pipes to contain them is very expensive. Pressures much higher than 200 atmospheres cost more to generate than you would get back in the value of the extra ammonia produced.

The catalyst

The iron catalyst speeds the reaction up, but has no effect on the proportion of ammonia in the equilibrium mixture. If the catalyst wasn't used, the reaction would be so slow that virtually no ammonia would be produced.

Uses of ammonia

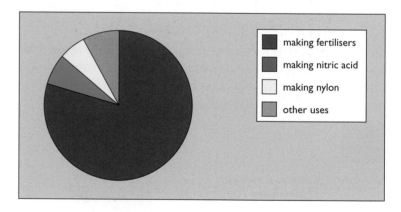

A high proportion of ammonia is used to make fertilisers. Most of the ammonia used to make nitric acid eventually ends up in fertilisers as well.

Making nitric acid

Ammonia is mixed with air and passed over a red-hot platinum–rhodium catalyst. The ammonia is oxidised by oxygen in the air to nitrogen monoxide, NO.

$$4NH_{3(g)} + 5O_{2(g)} \rightarrow 4NO_{(g)} + 6H_2O_{(g)}$$

On cooling, the nitrogen monoxide reacts with more oxygen from the air and is oxidised further to nitrogen dioxide, NO_2.

$$2NO_{(g)} + O_{2(g)} \rightarrow 2NO_{2(g)}$$

This is finally absorbed in water in the presence of still more oxygen to give nitric acid.

$$2H_2O_{(l)} + 4NO_{2(g)} + O_{2(g)} \rightarrow 4HNO_{3(aq)}$$

If you aren't sure about this, re-read pages 167–169.

Making fertilisers

Common nitrogen-containing fertilisers include ammonium nitrate and ammonium sulphate. These can be made by neutralisation reactions involving ammonia and either nitric acid or sulphuric acid.

$$NH_{3(g)} + HNO_{3(aq)} \rightarrow NH_4NO_{3(aq)}$$

$$2NH_{3(g)} + H_2SO_{4(aq)} \rightarrow (NH_4)_2SO_{4(aq)}$$

Using fertilisers

Plants use carbon dioxide from the air and water from the soil to produce the carbohydrates which make up much of their bulk. But they also need other elements like nitrogen to produce proteins, or phosphorus to make DNA, and they have to get these from compounds in the soil.

These nutrients are removed from the soil by the plants, and are also washed away by heavy rain. They have to be regularly replaced to keep the soil fertile.

Inorganic versus organic fertilisers

Fertilisers produced chemically are known as **inorganic fertilisers**. Manure and compost are **organic fertilisers**. Each can provide the nutrients that plants need to grow successfully.

The fertiliser in the photograph had this composition:

Typical Analysis	
Nitrogen as N	4.5%
Phosphorus as P_2O_5	3.5%
Potassium as K_2O	2.5%
Magnesium as MgO	1.0%
Sulphur as SO_3	0.5%
Calcium as CaO	9.0%

Plus full range of trace elements including Iron, Magnesium, Copper, Molybdenum and Zinc.

Notice the strange way that the percentages of the various elements are typically shown in fertilisers. Fertilisers don't contain P_2O_5, K_2O, etc. In reality they contain compounds like ammonium phosphate and potassium nitrate.

Bulky organic fertilisers have the advantage that they improve soil structure as well as adding nutrients. They improve the moisture-holding capacity of sandy soils, and break up heavy clay soils.

The nutrients are present as complicated molecules which have to be broken down by soil bacteria. This means that they are released gradually over a long period of time, and are less likely than inorganic fertilisers to be washed out of the soil by rain.

You need large quantities of bulky organic fertilisers to provide enough nutrients. It is easy to see how small mixed farms with animals as well as crops could use them. It isn't so easy to see how a huge farm growing only grain could do the same thing in a way that made economic sense.

Crops like wheat need a high input of fertilisers to give a good yield.

Organic gardening using poultry manure.

Muck spreading.

The chart shows the results of research carried out at the Rothamsted Experimental Station over a 5 year period. You can see that the unfertilised plot has very low yields compared with the others.

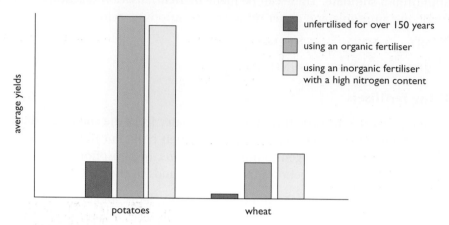

The wheat did better with the inorganic fertiliser – the potatoes with the organic one.

The advantages of inorganic fertilisers are that they are cheap, easy to use and very fast acting. For example, plants take up nitrogen as nitrate ions in solution, $NO_3^-{}_{(aq)}$. Nitrate fertilisers provide that directly. Ammonium salts are rapidly converted into nitrate ions by soil bacteria.

Inorganic fertilisers do, however, have major environmental disadvantages.

- Over the long term, if little or no organic matter is returned to the soil, the soil structure will break down with the risk of wind or water erosion.

- Inorganic fertilisers use large amounts of energy, both in their manufacture and transport. This can lead to pollution and global warming.

- Inorganic fertilisers are very soluble in water and can pollute water courses. Phosphates and nitrates from fertilisers cause very rapid growth of water plants (**eutrophication**). When these plants die, their decay uses up all the oxygen dissolved in the water, causing the death of all the other water life.

- Nitrates can also pollute water supplies. Because all nitrates are soluble, it is very difficult to remove them from drinking water. There is an association between a rare disease in babies ("**blue baby syndrome**") and nitrates in drinking water, and there is a theoretical (but unproven) possibility of a link between high nitrate levels and some cancers.

The Manufacture and Uses of Sulphuric Acid

The Contact Process

At the heart of this process is a reversible reaction in which sulphur dioxide is converted to sulphur trioxide, but first you have to produce sulphur dioxide.

Stage 1: making sulphur dioxide

Either burn sulphur in air:

$$S_{(s)} + O_{2(g)} \rightarrow SO_{2(g)}$$

or heat sulphide ores strongly in air:

$$4FeS_{2(s)} + 11O_{2(g)} \rightarrow 2Fe_2O_{3(s)} + 8SO_{2(g)}$$

FeS$_2$ is *pyrite* or *iron pyrites*. You will find a photograph of some crystals of pyrite on page 145.

Stage 2: making sulphur trioxide

Now the sulphur dioxide is converted into sulphur trioxide using an excess of air from the previous processes.

$$2SO_{2(g)} + O_{2(g)} \rightleftharpoons 2SO_{3(g)} \qquad \Delta H = -196 \text{ kJ mol}^{-1}$$

Notice that the forward reaction is an exothermic change.

```
+-------------------------+        +------------------+
| sulphur dioxide         |        | 450°C            |
|        and              | -----> | 1-2 atm          |
|      oxygen             |        | V2O5 catalyst    |
|    (excess air)         |        +------------------+
+-------------------------+                 |
                                            v
                                 +------------------+
                                 | sulphur trioxide |
                                 +------------------+
                                            |
                                            v
                                    further reactions
```

In this reaction an excess of oxygen is used, because it is important to make sure that as much sulphur dioxide as possible is converted into sulphur trioxide. Having sulphur dioxide left over at the end of the reaction is wasteful, and could cause possibly dangerous pollution.

Because the forward reaction is exothermic, there would be a higher percentage conversion of sulphur dioxide into sulphur trioxide at a low temperature. However, at a low temperature the rate of reaction would be very slow. 450°C is a compromise. Even so, there is about a 99.5% conversion.

If you have come to this via the index rather than reading the whole chapter, you would find it useful to read pages 167–169 before you go on.

There are 3 gas molecules on the left-hand side of the equation, but only 2 on the right. Reactions in which the numbers of gas molecules decrease are favoured by high pressures. In this case, though, the conversion is so good at low pressures that it isn't economically worthwhile to use higher ones.

The catalyst, vanadium(V) oxide, has no effect on the percentage conversion, but helps to speed up the reaction. Without the catalyst the reaction would be extremely slow.

Stage 3: making the sulphuric acid

In principle, you can react sulphur trioxide with water to make sulphuric acid. In practice, this produces an uncontrollable fog of concentrated sulphuric acid. Instead, the sulphur trioxide is absorbed in concentrated sulphuric acid to give **fuming sulphuric acid** (also called **oleum**).

$$H_2SO_{4(l)} + SO_{3(g)} \rightarrow H_2S_2O_{7(l)}$$

This is converted into twice as much concentrated sulphuric acid by careful addition of water.

$$H_2S_2O_{7(l)} + H_2O_{(l)} \rightarrow 2H_2SO_{4(l)}$$

Uses of sulphuric acid

Sulphuric acid has a wide range of uses throughout the chemical industry. The highest single use is in making fertilisers (including ammonium sulphate and "superphosphate" – essentially a mixture of calcium phosphate and calcium sulphate).

It is also used in the manufacture of detergents, and as car battery acid.

Concentrated sulphuric acid as a dehydrating agent

With copper(II) sulphate crystals

Concentrated sulphuric acid removes the water of crystallisation from copper(II) sulphate crystals. This is an example of dehydration. The crystals turn white as anhydrous copper(II) sulphate is formed.

$$CuSO_4 \cdot 5H_2O_{(s)} \rightarrow CuSO_{4(s)} + 5H_2O_{(l)}$$

The sulphuric acid can be thought of as being slightly diluted by the reaction, and therefore chemically unchanged. It isn't usually written into the equation. If you like, you can draw a longer arrow, and write "concentrated H_2SO_4" over the top of it.

With sugar (sucrose)

Dehydration isn't just the removal of whole water molecules from a substance. In most cases it involves removing separate hydrogen atoms and oxygen atoms from the molecule, but in the ratio of 2:1 – the same as they are in water.

For example, if concentrated sulphuric acid is added to sugar, the sugar swells up into a large mass of spongy carbon in a vigorously exothermic change.

$$C_{12}H_{22}O_{11(s)} \rightarrow 12C_{(s)} + 11H_2O_{(l)}$$

Again, if you want to, you could write "concentrated H_2SO_4" over the arrow.

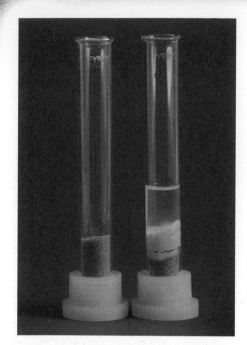

Concentrated sulphuric acid removes the water of crystallisation from copper(II) sulphate crystals.

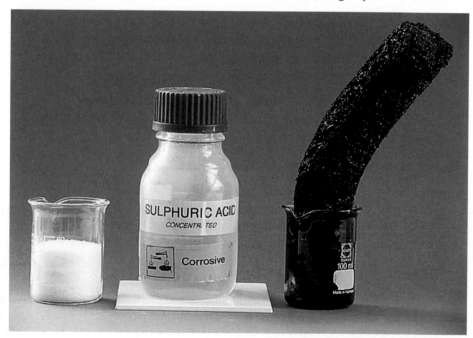

Dehydrating sugar with concentrated sulphuric acid.

End of Chapter Checklist

If you haven't got a copy of your specification, read the introduction on page vi.

You will need to be able to do some or all of the following. Check your Awarding Body's specification (syllabus) to find out exactly what you need to know.

- Understand what is meant by a reversible reaction using simple examples like the effect of heat on copper(II) sulphate crystals or ammonium chloride.

- Understand that a reversible reaction in a closed system can reach a state of dynamic equilibrium.

- Using Le Chatelier's Principle (or otherwise) state what happens to the position of equilibrium if you add or subtract substances, change the temperature, or change the pressure.

- Know that a catalyst has no effect on the position of equilibrium.

- Know the conditions for the Haber Process, and be able to explain the choice of conditions in terms of the percentage of ammonia in the equilibrium mixture, the rate of the reaction, and the economics of the process.

- Know how ammonia is converted into nitric acid, including equations.

- Know that ammonia and nitric acid are used to make fertilisers.

- Know the difference between organic and inorganic fertilisers and discuss the advantages and disadvantages of each.

- Describe the manufacture of sulphuric acid by the Contact Process, including reasons for the choice of conditions.

- Give some uses of sulphuric acid.

- Understand that concentrated sulphuric acid is a dehydrating agent, and describe and explain the dehydration of copper(II) sulphate crystals and sugar (sucrose).

Questions

More questions on reversible reactions can be found at the end of Section C on page 177.

1 **a)** Describe, including equations and essential conditions, the production of sulphuric acid starting from sulphur.

   b) Concentrated sulphuric acid is a *dehydrating agent*. Explain what that means and describe the addition of concentrated sulphuric acid to *either* copper(II) sulphate crystals *or* sugar (sucrose). Write an equation for the reaction you describe.

   c) At a temperature of about 170°C, concentrated sulphuric acid dehydrates ethanol, C_2H_5OH, to give a gas, X. Suggest a formula for X.

2 **a)** Given solutions of ammonia and dilute sulphuric acid and any indicator of your choice, describe how you would make a pure solution of ammonium sulphate.

   b) Design a simple experiment to show that your ammonium sulphate solution could function as a fertiliser.

3 Environmental groups like The Soil Association or Friends of the Earth are keen to promote organic farming, which means that no artificial fertilisers, pesticides or weedkillers should be use to produce crops. Describe the advantages and disadvantages of using **a)** organic fertilisers, **b)** inorganic fertilisers. Will the same arguments apply in rich countries like Britain and poor countries like Ethiopia? Explain your answer.

4 The reaction at the centre of the Haber Process for the manufacture of ammonia is:

$$N_{2(g)} + 3H_{2(g)} \rightleftharpoons 2NH_{3(g)} \quad \Delta H = -92 \text{ kJ mol}^{-1}$$

a) What would happen to the percentage of ammonia in the equilibrium mixture and to the rate of the reaction if you:

   i) increased the temperature

   ii) increased the pressure

   iii) added a catalyst?

b) In the light of your answer to *aii)*, explain why ammonia plants usually operate with pressures of about 200 atmospheres.

c) State the sources of the nitrogen and the hydrogen used in the Haber Process.

5 N_2O_4 molecules are pale yellow, whereas NO_2 molecules are dark brown. At room temperature, they are both present in an equilibrium mixture given by this equation:

$$N_2O_{4(g)} \rightleftharpoons 2NO_{2(g)} \quad \Delta H = +58 \text{ kJ mol}^{-1}$$

The forward reaction is an endothermic change, and the mixture is a medium brown colour at room temperature.

State and explain what would happen to the colour of the mixture if you:

a) heated the mixture to 150°C, keeping the pressure constant

b) increased the pressure on the mixture at room temperature.

6 The indicator litmus is a weak acid. It is a complicated molecule containing a hydrogen atom which can break free to form an ion. In the equation, Lit represents the rest of the molecule.

$$HLit_{(aq)} \rightleftharpoons H^+_{(aq)} + Lit^-_{(aq)}$$

In litmus solution there is a *dynamic equilibrium* involving the HLit molecules and the two ions. Because litmus is a weak acid, the *position of equilibrium* lies well to the left.

a) By reference to this example, explain what is meant by the terms *dynamic equilibrium* and *position of equilibrium*.

b) HLit molecules are red and Lit⁻ ions are blue. Explain why:

   i) litmus turns red when you add an acid (a source of $H^+_{(aq)}$) to it

   ii) litmus turns blue when you add an alkali (a source of $OH^-_{(aq)}$) to it.

End of Section Questions

You may need to refer to the Periodic Table on page 306.

1 Iron is produced in a blast furnace by the reduction of its ore, haematite, Fe_2O_3.

a) What do you understand by the term *reduction*?
(1 mark)

b) Give the proper chemical name for haematite.
(1 mark)

c) The main heat source in the furnace is provided by burning coke in air.

$$C_{(s)} + O_{2(g)} \rightarrow CO_{2(g)}$$

What name is given to a reaction which produces heat?
(1 mark)

d) The main reducing agent in the furnace is carbon monoxide. Write an equation to show its formation.
(1 mark)

e) Balance the equation:

$$Fe_2O_{3(s)} + CO_{(g)} \rightarrow Fe_{(l)} + CO_{2(g)}$$
(1 mark)

f) Limestone is added to the furnace to help in the removal of impurities in the ore like silicon dioxide, SiO_2. Explain the chemistry of this.
(3 marks)

g) Give one use for slag.
(1 mark)

h) The impure iron from the blast furnace can be used to make cast iron, but most is converted into various steels.

	cast iron	mild steel	high carbon steel	stainless steel
contains	4% C	0.25% C	1.5% C	18% Cr 8% Ni

i) Give one use in each case for cast iron, mild steel, high carbon steel and stainless steel.
(4 marks)

ii) Give two effects of increasing the proportion of carbon mixed with the iron.
(2 marks)

Total: 15 marks

2 *a)* Aluminium is manufactured by the electrolysis of aluminium oxide dissolved in molten cryolite.

i) Name the ore from which aluminium oxide is obtained.
(1 mark)

ii) At which electrode is the aluminium produced?
(1 mark)

iii) Oxygen is released at the other electrode. Explain why that creates a problem.
(2 marks)

iv) Aluminium is the commonest metal in the Earth's crust, and yet it is relatively expensive because its extraction is expensive. Why is the extraction expensive?
(1 mark)

b) Aluminium is already fairly resistant to corrosion, but its corrosion resistance can be increased by *anodising* the aluminium.

i) What happens to the aluminium when it is anodised?
(2 marks)

ii) Describe how you would anodise a piece of aluminium cut from a drinks can.
(3 marks)

c) High voltage overhead electricity cables are made of aluminium with a steel core, supported on galvanised steel pylons.

i) Aluminium is not such a good conductor of electricity as copper. Why is aluminium used for overhead power cables instead of copper?
(1 mark)

ii) Iron is a less good conductor of electricity than aluminium. Why are the cables constructed with a steel core?
(1 mark)

iii) Suggest two reasons why the pylons are made of steel rather than aluminium.
(2 marks)

iv) What is *galvanised* steel?
(1 mark)

v) Explain how galvanising iron helps to prevent it from rusting.
(2 marks)

Total 17 marks

3 *a)* Zinc occurs naturally as sphalerite (zinc blende), ZnS. The zinc sulphide is heated strongly in air to produce zinc oxide.

$$2ZnS_{(s)} + 3O_{2(g)} \rightarrow 2ZnO_{(s)} + 2SO_{2(g)}$$

The zinc oxide can be reduced to zinc in two ways. In the first method, it is heated with carbon in a blast furnace at a temperature in excess of 1000°C. Zinc boils at 907°C, and so is produced as a vapour which can be condensed. In the second method, the zinc oxide is converted into zinc sulphate solution which is then electrolysed.

i) Suggest a use for the sulphur dioxide produced during the formation of the zinc oxide. *(1 mark)*

ii) Reduction of zinc oxide by carbon produces zinc and carbon monoxide. Write the equation for the reaction.
(1 mark)

iii) What would you add to zinc oxide to produce a solution of zinc sulphate? Write an equation for the reaction involved. *(2 marks)*

iv) At which electrode would the zinc be formed during the electrolytic extraction? Write the equation for the reaction occurring at that electrode.
(2 marks)

b) Zinc is used to make the alloy brass.

i) Which metal is mixed with zinc to produce brass?
(1 mark)

ii) Give two differences in physical properties between pure zinc and brass. *(2 marks)*

Total 9 marks

4 In Britain, the chlor-alkali industry is centred around Cheshire because of the large underground salt deposits found in that county. The salt is extracted by solution mining, and the salt solution is electrolysed to produce three important chemicals.

a) What is the origin of the salt deposits in Cheshire? *(1 mark)*

b) Name the three chemicals produced by the electrolysis of salt solution, and give a use for each of them. *(6 marks)*

c) Which of the three is produced at the anode during the electrolysis? Write the equation for the reaction occurring at the anode. *(2 marks)*

Total 9 marks

5 a) In the production of quicklime (calcium oxide) from limestone, the limestone is heated in a lime kiln. The reaction occurring is:

$$CaCO_{3(s)} \rightleftharpoons CaO_{(s)} + CO_{2(g)}$$

i) What is the source of heat in the lime kiln? *(2 marks)*

ii) Explain why it is essential to sweep away the carbon dioxide as soon as it is formed. *(2 marks)*

iii) How is quicklime converted into slaked lime (calcium hydroxide)? *(1 mark)*

iv) Write the equation for the formation of slaked lime from quicklime. *(1 mark)*

v) State one use for slaked lime. *(1 mark)*

b) Give one possible advantage and one possible disadvantage to people living near a limestone quarry. *(2 marks)*

c) Describe briefly how cement is made from limestone. *(3 marks)*

Total 12 marks

6 The hydrogen for the Haber Process for the manufacture of ammonia is produced by reacting methane (natural gas) with steam. The reaction is carried out at a temperature of 850°C, a pressure of 30 atmospheres and in the presence of a nickel catalyst.

$$CH_{4(g)} + H_2O_{(g)} \rightleftharpoons CO_{(g)} + 3H_{2(g)} \quad \Delta H = +210 \text{ kJ mol}^{-1}$$

The forward reaction is endothermic. The carbon monoxide is removed from the hydrogen by further reactions.

a) Explain why a high temperature is beneficial to both the rate of the reaction and the percentage of hydrogen in the equilibrium mixture. *(4 marks)*

b) Suggest a reason that even higher temperatures aren't used. *(1 mark)*

c) Explain why the pressure used (30 atmospheres) is unexpectedly large for this reaction. *(3 marks)*

d) What effect does the nickel catalyst have on:

i) the rate at which the equilibrium mixture is formed *(1 mark)*

ii) the percentage of hydrogen in the equilibrium mixture? *(1 mark)*

Total 10 marks

7 Fritz Haber, a German chemist, discovered how to convert nitrogen in the air into ammonia in 1913. TNT (trinitrotoluene) uses nitric acid in its manufacture, and Haber's discovery enabled Germany to continue to produce explosives during the First World War, despite a blockade which stopped supplies of Chilean saltpetre (sodium nitrate), from which nitric acid was previously made. In more recent times, the ammonia from the Haber Process has mainly been used in the production of artificial fertilisers.

a) The conversion of ammonia into nitric acid involves the following stages:

A $\quad 4NH_{3(g)} + 5O_{2(g)} \rightarrow 4NO_{(g)} + 6H_2O_{(g)}$

B $\quad 2NO_{(g)} + O_{2(g)} \rightarrow 2NO_{2(g)}$

C $\quad H_2O_{(l)} + NO_{2(g)} + O_{2(g)} \rightarrow HNO_{3(aq)}$
$\qquad\qquad\qquad\qquad\qquad$ (unbalanced)

i) The catalyst for stage A is a mixture of very expensive metals. Name the catalyst and explain how the manufacturer can justify using such expensive materials. *(3 marks)*

ii) Balance the equation in stage C. *(1 mark)*

b) Nitric acid can be made from sodium nitrate by heating it with concentrated sulphuric acid. Suggest an equation for the reaction involved. *(2 marks)*

c) Explain briefly why it is necessary to add fertilisers to the soil. *(2 marks)*

d) Give one advantage of using inorganic fertilisers rather than organic ones. *(1 mark)*

e) A sample of an inorganic fertiliser claimed to contain ammonium sulphate. Describe tests which you could do to confirm that it contained:

i) ammonium ions *(3 marks)*

ii) sulphate ions. *(3 marks)*

f) Give one environmental disadvantage in using inorganic fertilisers like ammonium sulphate. *(2 marks)*

Total 17 marks

Section D: Air, Water and Earth

Chapter 17: The Air

> This chapter looks at the present composition of the atmosphere, the way it developed and some current environmental worries.

> It is important to realise that these figures apply only to *dry* air. Air can have anywhere between 0% and 4% of water vapour.

> A silica tube looks like glass, but won't melt, however strongly you heat it with a Bunsen burner.

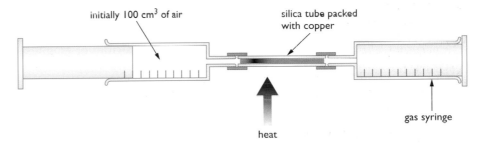

A modern volcano in Iceland.

The Present Atmosphere

Composition

The composition of the air is believed to have been largely unchanged for the past 200 million years. The approximate percentages (by volume) of the main gases present in *unpolluted, dry* air are:

nitrogen	78.1%	about 4/5
oxygen	21.0%	about 1/5
argon	0.9%	
carbon dioxide	0.03%	

There are also very small amounts of the other noble gases in air.

Showing that air contains about one fifth oxygen

This apparatus can be used to find the percentage of oxygen in the air:

initially 100 cm³ of air   silica tube packed with copper

heat

gas syringe

The apparatus initially contains 100 cm³ of air. This is pushed backwards and forwards over the heated copper, which turns black as copper(II) oxide is formed. The volume of air in the syringes falls as the oxygen is used up.

$$2Cu_{(s)} + O_{2(g)} \rightarrow 2CuO_{(s)}$$

As the copper reacts, the Bunsen burner is moved along the tube, so that it is always heating fresh copper.

Eventually, all the oxygen in the air is used up. The volume stops contracting and the copper stops turning black. On cooling, somewhere around 79 cm³ of gas is left in the syringes – 21% has been used up.

Therefore the air contained 21% of oxygen.

The Evolution of the Atmosphere

The first stages

More than 4000 million years ago a thin solid crust formed on the Earth. Molten rock kept breaking through the surface, producing volcanoes.

Volcanoes produce lots of gases as well as the more obvious rocks and lava. Primitive volcanoes produced large amounts of carbon dioxide and steam, together with smaller amounts of gases like ammonia (NH_3), methane (CH_4), nitrogen, hydrogen and carbon monoxide.

Warning! There are some differences between the various Awarding Bodies as to which minor gases were present in ancient volcanoes. You *must* check your specification to make sure you learn the right ones. If you don't have a copy of your specification, read the introduction on page vi.

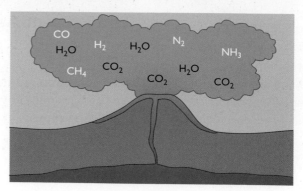

By about 4000 million years ago the Earth's surface had cooled to below 100°C, and the steam had condensed as liquid water to form the first oceans.

As the oceans cooled, carbon dioxide started to dissolve in the water. Gases are more soluble if they are under high pressure, and the pressure of carbon dioxide in the early atmosphere was much higher than it is today.

Some of the carbon dioxide stayed in solution, but some of it combined with the water to produce carbonate or hydrogencarbonate ions. Precipitation reactions involving calcium or magnesium ions dissolved from the rocks produced solid calcium or magnesium carbonate. The first limestones were being formed.

The arrival of life

You will find more about fermentation on page 232.

Primitive life would have used quite different biological processes from most life now, because there was no oxygen. The very first organisms probably got their energy from reactions rather like fermentation, and would have been found near volcanic openings in the deep oceans.

Eventually primitive plants developed which were capable of taking energy from sunlight using **photosynthesis**. Photosynthesis is an endothermic process which converts carbon dioxide and water into carbohydrates (such as glucose) and oxygen.

$$6CO_{2(g)} + 6H_2O_{(l)} \rightarrow C_6H_{12}O_{6(aq)} + 6O_{2(g)}$$

The reaction needs the presence of the green plant pigment, **chlorophyll**.

Ozone is a form of oxygen containing 3 atoms per molecule – O_3. It is formed in the high atmosphere by the effect of solar radiation on ordinary oxygen. There is more about the ozone layer on page 184.

Because there was no oxygen, there also wasn't any ozone layer. Ozone absorbs harmful ultra-violet radiation from the sun. There was therefore nothing apart from water to protect life from that damaging radiation.

Primitive plants would have lived just below the surface of the water, protected from the ultra-violet radiation, but with enough sunlight to allow photosynthesis. Gradually, enough oxygen would be formed to produce an ozone layer which would provide enough protection for plants to emerge onto dry land.

Over thousands of millions of years, plants have removed carbon dioxide from the atmosphere and replaced it by oxygen. The carbon is locked up in all existing life, in **fossil fuels** like coal and oil which are the remains of ancient life, in limestone, and dissolved in the oceans.

Carbon from the original atmosphere is now locked up in present-day living things (animal and vegetable),...

...in fossil fuels,...

...and in limestones like chalk.

Other changes to the atmosphere

There is virtually no free hydrogen in the atmosphere. Hydrogen molecules are so fast moving that they escape into space.

Ammonia in the early atmosphere was oxidised by the increasing levels of oxygen to give nitrogen, nitric acid and water. Methane was oxidised to carbon dioxide and water.

Ammonia can also be removed by **nitrifying bacteria** which convert it into nitrates in the soil. Other bacteria – **denitrifying bacteria** – will convert ammonia and nitrates into nitrogen gas. Because nitrogen is so unreactive, once it is formed and added to the atmosphere, it tends to stay there. That's the reason for the high proportion of nitrogen in the atmosphere today.

The carbon cycle

The level of carbon dioxide in the air at the beginning of the twentieth century was stable at about 0.03%, but since then it has risen by about a tenth, because it is being added to the atmosphere faster than it is being removed. This has important possible consequences in terms of the "greenhouse effect".

The greenhouse effect is explored on pages 182–183.

Processes removing carbon dioxide from the atmosphere

Carbon dioxide is removed by photosynthesis and by solution in water – both of which we have already discussed.

Processes adding carbon dioxide to the atmosphere

(1) Combustion of fuels

Coal, natural gas and products made from crude oil like petrol and diesel all contain carbon and all burn to form carbon dioxide. The increase in carbon dioxide in the atmosphere over the last century is largely due to increased burning of fossil fuels. Burning them returns carbon dioxide to the atmosphere which was originally removed millions of years ago.

(2) Volcanic activity

Just as they did in the early atmosphere, volcanoes still release carbon dioxide today.

Returning carbon dioxide to the atmosphere which was last there millions of years ago.

(3) Decay

When a plant grows, it removes carbon dioxide from the atmosphere by photosynthesis. When it dies and decays, carbon dioxide is returned to the atmosphere during the decay process.

(4) Respiration

Respiration isn't the same as breathing! Respiration is a process which goes on in cells – in both plants and animals – and provides the main energy source for the organism.

For example, if you were to burn glucose completely in air or oxygen, carbon dioxide and water would be formed.

$$C_6H_{12}O_{6(s)} + 6O_{2(g)} \rightarrow 6CO_{2(g)} + 6H_2O_{(l)}$$

Respiration has exactly the same overall equation (except that the glucose would be in solution), although it takes place in numerous small steps within the cell. It would also release exactly the same amount of energy, although in a slower and more controlled way.

Providing the fuel for respiration in cells!

ideas
evidence

Most scientists believe that global warming is happening and that it is caused by gases like carbon dioxide. But some reject this, and claim that the increases in average world temperatures are part of natural heating and cooling cycles. Views are likely to change during the lifetime of this book. Find out the current state of scientific thought by doing an internet search on **global warming**.

Some Environmental Problems

Global warming – the "greenhouse effect"

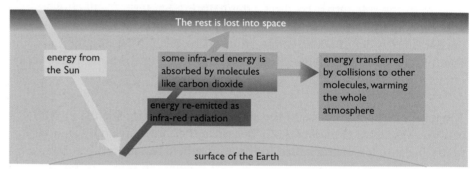

The rest is lost into space

energy from the Sun

some infra-red energy is absorbed by molecules like carbon dioxide

energy re-emitted as infra-red radiation

energy transferred by collisions to other molecules, warming the whole atmosphere

surface of the Earth

The absorption of infra-red radiation in this way is nothing new – it has always happened and is one of the processes which help to maintain the Earth at a temperature suitable for life. The worry is that increased amounts of carbon dioxide caused by burning fossil fuels are upsetting the balance and causing unusually rapid increases in global temperatures.

Shifting weather patterns will make droughts and humanitarian disasters even worse.

Rising sea levels (caused mainly by the expansion of water as it warms) and more extreme weather conditions will cause serious flooding.

The solution is to use less fossil fuels, by using alternative energy sources, and by using existing sources more efficiently...

...or by making changes to lifestyles to stop wasting fuel.

Acid rain

Rain is naturally slightly acidic because of dissolved carbon dioxide. Acid rain is even more acidic because of the presence of various pollutants.

Acid rain is caused when water and oxygen in the atmosphere react with sulphur dioxide to produce sulphuric acid, or with various oxides of nitrogen, NO_x, to give nitric acid. These come mainly from power stations and factories burning fossil fuels, or from motor vehicles.

Several oxides of nitrogen are involved, including NO and NO_2. They are often given the general formula NO_x.

Fossil fuels often contain sulphur compounds. Burning them produces sulphur dioxide.

Use of very low sulphur fuels limits the production of sulphur dioxide, but the spark in a petrol engine causes oxygen and nitrogen from the air to combine to make oxides of nitrogen, NO_x.

Limestone makes an attractive building stone, but because it is a carbonate...

...it reacts with the acid rain. Metals like iron are also affected by acid rain.

Trees dying from the effects of acid rain.

The ozone layer exists higher than the highest clouds.

Acid rain is mainly worrying for its devastating effect on trees, and on life in lakes. In some areas, a high proportion of trees are affected and are either sick or dying. In some lakes the water is so acidic that it won't support life.

The solution to the problem of acid rain involves removing sulphur from fuels, "scrubbing" the gases from power stations and factories to remove SO_2 and NO_x, and using catalytic converters in cars. The catalyst helps to convert oxides of nitrogen into harmless nitrogen gas, but has no effect on sulphur dioxide. Unfortunately, catalytic converters don't work properly until the catalyst becomes really hot, and so aren't effective on short journeys.

The holes in the ozone layer

In the stratosphere, 10 to 50 km above the Earth's surface, oxygen molecules absorb ultra-violet light. This breaks the covalent bonds and splits the molecules into individual atoms called **free radicals**. A free radical is any atom or group of atoms which has an unpaired electron, and is formed when a covalent bond splits evenly. The unpaired electron is often drawn as a dot beside the symbol for the atom.

$$O_{2(g)} \rightarrow O^\bullet_{(g)} + O^\bullet_{(g)}$$

Some of the oxygen radicals react with unchanged oxygen molecules to produce **ozone**, O_3.

$$O_{2(g)} + O^\bullet_{(g)} \rightarrow O_{3(g)}$$

Ozone itself can also absorb ultra-violet radiation to reverse the reaction.

$$O_{3(g)} \rightarrow O_{2(g)} + O^\bullet_{(g)}$$

The net effect of these reactions which absorb ultra-violet radiation is that 95–99% of all the ultra-violet from the Sun never gets through the **ozone layer** in the stratosphere. This is important because exposure to ultra-violet radiation leads to increased sunburn, increased ageing of the skin and, most importantly, increased incidence of skin cancer. It also has a damaging effect on crops, reducing yields.

The amount of ozone in the stratosphere drops quite dramatically at certain times of the year, particular over the Polar regions. This is partly caused by chlorine compounds released into the atmosphere over the past decades – particularly **chlorofluorocarbons**, CFCs. These are molecules containing chlorine and fluorine as well as carbon and possibly hydrogen, which used to be widely used as solvents, refrigerants and propellants in aerosol cans.

Sunlight can break the chlorine–carbon bonds in the CFCs to make chlorine free radicals; like all free radicals, these are extremely reactive. They react with the ozone in two steps to convert it into ordinary oxygen:

$$O_{3(g)} + Cl^\bullet_{(g)} \rightarrow O_{2(g)} + ClO^\bullet_{(g)}$$
$$ClO^\bullet_{(g)} + O^\bullet_{(g)} \rightarrow O_{2(g)} + Cl^\bullet_{(g)}$$

Notice that at the end of this process, another chlorine radical is produced which can go on to destroy more ozone – and so on and so on.

There are international agreements to stop the use of CFCs but, because this is a global problem, they only work if all countries follow them. Existing levels of CFCs in the atmosphere will go on destroying ozone for many years, and will be added to as CFCs leak from old abandoned refrigerators. There is no quick cure for the destruction of the ozone layer.

End of Chapter Checklist

If you haven't got a copy of your specification, read the introduction on page vi.

You will need to be able to do some or all of the following. Check your Awarding Body's specification (syllabus) to find out exactly what you need to know.

- Know the composition of the air, and describe a simple experiment to find the percentage of oxygen.

- Describe and explain the evolution of the atmosphere.

- Know the main processes which regulate the percentage of carbon dioxide in the atmosphere.

- Describe the causes, effects and possible cures relating to the following environmental problems: global warming, acid rain, depletion of the ozone layer.

Questions

More questions on the air can be found at the end of Section D on page 207.

1   When iron rusts it combines with water and oxygen in the air. Design a simple experiment to show that air is approximately one-fifth oxygen, using only a piece of iron wool, some water, and any glassware that you might normally have access to.

2   a)   The early atmosphere contained lots of carbon dioxide and hardly any oxygen. Name the main process which removed carbon dioxide from the atmosphere:

    i)   before the arrival of life

    ii)   after life had evolved.

    b)   Write an equation to show how oxygen in the atmosphere is produced.

    c)   The modern atmosphere contains large amounts of nitrogen. Explain briefly where this has come from.

    d)   Burning coal and burning wood both add carbon dioxide to the atmosphere. Using alcohol (made from sugar cane or sugar beet) instead of petrol to power a car produces carbon dioxide just as petrol does. Why do you think that an environmentalist might be less worried about burning wood or alcohol than about burning fossil fuels?

3   Three important environmental problems at the moment are acid rain, destruction of the ozone layer, and global warming.

    a)   State which of these problems each of the following substances makes an important contribution to:

    i)   carbon dioxide, CO_2

    ii)   chlorofluorocarbons, CFCs

    iii)   nitrogen oxides, NO_x

    iv)   sulphur dioxide, SO_2.

    b)   For each of the substances mentioned in part a):

    i)   explain what aspect of modern life has caused their concentrations in the atmosphere to increase

    ii)   suggest one way to prevent further increases.

4   Read the following passage and then answer the questions.

In 1895 an English physicist, Lord Rayleigh noticed a tiny difference between the density of nitrogen made from ammonia (1.2505 g/dm^3) and nitrogen obtained from the air (1.2572 g/dm^3). The nitrogen was obtained from the air by removing oxygen and carbon dioxide from it.

Sir William Ramsay passed nitrogen which had been obtained from the air over hot magnesium. Magnesium combines with nitrogen to make solid magnesium nitride, Mg_3N_2. At the end of this, he was left with a small amount of gas which wouldn't combine with the magnesium. This gas proved to be argon, contaminated with very small quantities of other noble gases.

    a)   Suggest one reaction you could use to remove oxygen from the air without replacing it by another gas. Write the equation for the reaction you give.

    b)   How might you remove carbon dioxide from the air?

    c)   Write a balanced equation for the reaction between hot magnesium and nitrogen. Don't forget the state symbols.

    d)   Name two gases which would remain mixed with the argon at the end of the process.

    e)   By considering the density figures given, is argon more dense or less dense than nitrogen? Explain your reasoning.

5 Rich countries like Britain and America produce far more carbon dioxide per head of population than poor countries do.

a) Explain briefly why the amount of carbon dioxide being produced matters to everybody on Earth.

b) The poorer countries very reasonably want to bring their living standards up to the same level as the richer ones, but that could result in a huge increase in the amount of carbon dioxide being produced. What do you think that the British government or British industry could usefully do to help developing countries improve their standards of living, without at the same time causing additional environmental problems?

6 By doing an internet (or other) search, find out why acid rain kills trees and fish. (Beware! If you do an internet search, go to at least three different sites to be sure that the information you find is reliable. Some quite reputable-looking sites have some obviously incorrect chemistry in them.)

7 A Friends of the Earth slogan is "Think globally; act locally". What can you *personally* do to help prevent global warming, acid rain and ozone depletion? Your suggestions should all be things which you and your family could put into effect as from tomorrow.

Chapter 18: Water

Water is the most abundant substance on the surface of the Earth. This chapter looks at the solubility of substances in water and some of the implications of this for water supplies.

Water as a Solvent

The oceans

About 4000 million years ago the surface temperature of the Earth fell below 100°C. Water vapour from early volcanoes condensed to a liquid, and the first oceans were formed.

Since then there has been a constant interchange of water between the atmosphere and the oceans via the **water cycle**.

Heat from the Sun evaporates sea water. The water vapour cools as it rises and condenses into water droplets in clouds. As the clouds rise, further cooling leads to rain.

Soluble substances from rocks dissolve in the rain water and end up in the oceans, together with other soluble compounds produced by underwater volcanic activity.

Some common soluble ions like sodium and chloride aren't removed by other processes, and so their levels increase – the sea gradually becomes more salty. The sodium chloride crystallises out to form salt deposits only when an inland sea dries up.

Other ions like calcium and carbonate are extracted from the water to form the shells and skeletons of sea creatures, or else produce a precipitate of calcium carbonate which settles to the bottom of the sea. Both processes lead eventually to the formation of limestone as the calcium carbonate deposits (either shells or precipitate) are covered by layers of other sediments.

Water is the most abundant substance on the surface of the Earth.

The solubility of gases in water

All gases dissolve in water to some extent, but their solubility is usually small unless they react with the water. For example, hydrogen chloride gas is very soluble in water, reacting to form ionic hydrochloric acid. This reaction between hydrogen chloride and water is explained in detail on page 71.

Gases become more soluble if their pressure increases. Carbonated drinks like fizzy lemonade are made by dissolving carbon dioxide in water under pressure. When the pressure is released by opening the bottle, the carbon dioxide becomes less soluble and forms bubbles.

The solubility of gases is also affected by changes in temperature. *As temperature increases, the solubility of gases falls.*

Power stations use large amounts of cooling water. If this is discharged into rivers, the water dissolves less oxygen because it is warm. That can make life difficult for fish and other aquatic creatures which depend on the dissolved oxygen.

You can see a similar effect on fish in a small pond during hot weather. The amount of dissolved oxygen falls, and the fish have to come to the surface to try to get oxygen directly from the air.

The fizzing happens because the dissolved carbon dioxide becomes less soluble when the pressure in the bottle drops as the cork is removed.

Fish need the oxygen dissolved in water in order to live.

Power station cooling towers. The hot water discharged from these will dissolve less oxygen than cold water.

The solubility of solids

Measuring solubility

The solubility of a solid at a particular temperature is usually defined as "the mass of solute which will saturate 100 g of water at that temperature".

A **saturated solution** is a solution which contains as much dissolved solute as possible at a particular temperature. There must be some undissolved solute present.

To measure the solubility of potassium nitrate at exactly 40°C, you start by making a saturated solution of potassium nitrate in water at a temperature just over 40°C.

> The **solute** is the substance which dissolves in the solvent. In this case the solvent is water. It is possible to get "supersaturated" solutions with some solutes. These contain more dissolved substance than you would expect at a particular temperature. If you add even one tiny crystal of solid to these, all the extra solute will crystallise out, and you are left with a normal saturated solution. You don't need to worry about this at GCSE. Having undissolved solid present when you make a saturated solution prevents supersaturated solutions forming.

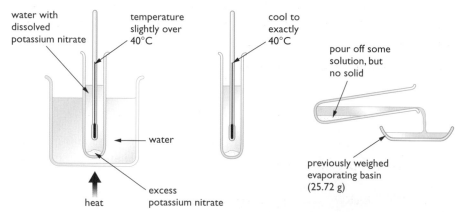

Now you have to find out how much water and solid there are in the solution.

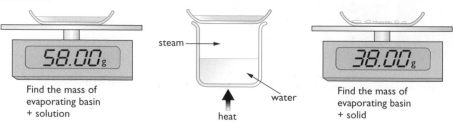

The solution is evaporated gently over a **steam bath** to avoid any solution spitting out of the basin. The best way to check that all the water has been evaporated is to weigh the basin when it *looks* dry and then re-heat it to make sure that there is no further loss of mass. This is called "heating to constant mass".

Using the results shown in the diagrams:

Mass of evaporating basin	= 25.72 g
Mass of evaporating basin + solution	= 58.00 g
Mass of evaporating basin + dry crystals	= 38.00 g
Mass of crystals	= 38.00 – 25.72 g
	= 12.28 g
Mass of water	= 58.00 – 38.00 g
	= 20.00 g

If 12.28 g of crystals saturate 20.00 g of water, you would need 5 times as much to saturate 100 g. That works out as 61.4 g. Therefore, the solubility of potassium nitrate at 40°C is 61.4 g per 100 g of water.

Solubility curves

The solubility of solids changes with temperature, and you can plot this on a **solubility curve**. Most solids have solubility curves like A or B in the sketch. Their solubility increases with temperature – either dramatically, or just a bit.

You can use solubility curves to work out what mass of crystals you would get if you cooled a saturated solution.

For example, if you cooled a saturated solution of A from, say, 80°C to 20°C, less of A will dissolve. The excess solid would precipitate out as crystals. You could use the curve to find out how much would dissolve in 100 g of water at 80°C and at 20°C. The difference would be the mass of crystals that you would obtain.

A worked example will make this clear.

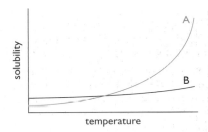

Typical solubility curves.

worked
  example

The table shows the solubility of potassium chloride at various temperatures.

Temperature (°C)	10	30	50	70	90
Solubility (g per 100 g of water)	31	37.5	43	48.5	54

(a) Plot a solubility curve for potassium chloride.

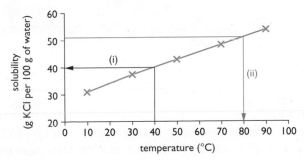

The lines marked (i) and (ii) on the graph come from answers to the next part of the question. It so happens that this solubility curve is virtually a straight line. You won't always be so lucky!

(b) Use the curve to find (i) the solubility of potassium chloride at 40°C, (ii) the temperature at which crystals will first appear if you cooled a hot solution containing 51 g of potassium chloride in 100 g of water.

Use the graph to find the information you want.

In part (i), the graph shows that at 40°C the solubility is 40 g per 100 g of water.

Part (ii): Solubility measures the mass of potassium chloride which will saturate 100 g of water at a particular temperature. In part (ii), the crystals will start to appear as soon as the solution becomes saturated. From the graph, potassium chloride would form a saturated solution containing 51 g of KCl per 100 g of water at 80°C.

Crystals would first appear at 80°C.

(c) What mass of potassium chloride would crystallise from the solution in (b)(ii) if the temperature fell to 10°C.

You can use the data in the table to find the solubility at 10°C. The value is 31 g per 100 g of water. That means that 31 g of potassium chloride will stay in solution. Since you started with 51 g, the rest must have formed crystals.

Mass of crystals = 51 − 31 g = 20 g

20 g of potassium chloride will crystallise out.

> In different circumstances, you might have to find the 10°C figure from the graph as well.

Hard Water

Hard water is water that doesn't easily form a lather with soap. You need to use large amounts of soap to get anything more than a white scum. By contrast, with soft water you can get lots of bubbles with a minimum amount of soap.

Hard water and soap

How does soap work?

Soap is the sodium or potassium salt of one of a number of large organic acids, of which stearic acid, $C_{17}H_{35}COOH$, is typical. Soap could therefore contain sodium stearate, $C_{17}H_{35}COONa$. Like all sodium compounds, this is ionic: $C_{17}H_{35}COO^-Na^+$.

The important bit is the stearate ion. A model shows that it is rather tadpole-like. The tail of the tadpole is a long chain of carbon atoms with hydrogens attached. The head, where the oxygens are, carries a negative charge.

The left-hand tube shows the effect of soap solution on hard water. In the right-hand tube, the soap solution was shaken with soft water.

A stearate ion.

> A hydrocarbon is a compound which consists of carbon and hydrogen only.

The long hydrocarbon tail is soluble in other hydrocarbons, including grease and oil. The negatively charged head is attracted to water molecules. Grease doesn't dissolve in water because there isn't enough attraction between the hydrocarbon molecules and water molecules. Soap overcomes that difficulty.

The hydrocarbon tails dissolve in the grease, leaving the ionic heads sticking out into the water and attracted to the water molecules.

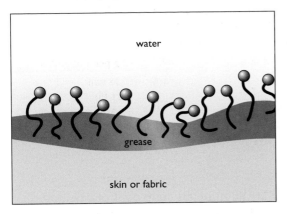

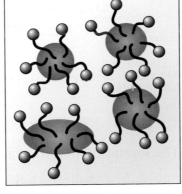

The stearate ions penetrate the grease...

...and surround it when it breaks up into tiny droplets.

If you wash your hands using soap and hot water, the grease breaks up into tiny droplets, each surrounded by negative charges.

The negative charges around each droplet repel the similar charges around neighbouring droplets. The droplets never come back together again into a single large piece of grease.

This isn't a true *solution* of grease in water. Each grease droplet is made up of lots of molecules and not just one. An **emulsion** is formed – small droplets of one liquid dispersed throughout a second one.

Why is some water hard?

Soap will only work if the stearate ions are in solution. Calcium or magnesium ions present in hard water remove the stearate ions in a precipitation reaction. White calcium or magnesium stearate is formed. This is **scum**. For example:

$$Ca^{2+}{}_{(aq)} + 2C_{17}H_{35}COO^-{}_{(aq)} \rightarrow (C_{17}H_{35}COO)_2Ca_{(s)}$$

This equation looks quite daunting because of the complicated formula of the stearate ion. Stearate ions are often given the simplified formula, St^-, and the equation then becomes much easier.

$$Ca^{2+}{}_{(aq)} + 2St^-{}_{(aq)} \rightarrow CaSt_{2(s)} \qquad \text{(where St = stearate)}$$

A lather won't form until all the calcium ions present have reacted with stearate ions to form scum. In hard water areas, this involves wasting lots of soap as well as the unpleasantness of washing in water with a thick layer of scum on it.

Comparing the hardness of different samples of water

Soap solution is added a little at a time to a known volume of water – whether hard or soft. The water is shaken vigorously after each addition to see if a permanent lather will form. A permanent lather is one which will last for a minute or so without the bubbles collapsing.

The experiment is repeated using the same volume of different water samples. Pure distilled water would be used as a control.

In a true solution, you have individual particles of the solute present – whether ions or molecules.

Important! From now on, we shall often seem to ignore the magnesium case. This is to save space. Anything said about calcium ions is equally true of magnesium ions. All you have to do is mentally swap the name or symbol.

This is the form of the equation that you will probably want to use in an exam. You should, however, add the phrase "(where St = stearate)".

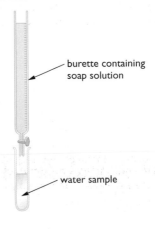

burette containing soap solution

water sample

The table shows some possible results for the volumes of soap solution added to equal volumes of different water samples.

Type of water	Volume (cm³)
distilled water	0.3
soft water	0.7
hard water A	4.5
hard water B	6.3

The more soap solution needed, the harder the water.

The formation of hard water

Anhydrite, gypsum and alabaster are all minerals composed of calcium sulphate. Calcium sulphate is very slightly soluble in water, and so some calcium ions dissolve in rain water. Other soluble calcium and magnesium salts in rocks dissolve in the same way.

But most calcium and magnesium ions get into the water via a chemical reaction. Rain water containing dissolved carbon dioxide is slightly acidic and reacts with limestone. This could be normal limestone, including chalk ($CaCO_3$), or a magnesium limestone ($MgCO_3$).

The product is calcium (or magnesium) hydrogencarbonate solution.

$$CaCO_{3(s)} + CO_{2(aq)} + H_2O_{(l)} \rightarrow Ca(HCO_3)_{2(aq)}$$

or

$$MgCO_{3(s)} + CO_{2(aq)} + H_2O_{(l)} \rightarrow Mg(HCO_3)_{2(aq)}$$

Caves at Nerja, southern Spain. These were produced by carbon dioxide dissolved in rain water reacting with the limestone.

These reactions are responsible for the formation of cave systems in areas with limestone rocks.

The solutions formed contain calcium or magnesium ions ($Ca^{2+}_{(aq)}$ or $Mg^{2+}_{(aq)}$), together with hydrogencarbonate ions, $HCO_3^-_{(aq)}$. It is the metal ions which cause the water to be hard.

Hard water – good or bad?

The formation of scum

We've already seen that the stearate ions in soap are precipitated by calcium or magnesium ions as scum. You need to use more soap to compensate for this.

You can avoid the problem by replacing soap by another detergent. Washing-up liquid and shampoo are both detergents, and aren't affected by hard water. They don't react with calcium or magnesium ions and so don't produce scum.

The formation of limescale

Limescale is formed when hard water containing calcium (or magnesium) hydrogencarbonate is heated. If you live in a hard water area, you will be familiar with the white "fur" that forms in electric kettles.

Heating produces solid calcium carbonate – limescale.

$$Ca(HCO_3)_{2(aq)} \rightarrow CaCO_{3(s)} + CO_{2(g)} + H_2O_{(l)}$$

This reaction is simply the reverse of the one which produced the calcium

Heating hard water containing calcium hydrogencarbonate produces limescale.

Technically, soap is a detergent. The word describes the effect the soap has on grease. Other detergents are known as "synthetic detergents" or "soapless detergents".

hydrogencarbonate solution in the first place.

Limescale can eventually block pipes. It also leads to waste of energy. Limescale coating an electric kettle element slows the transfer of heat to the water, for example. A copper boiler with a thick layer of limescale on the inside may eventually melt because the heat isn't being transferred to the water properly.

Limescale in a kettle can be removed by using solutions of acids. Commercial products are known as **descalers**, but you could equally well use vinegar (containing ethanoic acid). The acid reacts with the carbonate to form carbon dioxide and water.

$$CO_3^{2-}{}_{(s)} + 2H^+{}_{(aq)} \rightarrow CO_{2(g)} + H_2O_{(l)}$$

If you aren't happy with the ionic equation, you might like to look at pages 69 and 237. The ionic equation is given because all sorts of acids might be used. Some of them have complicated formulae.

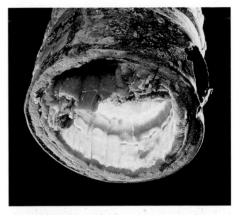

More limescale.

Hard water isn't all bad

In hard water areas, even cold water pipes get coated with a layer of calcium carbonate. This helps to stop the pipe corroding and releasing poisonous copper or lead ions into the water.

Calcium ions in hard water help to build strong teeth and bones, and also help to prevent heart diseases by reducing blood pressure.

Temporary and permanent hardness

Hardness of water due to calcium (or magnesium) hydrogencarbonate is known as **temporary hardness**. Once the water has been boiled, the calcium ions are all precipitated as insoluble calcium carbonate. If the calcium ions aren't in solution any longer, they can't have any effect on soap.

- Temporary hardness is removed by boiling the water.

Other calcium and magnesium salts in the water aren't affected by heating. For example, if the hardness is due to dissolved calcium sulphate, you wouldn't see any change when it was boiled. This is known as **permanent hardness**. Permanently hard water doesn't produce limescale.

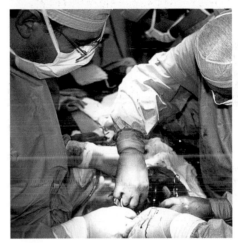

Studies show that calcium ions reduce blood pressure and may help to reduce heart disease.

- Permanent hardness isn't removed by boiling.

Removing hardness from water

By boiling

Simple boiling would only remove temporary hardness, and energy costs would make it uneconomic. Distillation would, of course, remove all dissolved material from the water, but would normally be much too expensive.

By precipitating the calcium and magnesium ions

Both types of hard water can be softened by adding sodium carbonate. The calcium or magnesium ions are precipitated out as insoluble calcium or magnesium carbonate.

$$Ca^{2+}{}_{(aq)} + CO_3^{2-}{}_{(aq)} \rightarrow CaCO_{3(s)}$$

Sodium carbonate is sometimes known as "washing soda", and is the major component of "bath salts".

Using an ion exchange resin

Domestic water softeners consist of small beads of an ion exchange resin packed into a container. The hard water flows through this and comes out at the other end as soft water.

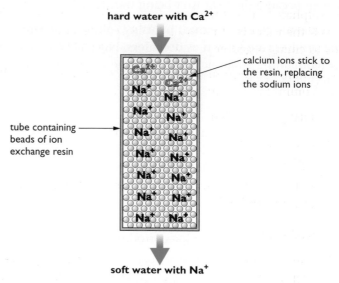

hard water with Ca^{2+}

calcium ions stick to the resin, replacing the sodium ions

tube containing beads of ion exchange resin

soft water with Na^+

The ion exchange resin does exactly what it says. It exchanges other positive ions for sodium ions. The resin is a complex lattice with sodium ions attached to it. Calcium and other positive ions stick to the lattice and the sodium ions are washed off. The sodium ions have no effect on soap.

$$2Na^+_{\text{(on resin)}} + Ca^{2+}_{\text{(aq)}} \rightarrow 2Na^+_{\text{(aq)}} + Ca^{2+}_{\text{(on resin)}}$$

Eventually all the sodium ions will have been exchanged and the resin will stop working. To regenerate it, all you need to do is to pour concentrated salt solution (sodium chloride solution) through it. All the trapped positive ions are washed off and replaced by sodium ions again.

Some ion exchange resins remove all ions – positive and negative. These replace all the positive ions present by hydrogen ions, $H^+_{\text{(aq)}}$, and all the negative ones by hydroxide ions, $OH^-_{\text{(aq)}}$. The hydrogen ions and hydroxide ions then react with each other to make water.

$$H^+_{\text{(aq)}} + OH^-_{\text{(aq)}} \rightarrow H_2O_{\text{(l)}}$$

The water emerging from the resin is known as **deionised water**. These resins would be too expensive to use domestically because there is no easy way to regenerate them when all the hydrogen ions and hydroxide ions have been washed off.

> Notice that each calcium ion replaces two sodium ions. This is to maintain a proper balance of charges in the resin.

Water Treatment

Water will be made fit for drinking using some or all of the following processes, depending on what has to be removed.

Screening

Water from a river or reservoir is passed through grids made of metal bars to remove large bits of solid matter like leaves, twigs and fish.

Sedimentation and filtration

After screening, the water is likely to contain fine solids, some material present as colloids, and some in solution. (You will find more about colloids on pages 196–197.)

Aluminium sulphate is added to the water, together with calcium hydroxide (slaked lime) if the water isn't naturally alkaline. The aluminium ions cause any colloids to clump together into solids (see page 197).

A coarse precipitate of aluminium hydroxide is formed from reaction between the aluminium ions and the hydroxide ions:

$$Al^{3+}_{(aq)} + 3OH^-_{(aq)} \rightarrow Al(OH)_{3(s)}$$

This precipitate tends to absorb some of the organic molecules which cause colour or unpleasant tastes in the water.

Most of the solids now settle out in **sedimentation** tanks, and the rest are removed by **filtration** through sand and gravel.

Use of activated carbon

Occasionally water is quite brown in colour or contains lots of organic molecules, giving it strange tastes or smells. These impurities can be removed by treating the water with **activated carbon**. Activated carbon is a specially treated porous form of carbon which is very absorbent.

pH adjustment

If the water is naturally acidic, enough calcium hydroxide is added to make it slightly alkaline. This helps to prevent pipes corroding and the release of poisonous ions like copper and lead into the water supply.

Chlorination

Chlorine is added to water to kill bacteria. Some water treatment companies also add it at the beginning of the process to prevent growth of algae in the treatment works. Any excess chlorine added can be removed by reacting it with sulphur dioxide.

$$Cl_{2(aq)} + SO_{2(aq)} + 2H_2O_{(l)} \rightarrow 2HCl_{(aq)} + H_2SO_{4(aq)}$$

Very dilute solutions of hydrochloric acid and sulphuric acid are formed. Obviously if addition of chlorine is necessary, it would have to be done before the pH is adjusted.

Problems after treatment

Even if the water leaving the treatment plant is pure, it may get contaminated on the way to the customer. Old iron pipes corrode, producing iron(III) ions which react with hydroxide ions in the slightly alkaline water. A brown precipitate of iron(III) hydroxide is formed.

$$Fe^{3+}_{(aq)} + 3OH^-_{(aq)} \rightarrow Fe(OH)_{3(s)}$$

If this precipitate is disturbed, it can find its way into the home as brown-coloured water. This leaves rusty stains on clothes washed in it, turns vegetables brown if they are cooked in it, and gives tea an inky colour and a bitter taste.

Water from a reservoir on Dartmoor will need to be treated differently from...

...water taken from the River Thames.

Ozone, O_3, is sometimes used to sterilise the water instead of chlorine. Knowledge of this isn't currently required by any GCSE specification, however.

In Britain we take clean drinking water for granted and worry if it is discoloured. Millions of people aren't so fortunate.

Colloids

You only need to remember details from this section if you are doing the OCR specification with Extension Block B.

In a **solution**, the individual particles (ions or molecules) of the solute are randomly mixed with the solvent. In a **suspension**, the particles are clumped together in a form which will separate from the liquid given time. For example, even a fine precipitate will eventually settle to the bottom. In full fat milk, cream soon separates and floats to the top.

A colloid is a half-way stage between these two. If you have read the whole of this chapter you have already come across a description of one particular colloid. When soap helps to "dissolve" grease, an **emulsion** is formed. This is a type of colloid. Soap is described as an **emulsifying agent**.

Emulsions

It would be helpful if you read pages 190–191 before you go on.

Emulsions are formed when tiny drops of one liquid are dispersed in a second liquid.

- The droplets are described as the **disperse phase**.

- The liquid in which they are dispersed is the **continuous phase**.

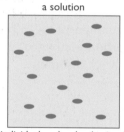
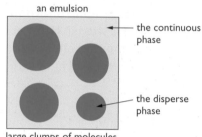

a solution

individual molecules (or ions) scattered throughout a solvent

an emulsion

the continuous phase

the disperse phase

large clumps of molecules, but prevented from getting any larger so that they settle out

This is intended to be a general case – not specifically the case of soap and grease. That's why no charges have been shown on the disperse phase. Some emulsions may form in a different way without the presence of these charges.

Semi-skimmed milk is a simple example of an emulsion. Droplets of fat are dispersed in water. Full fat milk contains this emulsion as well, but also has an excess of fat present as a simple suspension. That excess fat separates out on the top as cream.

Foams

A **foam** has a gas dispersed in a liquid. Foams can be as varied as whipped cream (air dispersed in cream) or the specialist foams used by airport fire crews. It is also possible to get solid foams in which a gas is dispersed in a solid. Expanded polystyrene is a simple example of a solid foam.

Sols

In a sol, a solid is dispersed in a liquid. A simple example involves the tiny particles of clay which are present in river water. They have to be removed during the water treatment process.

A fire-fighting foam.

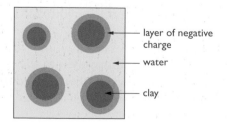

layer of negative charge

water

clay

196

Each clay particle has a number of negative ions stuck to its surface. The small particles don't clump together and form a precipitate because the negative charges repel each other.

In water treatment, the individual clay particles are made to coagulate (clump together) by adding aluminium sulphate solution. The Al^{3+} ions present stick to the surface of the clay particles and neutralise the negative charge. Aluminium ions are chosen because they each carry three positive charges and are particularly effective at neutralising negative ones.

A positive ion with only one charge (like Na^+) is less effective, but will do the job if the concentration is high enough. For example, the colloidal clay carried by rivers is precipitated out when the water comes into contact with the sodium and other positive ions present in the sea. Larger solid clay particles in suspension in the water also settle out at the same time. A combination of the two effects produces the muddy deltas typical of large rivers.

Colloidal clay is precipitated when it comes into contact with sea water.

End of Chapter Checklist

If you haven't got a copy of your specification, read the introduction on page vi.

You will need to be able to do some or all of the following. Check your Awarding Body's specification (syllabus) to find out exactly what you need to know.

- Explain the origin of the oceans.

- Know how the solubility of gases changes with temperature and pressure.

- Understand how the solubility of a solid is measured and how to construct and interpret solubility curves.

- Understand what makes water hard, and explain the effect of hard water on soap.

- Give advantages and disadvantages of hard water.

- Understand the difference between temporary and permanent hardness, and describe and explain the effect of boiling on temporarily hard water.

- Know how water can be softened by addition of sodium carbonate and by use of ion exchange resins.

- Understand the main processes involved in treating water for domestic use.

- Understand what is meant by a colloid and give examples restricted to an emulsion, a foam and a sol.

- Understand the terms *disperse phase* and *continuous phase* as applied to a colloid.

- Know that colloidal clay particles are prevented from coagulating because they carry a negative charge, and understand the effect of positive ions on the colloid.

Questions

More questions on water can be found at the end of Section D on page 207.

1 The solubility of sodium chlorate in water was measured at a number of different temperatures.

Temperature (°C)	0	20	40	60	80	100
Solubility (g per 100 g of water)	3	8	14	23	38	55

a) Use these figures to plot a solubility curve, with the temperature on the horizontal axis, and solubility on the vertical one.

b) Use your graph to find the solubility of sodium chlorate at 50°C.

c) 20 g of sodium chlorate was added to 100 g of water and the mixture heated to about 70°C. It was then left to cool with the thermometer in the solution. Use your graph to answer the following questions. (You can assume that sodium chlorate doesn't form supersaturated solutions.)

i) At what temperature would crystals first appear in the solution?

ii) If the solution was cooled to 17°C, work out the total mass of crystals formed.

2 A bottle of spring water from Buxton in Derbyshire had the following analysis. The figures are for ion concentrations in milligrams per litre.

Ca^{2+}	55	HCO_3^-	248
Mg^{2+}	19	Cl^-	42
Na^+	25	SO_4^{2-}	23
K^+	1	NO_3^-	less than 0.1

a) Name *two* compounds which are present in the water.

b) Write the symbols for *two* ions which make this water hard.

c) Give the mineral (*not* the chemical) names for any two rocks that this water might have been in contact with.

d) Hard water is water which doesn't readily form a lather with soap. Explain why this sample of water would have this effect on soap. You can assume soap to be sodium stearate, and use NaSt as its formula.

e) Describe and explain any method (apart from distillation) which would remove *all* the hardness from this sample.

3 A reservoir was fed by streams flowing over moorland grazed by sheep. The water had a pH of about 5 and was rather brown in colour because of the peat in the area. Describe and explain any *three* things that the local water company would have to do to the water to make it fit to drink. You should make clear in each case what the disadvantage to the consumer would be if the treatment wasn't carried out.

4 a) Explain what is meant by

    i) temporary hardness,

    ii) permanent hardness.

b) You are provided with a sample of hard water which is known to contain both calcium hydrogencarbonate (responsible for temporary hardness) and calcium sulphate (causing permanent hardness). You also have access to distilled water and some soap solution whose concentration is unknown to you, together with normal lab glassware. Design an experiment to find the *percentage* of the hardness in the water sample which is permanent. You should state clearly what preliminary experiments you would have to do to find out what volume of hard water to use, and whether the soap solution needed diluting.

5 A pair of students measured the solubility of a compound in water in the following way.

They placed 30.0 g of the solid in a small flask and added 20.0 cm³ of pure water. They heated the flask in a beaker of boiling water until all the solid had dissolved. Then they removed the flask from the hot water, placed a thermometer in the solution, and watched carefully for the first trace of crystals to appear. That shows that the solution had become saturated. They measured the temperature at that point.

Next they added some more water, and repeated the experiment. They continued to do this until they had the set of results below.

Mass of solid used = 30.0 g

Total volume of water used (cm³)	Temperature at which crystals appeared (°C)
20.0	92.5
30.0	80.5
40.0	72.5
60.0	56.0
80.0	40.0
100.0	25.5

a) Calculate the solubility of the solid at each of the temperatures in grams of solid per 100 g of water. Take the density of water to be 1 g/cm³.

b) Plot a graph of solubility on the vertical (y) axis against temperature on the horizontal (x) axis.

c) Use your graph to find the solubility of the solid at 50°C.

6 100 g of a mixture of two soluble compounds **A** and **B** contained 95 g of **A** and 5 g of **B**. The solubility curves for the two compounds are shown in the diagram.

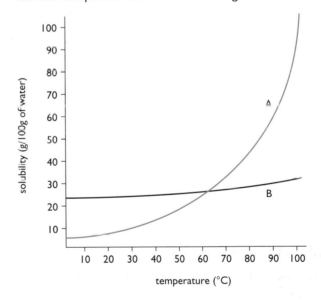

The mixture was dissolved in 100 g of water at 100°C, and then cooled to 0°C by surrounding it with ice.

This technique is known as "recrystallisation" and can be used to obtain pure crystals of **A** from the mixture.

a) Explain why the technique results in pure crystals of **A**. (Hint: Consider what happens to **A** and to **B** during the cooling.)

b) Suppose **A** was contaminated by an insoluble impurity **C** as well as with **B**. How would you change the experiment to remove **C** as well as **B**?

7 a) Colloids consist of a *disperse phase* and a *continuous phase*. Give an example of *i)* an emulsion, *ii)* a foam – in each case, naming the substances making up the disperse and continuous phases.

b) Rivers carry clay as a *sol* – tiny clay particles dispersed in water.

    i) Explain why the clay particles don't clump together and form a sediment.

    ii) The clay particles do clump together if solutions containing positive ions are added to the water. Explain why.

    iii) Suppose you were given separate solutions containing sodium, magnesium, and iron(III) ions in equal concentrations. Rank these in order of their effectiveness in precipitating colloidal clay from water. Explain your ranking.

8 In this question you will be asked to explain various things. Your explanations should use scientific terms, and include equations for any reactions you mention.

    a) Explain why the first oceans weren't formed until about 4000 million years ago, although there were large quantities of water vapour in the atmosphere for millions of years before that.

    b) The oceans are constantly fed by river water. Explain why the sea contains a much higher percentage of sodium chloride than river water does.

    c) Two clear plastic bottles both contained mineral water. One was labelled as "still", the other as "sparkling". There is no obvious difference in appearance between the two until you take the top off the bottles. Then the sparkling one fizzes. Explain what is happening.

    d) Explain why rain water flowing through limestone sometimes produces large caves.

    e) If you fill a clean vacuum flask with boiling water in London (a hard water area), by the time you come to use it a couple of hours later the flask has a white sediment in it. Explain where the sediment comes from.

    f) Limestone caves often have stalactites (icicle-like rock formations which hang down from the roof of the cave) and stalagmites (similar structures which grow up from the floor). From your knowledge of chemistry, suggest how these might be produced.

Chapter 19: Rocks

This chapter looks at the three main types of rock (igneous, sedimentary and metamorphic) and the processes which create them.

New rock being formed as magma oozes out into sea water at a plate boundary.

The large crystals in this granite show that it cooled and crystallised slowly.

It's hard to believe now that this granite was originally formed deep inside a mountain.

Background

Plate tectonics

The surface of the Earth is broken up into a series of large rigid pieces called **plates**, floating on a layer of partly molten rock. These plates move around relative to each other at a rate of a few centimetres per year. The word *tectonics* comes from a Greek word *tekton* which means "a builder". The movement of the plates is responsible for building mountains.

When plates move apart from each other, molten rock called **magma** wells up towards the surface and crystallises to form new solid rock.

Where plates collide with each other, violent earthquakes occur, volcanoes form, and existing rock is subjected to huge forces. These forces can cause the rocks to distort, and can lift them to build huge mountain ranges. These processes happen over millions of years.

Erosion

Over more millions of years, mountains are worn away again. **Weathering** breaks up rock, either by chemical reactions or by the action of wind, water and ice. **Transportation** then moves the broken-up rock fragments elsewhere. The wearing away of the land is known as **erosion**.

Rock Types

Igneous rocks

Igneous rocks are formed when magma cools and crystallises. If it cools very quickly, lots of tiny crystals are formed. On the other hand, if it crystallises slowly there is time for large crystals to be formed.

Granite

Granite was originally formed deep under the surface of the Earth. It is an **intrusive rock** because it was formed by molten magma forcing its way into existing rocks, lifting them to form mountains. Because the granite was formed deep beneath the surface, it cooled slowly and formed large crystals of the various minerals that it contains. In the photograph, the largest crystals are of a mineral called feldspar.

Don't be misled by the fact that a lot of granite now appears on the surface of the Earth. There were originally many layers of rock on top of it which have since been removed by weathering and erosion. All the granite that you now see was originally formed beneath high mountains.

Basalt

Basalt is a fine-grained rock produced when magma cools very quickly. It is produced as plates under the oceans move apart or from lava flows. Basalt is an **extrusive rock**. It was formed on the surface, not deep in the Earth.

As this lava cools, basalt will be formed.

The Giant's Causeway in Northern Ireland is made of basalt from an ancient lava flow.

The ash and "bombs" (large bits of rock) thrown out of volcanoes are also igneous, but obviously in fragments. Rocks formed from these are known as **pyroclastic rocks** (from two Greek words meaning "fire" and "broken").

Sedimentary rocks

When rocks (of whatever type) are weathered and eroded, the bits broken off are carried to other places as mud or sand. Over millions of years, these build up to very thick layers. Because of changes in sea levels, different sediments build up on top of the original ones.

The sediments are compressed by the pressure of the layers on top of them. The individual fragments are cemented together to form rocks by minerals dissolved in the water surrounding them.

Overall, this process is known as **lithification** (literally "stone making").

The layering of sedimentary rocks is clearly shown in the Grand Canyon in America. The Grand Canyon was formed as the Colorado River cut as much as 1.6 km deep through layer upon layer of sedimentary rocks.

The rocks at the very bottom of the Canyon are the remains of an old mountain range. The higher up you go in the Canyon, the younger the rocks get as each new layer was formed on the layers underneath. This is confirmed by the fossil record.

You might be interested in doing an internet search on **grand canyon**. There are some excellent sites, including several controversial ones produced by groups who try to use the fossil evidence to disprove the theory of evolution.

As a general rule, younger rocks are found on top of older ones. But very occasionally, the rocks get so distorted and folded because of movements in the Earth's crust that the layers can get inverted. In the Grand Canyon, the layers are nearly horizontal and so it is easy to follow the sequence from younger to older rocks as you go deeper.

The Grand Canyon – a slice through history.

Sedimentary rocks often contain **fossils** – the remains of ancient life preserved in stone. Fossils are obviously missing from igneous rocks, which were formed as molten magma crystallised.

The oldest sedimentary rock (found towards the bottom of the Grand Canyon) is about 1250 million years old. The rock is so old that only fossil algae are found in it.

As you go higher, the rocks get younger, and contain more and more fossils. This is because there were more and more forms of life evolving whose remains could be trapped in the sediments. The top layer (about 250 million years old) contains numerous fossils of sea creatures. The presence of particular fossils is a useful guide to the relative age of a rock.

The Grand Canyon has a good cross-section of sedimentary rocks, including shale, sandstone and limestone. There is also some basalt, showing that there was volcanic activity at some point.

Shale is a fine-grained rock formed from the compression of mud and silt. It was originally laid down around old river estuaries.

Sandstone is formed from compacted and cemented layers of sand. You can see the original grains of sand, with particles up to about 2 mm. There may be ripple marks in it – signs of patterns in the original sand. Sandstone might have been formed from sand in the sea, in rivers or in deserts.

Limestone

Limestone is calcium carbonate, and is mainly formed from the shells and skeletons of sea creatures. Huge layers of these accumulated at the bottom of oceans and were compressed and cemented into limestone over millions of years. There is more about limestone on pages 161–163.

Fossilised sea creatures in limestone.

Tilting and folding of sedimentary rocks

Movements of the Earth's crust can subject rocks to such large forces that they are tilted and folded. By looking at the way the rocks have been folded, you can estimate the direction and size of the forces involved.

Earth movements can also produce **faults**, in which layers of rock get broken and displaced – moving up or down or sideways relative to each other.

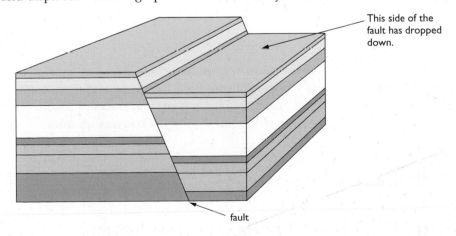

This side of the fault has dropped down.

fault

These rocks were originally laid down in horizontal layers.

You can see in the diagram that at some stage the right-hand side dropped downwards relative to the left-hand side.

Metamorphic rocks

Metamorphic rocks are produced by the action of high temperatures and pressures on other rocks over a long time scale. The original sedimentary rocks are changed into new rocks, but without actually melting in the process – otherwise you would form new igneous rocks.

When granite forms, the rocks in contact with the hot magma will be changed. This is known as **contact metamorphism**. Any other **igneous intrusion** (molten rock being forced into existing solid rock) will have the same effect.

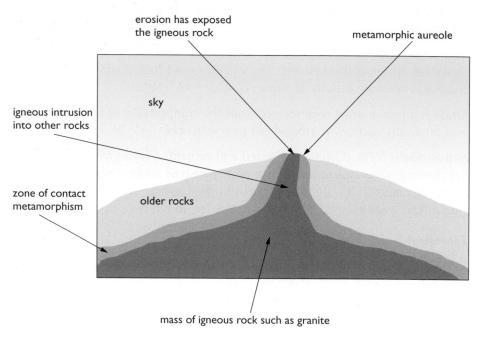

The original rocks nearest to the igneous rock will be changed most. As the temperature falls as you move away from the intrusion, the degree of change will get less and less. Notice that an intrusion must obviously be a younger rock than those that it has forced its way through.

Metamorphic rocks also form because of the stresses and high temperatures associated with tectonic plate movements and mountain building. This produces **regional metamorphism** extending over wide areas.

Marble

Marble is a metamorphic rock formed from limestone. It can be produced by either contact or regional metamorphism. Both limestone and marble are made of calcium carbonate. Some marbles still contain traces of fossils, but in others the fossils have been destroyed during the conversion from limestone.

Slate

Slate is formed by **low grade metamorphism** of shale. It is produced by relatively low temperatures and pressures and is the first stage in the formation of a sequence of other less familiar rocks. **High grade metamorphism** of shale (at higher temperatures and pressures) leads to rocks like schist and gneiss.

The Taj Mahal is built of white marble.

End of Chapter Checklist

If you haven't got a copy of your specification, read the introduction on page vi.

You will need to be able to do some or all of the following. Check your Awarding Body's specification (syllabus) to find out exactly what you need to know.

- Know what is meant by an igneous rock and give examples.
- Understand that the crystal sizes in igneous rocks depend on the rate of cooling.
- Give examples of the formation of sedimentary rocks, and understand that younger rocks are normally found on top of older ones.
- Understand that fossils found in sedimentary rocks can help to date the rocks.
- Understand how metamorphic rocks are formed in contact or regional metamorphism.
- Know that marble is formed from limestone.

Questions

More questions on rocks can be found at the end of Section D on page 207.

1 Granite and basalt are both igneous rocks.

a) What do you understand by the term *igneous rock*?

b) Granite consists of large visible crystals of the various minerals that it contains. Basalt is made up of very tiny crystals. Describe how granite and basalt are formed, explaining why that results in the different crystal sizes.

2 The top four layers of rock in a part of the Grand Canyon are shown in the diagram.

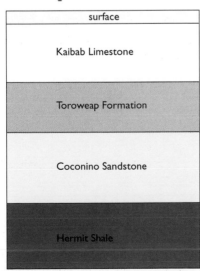

surface
Kaibab Limestone
Toroweap Formation
Coconino Sandstone
Hermit Shale

The Toroweap Formation consists of sandstone and limestone. All the layers are made of sedimentary rocks.

a) What do you understand by the term *sedimentary rock*?

b) Which of these rocks was formed most recently? Explain your answer.

c) How could you show, using simple chemical tests, that the top layer was limestone?

d) The Kaibab Limestone contains plentiful fossils of marine organisms. Explain briefly the origin of this limestone.

e) The Hermit Shale contains fossils of land plants. Explain how shale is formed.

f) Underlying the sedimentary rocks in the Grand Canyon is a thick layer of schist. Schist is a metamorphic rock. What do you understand by a *metamorphic rock*?

3 The Andes in South America are highly folded mountains consisting of limestone, sandstone, slate and granite. There are also large amounts of lava due to volcanic activity.

a) From the list of rocks mentioned, name *i)* two igneous rocks, *ii)* two sedimentary rocks, *iii)* one metamorphic rock.

b) The sedimentary rocks would originally have been formed at or below sea level. Explain why they are now found at the top of high mountains.

c) What is meant by *folded* rocks? How does folding give evidence for plate movements?

d) Granite is formed deep underneath mountains. Explain why some granite is found on the surface in the Andes.

4 The diagram shows some sedimentary rocks with an intrusion of igneous rock.

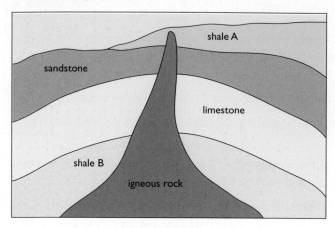

a) List the rocks in order of age, starting with the oldest and ending with the youngest.

b) Marble is a metamorphic rock. Where in these rocks would you expect to find some marble? Explain your reasoning.

End of Section Questions

1 Limestone and marble are both calcium carbonate. Limestone is a sedimentary rock which often contains fossils of sea creatures. Marble is a crystalline rock which rarely contains any traces of fossils.

a) Explain how limestone is formed. *(3 marks)*

b) Explain how marble is formed. *(2 marks)*

c) Describe a chemical test to show that both limestone and marble contain carbonate ions. *(2 marks)*

When water flows over limestone, the following reaction occurs:

$$CaCO_{3(s)} + CO_{2(g)} + H_2O_{(l)} \rightarrow Ca(HCO_3)_{2(aq)}$$

The water becomes *hard*.

d) What is meant by *hard water*? *(1 mark)*

e) Explain the formation of limescale when water containing dissolved calcium hydrogencarbonate is boiled. *(2 marks)*

f) Name one substance which might make water hard and which would not be affected by boiling. *(1 mark)*

g) Give one advantage of hard water over soft water. *(1 mark)*

h) Explain, with the help of an equation, why hard water can be softened by adding sodium carbonate. *(3 marks)*

Total 15 marks

2 The diagram shows a sequence of rocks. The limestone is at the surface of the Earth.

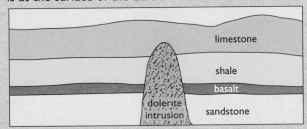

a) Dolerite is an *igneous* rock. Explain what is meant by the term igneous. *(1 mark)*

b) Give the name of another igneous rock shown in the diagram, and explain briefly how it might have been formed. *(3 marks)*

c) Which is the oldest sedimentary rock shown in the diagram? *(1 mark)*

d) Which is the youngest rock shown? Explain your reasoning. *(3 marks)*

e) In this example the layers of rock are fairly horizontal, but in many places, the layers of rock are highly folded and faulted. What evidence does such folding and faulting give about the Earth's crust in that region? *(2 marks)*

Total 10 marks

3 A student determined the solubility of a salt (not sodium chloride) in water at 20°C by the following procedure. She added some of the salt to some slightly warmed water in a beaker until no more salt would dissolve. The undissolved solid was allowed to settle and the solution was cooled back to 20°C. She quickly poured some of the solution into a previously weighed evaporating basin, reweighing it with the solution before evaporating it gently to dryness. She then weighed the basin containing the dry solid.

Results: Mass of basin + solution   = 60.0 g
           Mass of basin + solid   = 40.0 g
           Mass of empty basin   = 37.0 g

a) Suggest one reason why it is important to evaporate the solution *gently*. *(2 marks)*

b) How could the student be sure that all the water had been evaporated off? *(2 marks)*

c) Work out the mass of water lost from the evaporating basin. *(1 mark)*

d) Work out the mass of solid left after evaporation. *(1 mark)*

e) Work out the solubility of the salt in grams per 100 g of water. *(1 mark)*

The student repeated the experiment at several different temperatures and obtained these results.

Temperature (°C)	20	30	40	50	60	70	80
Solubility (g per 100 g of water)		19.5	26.0	35.0	47.0	63.5	85.0

f) Plot a graph of solubility on the vertical (y) axis against temperature on the horizontal (x) axis. Use your result from part **e)** for the solubility at 20°C. *(4 marks)*

g) Use your graph to find the temperature at which the solubility becomes 40 g of salt per 100 g of water. *(1 mark)*

h) The student dissolved 50 g of the salt in 100 g of water at 80°C, and then cooled the solution to 35°C. What mass of salt would crystallise out as a result of this temperature change? *(2 marks)*

Total 14 marks

4 About 4000 million years ago, when the Earth's atmosphere contained large amounts of carbon dioxide and steam, together with smaller amounts of gases like ammonia (NH_3), methane (CH_4), nitrogen, hydrogen and carbon monoxide, the first oceans formed. The carbon dioxide level in the atmosphere was reduced by solution in the oceans, the formation of carbonate rocks, and the formation of fossil fuels.

a) Explain why the first oceans didn't form until about 4000 million years ago. *(2 marks)*

b) Name a carbonate rock. *(1 mark)*

c) Name a fossil fuel. *(1 mark)*

d) The primitive oceans were hotter than the present oceans, and the carbon dioxide was at a higher pressure. What effect would there be on the solubility of the carbon dioxide because of *i)* the higher temperature, *ii)* the higher pressure? *(2 marks)*

e) Explain the effects of the evolution of plants on the composition of the atmosphere. *(3 marks)*

f) Over the past century or so, the levels of carbon dioxide in the atmosphere have been rising.

  i) Give one reason why the levels have been rising. *(1 mark)*

  ii) Give one environmental problem that may be caused because of increasing amounts of carbon dioxide in the atmosphere. *(1 mark)*

  iii) Suggest one measure that might be taken to prevent that problem. *(1 mark)*

g) Environmental problems are also caused by the presence of sulphur dioxide and nitrogen oxides in the air. Describe briefly the problem associated with these gases. You should state the source of the pollution, the environmental effect that it has, and how that effect might be overcome. *(4 marks)*

Total 16 marks

5 In 1988 a delivery driver emptied 20 tonnes of poisonous aluminium sulphate into the wrong tank at a water treatment works at Camelford in Cornwall. The result was a lot of ill-health in the local population. Water companies also add other poisonous substances to the water. Chlorine is routinely added and a few companies add fluorides such as sodium fluoride.

a) Why are small amounts of chlorine gas added to drinking water? *(1 mark)*

b) Decreasing numbers of water companies add fluorides to their water. Explain:

  i) the advantage of adding the sodium fluoride

  ii) why the number of companies adding it is decreasing. *(2 marks)*

c) Explain the importance of adding aluminium sulphate in the treatment of domestic water supplies. You should explain the reason for adding it, and (briefly) how it achieves its effect. *(5 marks)*

Total 8 marks

6 The table shows how the solubility of ammonia gas, $NH_{3(g)}$, varies with temperature.

Temperature (°C)	0	10	20	30	40
Solubility of ammonia (g per 100 g water)	90.0	69.0	53.0	41.0	31.0

a) Plot a graph of the solubility of ammonia (on the vertical (y) axis) against temperature (on the horizontal (x) axis). *(3 marks)*

b) Use your graph to find the solubility of ammonia at 15°C. *(1 mark)*

c) What does the graph show about the effect of temperature on the solubility of ammonia? *(1 mark)*

d) How would you expect the solubility of ammonia to change if you increased the pressure of the ammonia gas in contact with the water at a particular temperature? *(1 mark)*

e) Ammonia reacts with water to give a solution containing ammonium ions and hydroxide ions. Write an ionic equation for the reaction. *(2 marks)*

f) Oxygen is more soluble in water than nitrogen is. When air is in contact with water, the dissolved gases contain about 33% of oxygen rather than the 21% present in air. You can show this by boiling the water to collect the dissolved gases, and then finding the percentage of oxygen in the mixture.

Describe a simple experiment you could use to show that the mixture of gases formed contains about 33% of oxygen. *(5 marks)*

Total 13 marks

Section E: Organic Chemistry

Chapter 20: Useful Products from Crude Oil

The oil industry is at the very heart of modern life – providing fuels, plastics and the organic chemicals which go to make things as different as solvents, drugs, dyes and explosives. This chapter explores the way that an unappealing sticky black liquid is converted into useful things.

This sticky black liquid underpins modern life.

What is Crude Oil (Petroleum)?

The origin of crude oil

Millions of years ago plants and animals living in the sea died and fell to the bottom. Layers of sediment formed on top of them. Their shells and skeletons formed limestone. The soft tissue was gradually changed by high temperatures and pressures into crude oil. Crude oil is a **finite, non-renewable resource**. Once all the existing supplies have been used they won't be replaced – or at least, not for many millions of years.

Crude oil contains hydrocarbons

Crude oil is a mixture of **hydrocarbons** – compounds containing carbon and hydrogen only. Hydrocarbons are **organic** molecules. The term *organic* was originally used because it was believed that organic compounds could only come from living things. Now it is used for any carbon compound except for the very simplest (like carbon dioxide and the carbonates).

Hydrocarbons can exist as chains, branched chains or rings of carbon atoms with hydrogens attached.

Hydrocarbons can occur as...

chains...   branched chains...   rings...   or a combination

How the properties of hydrocarbons change with size of molecule

As the number of carbon atoms in the molecules increases, several properties of the compounds change in a regular pattern. Most of these changes are the result of increasing attractions between neighbouring molecules. As the molecules get bigger, these **intermolecular attractions** increase and it gets more difficult to pull one molecule away from its neighbours.

As the molecules get bigger:

- Boiling point increases – the larger the molecule, the higher the boiling point. This is because large molecules are attracted to each other more strongly than smaller ones. More heat is needed to break these stronger attractions to produce the widely separated molecules in the gas.

- The liquids become less volatile. The bigger the hydrocarbon, the more slowly it evaporates at room temperature. This is again because the bigger molecules are more strongly attracted to their neighbours and so don't turn to a gas so easily.

Intermolecular attractions are explained on pages 18–19.

We usually count a substance as being volatile if it turns to a vapour easily at room temperature. This means that it will evaporate quickly at that temperature.

Chapter 20: Useful Products from Crude Oil

209

- The liquids flow less easily (they become more viscous). Liquids containing small hydrocarbon molecules are runny. Those containing large molecules are much stickier because of the greater attractions between their molecules.

- Bigger hydrocarbons do not burn as easily as smaller ones. This limits the use of the bigger ones as fuels.

Separating the Crude Oil

Fractional distillation

The crude oil is heated and passed into a **fractionating column** which is cooler at the top and hotter at the bottom. The crude oil is split into various **fractions**.

Warning! You will find a lot of disagreement about exactly what fractions are produced in this first distillation. Don't worry about this. It's never a problem in exam questions.

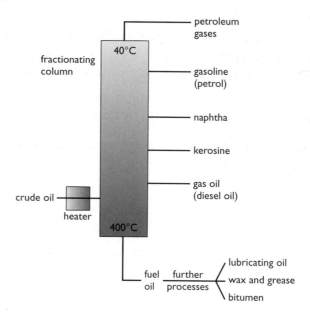

How far up the column a particular hydrocarbon gets depends on its boiling point. Suppose a hydrocarbon boils at 120°C. At the bottom of the column, the temperature is much higher than 120°C and so the hydrocarbon remains as a gas. As it travels up through the column, the temperature gets lower. When the temperature falls to 120°C, that hydrocarbon will start to turn to a liquid. It condenses and can be tapped off.

The hydrocarbons in the petroleum gases have boiling points which are so low that the temperature of the column never falls low enough for them to condense to liquids.

The temperature of the column isn't hot enough to boil the large hydrocarbons found in the fuel oil and this remains as a liquid. Some of the fuel oil is fractionally distilled under reduced pressure. The residue at the end of all this is **bitumen**, which is used in road making.

Things boil at lower temperatures if you reduce the pressure. Distillation under reduced pressure prevents the large molecules breaking up as a result of the high temperatures.

Uses of the fractions

All hydrocarbons burn in air (oxygen) to form carbon dioxide and water and release lots of heat in the process. They can therefore be used as **fuels**.

For example, burning methane (the major constituent of natural gas):

$$CH_{4(g)} + 2O_{2(g)} \rightarrow CO_{2(g)} + 2H_2O_{(l)}$$

...or burning octane (one of the hydrocarbons present in petrol):

$$2C_8H_{18(l)} + 25O_{2(g)} \rightarrow 16CO_{2(g)} + 18H_2O_{(l)}$$

If there isn't enough air (or oxygen) you get **incomplete combustion**. This leads to the formation of carbon (soot) or carbon monoxide instead of carbon dioxide. Carbon monoxide is colourless and odourless and very poisonous.

Carbon monoxide is poisonous because it combines with **haemoglobin** (the molecule which carries oxygen in the bloodstream), preventing it from carrying the oxygen. You are made ill, or even die, because not enough oxygen gets to the cells in your body.

Petroleum gases

Petroleum gases are a mixture of methane, ethane, propane and butane which can be separated into individual gases if required. These gases are commonly used as LPG (liquefied petroleum gas) for domestic heating and cooking.

Gasoline (petrol)

As with all the other fractions, petrol is a mixture of hydrocarbons with similar boiling points. Its use is fairly obvious!

Naphtha

Naphtha is used as a source of organic chemicals for industry as well as a constituent of petrol. Useful molecules like ethene and propene can be made by **cracking** the naphtha. You will find more about cracking on pages 212–213.

Kerosine

Kerosine is used as fuel for jet aircraft, as domestic heating oil and as "paraffin" for small heaters and lamps.

Gas oil (diesel oil)

This is used for buses, lorries, some cars, and railway engines where the line hasn't been electrified. Some is also cracked to make other organic chemicals and produce more petrol.

Don't try to learn these equations – there are too many possible hydrocarbons you could be asked about. Provided you know (or are told) the formula, they are easy to balance.

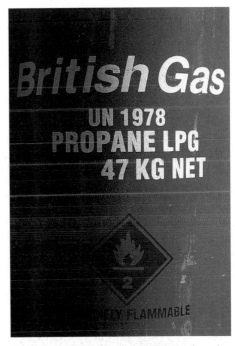

Propane is liquefied under pressure and used for domestic heating.

Kerosine is used as aviation fuel.

A train powered by diesel oil.

Chapter 20: Useful Products from Crude Oil

211

Fuel oil

This is used for ships' boilers and for industrial heating.

Some of the fuel oil is also distilled again, under reduced pressure, to make lubricating oil, grease, wax (for candles) and bitumen.

Bitumen

Bitumen is a thick black material which is melted and mixed with rock chippings to make the top surfaces of roads.

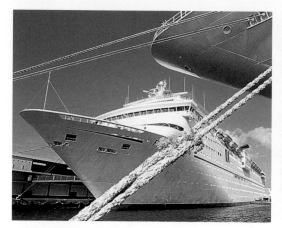

Ships' boilers burn fuel oil.

Bitumen is used in road construction.

Cracking

Although the fractions from crude oil distillation are useful fuels, there are two problems:

- The amounts of each fraction you get will depend on the proportions of the various hydrocarbons in the original crude oil, not the amounts in which they are needed. Far more petrol is needed, for example, than is found in crude oil.

- Apart from burning, the hydrocarbons in crude oil are fairly unreactive. To make other organic chemicals from them they must first be converted into something more reactive.

Cracking is a useful process in which large hydrocarbon molecules are broken into smaller ones. The big hydrocarbon molecules in gas oil, for example, can be broken down into the smaller ones needed for petrol.

The majority of the hydrocarbons found in crude oil have single bonds between the carbon atoms. During the cracking process, molecules are also formed which have double bonds between carbon atoms. These new molecules are much more reactive and can be used to make lots of other things.

How cracking works

The conditions

The naphtha or gas oil fraction is heated to give a gas and then passed over a catalyst of mixed silicon dioxide and aluminium oxide at about 500°C. Cracking can also be carried out at higher temperatures without a catalyst.

The reactions

Cracking is just an example of thermal decomposition – a big molecule splitting into smaller ones. The molecules are broken up in a fairly random way. This is just one possibility:

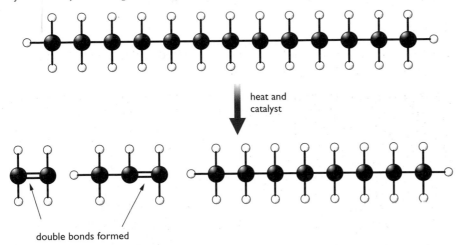

heat and
catalyst

double bonds formed

The molecules have been drawn to show the various covalent bonds. They have also been "straightened out". The real molecules are much more "worm-like"! Organic molecules are almost always drawn in this simplified form.

As an equation, this would read:

$$C_{13}H_{28(l)} \rightarrow C_2H_{4(g)} + C_3H_{6(g)} + C_8H_{18(l)}$$

Cracking produces a mixture of **alkanes** and **alkenes**. Alkanes are hydrocarbons in which all the carbon atoms are joined by single bonds – like $C_{13}H_{28}$ and C_8H_{18}. An alkene contains at least one carbon–carbon double bond – like C_2H_4 and C_3H_6.

Don't worry for now if you don't know which are gases and which are liquids. You will find out how to work this out in the next chapter.

The molecule might have split quite differently. All sorts of reactions are going on in a catalytic cracker. Two other possibilities might be:

$$C_{13}H_{28(l)} \rightarrow 2C_2H_{4(g)} + C_9H_{20(l)}$$

$$C_{13}H_{28(l)} \rightarrow 2C_2H_{4(g)} + C_3H_{6(g)} + C_6H_{14(l)}$$

Some reactions might even produce a small percentage of free hydrogen. For example:

$$C_{13}H_{28(l)} \rightarrow 2C_2H_{4(g)} + C_3H_{6(g)} + C_6H_{12(l)} + H_{2(g)}$$

In this case, all the hydrocarbons formed will have double bonds. They are all alkenes.

It is essential that you don't try to learn these equations. In an exam, you could be given one of a wide range of possible hydrocarbons to start with. You need to understand what is going on so that you can be adaptable. For a typical question on cracking, see Question 3 at the end of this chapter.

Polymerisation

The polymerisation of ethene

Ethene is one of the alkenes produced by cracking. It is the smallest hydrocarbon containing a carbon–carbon double bond. All of the following are different ways of writing or drawing an ethene molecule.

$$\begin{array}{cc} H & H \\ | & | \\ C = C \\ | & | \\ H & H \end{array}$$

$CH_2{=}CH_2$

Under the right conditions, molecules containing carbon–carbon double bonds can join together to produce very long chains. Part of the double bond is broken, and the electrons in it are used to join to neighbouring molecules. This is called **addition polymerisation**.

Polymerisation is the joining up of lots of little molecules (the **monomers**) to make one big molecule (the **polymer**). In the case of ethene, lots of ethene molecules join together to make **poly(ethene)** – more usually called polythene.

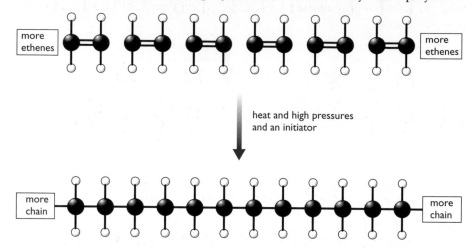

The chain length can vary from about 4000 to 40,000 carbon atoms.

For normal purposes, this is written using **displayed formulae**. A displayed formula is rather like the models drawn here, but with symbols for the atoms rather than circles.

For exam purposes, the acceptable form is:

$$n \quad \begin{array}{c} H \quad H \\ | \quad | \\ C = C \\ | \quad | \\ H \quad H \end{array} \quad \longrightarrow \quad \left(\begin{array}{c} H \quad H \\ | \quad | \\ C - C \\ | \quad | \\ H \quad H \end{array} \right)_n$$

Uses of poly(ethene)

Poly(ethene) comes in two types – low density poly(ethene) (LDPE) and high density poly(ethene) (HDPE).

Low density poly(ethene) is mainly used as a thin film to make polythene bags. It is very flexible and not very strong.

High density poly(ethene) is used where greater strength and rigidity is needed – for example, to make plastic milk bottles. If you look underneath a plastic milk bottle, you will probably find the letters HDPE next to a recycling symbol.

The polymerisation of propene

Propene is another alkene – this time with three carbon atoms in each molecule. Its formula is normally written as $CH_3CH=CH_2$. Think of it as a modified ethene molecule – with a CH_3 group attached in place of one of the hydrogen atoms.

An **initiator** is used to get the process started. You mustn't call it a catalyst, because it gets used up in the reaction. People occasionally wonder what happens at the ends of the chains. They don't end tidily! Bits of the initiator are bonded on at either end. You don't need to worry about that for GCSE.

In this structure for poly(ethene), "n" represents a large but variable number. It simply means that the structure in the brackets repeats itself lots and lots of times in the molecule.

This is the real shape…

…but you'll make life much easier for yourself if you think of it like this

When propene is polymerised you get **poly(propene)**. This used to be called polypropylene.

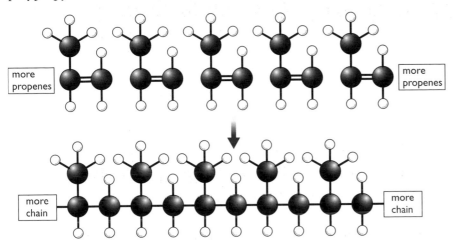

more propenes

more propenes

more chain

more chain

You will find that it is much easier to work out the structure of its polymer if you keep the CH_3 group tucked up out of the way when you draw it.

Write this as:

$$n \quad \begin{array}{c} CH_3 \quad H \\ | \quad\quad | \\ C = C \\ | \quad\quad | \\ H \quad\quad H \end{array} \longrightarrow \left(\begin{array}{c} CH_3 \quad H \\ | \quad\quad | \\ C - C \\ | \quad\quad | \\ H \quad\quad H \end{array} \right)_n$$

Uses of poly(propene)

Poly(propene) is somewhat stronger than poly(ethene). It is used to make ropes and crates (among many other things). If an item has a recycling mark with PP inside it or near it, it is made of poly(propene).

Poly(propene) is used to make crates…

…and ropes.

The polymerisation of chloroethene

Chloroethene is an ethene molecule in which one of the hydrogen atoms is replaced by a chlorine. Its formula is $CH_2=CHCl$. It used to be called vinyl chloride. Polymerising chloroethene gives you poly(chloroethene). This is usually known by its old name, polyvinylchloride or PVC.

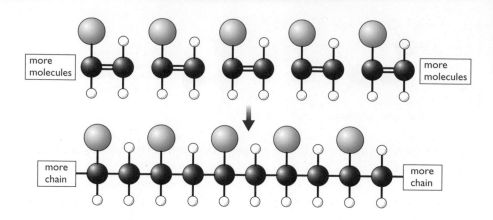

It doesn't matter when you draw this whether you put the chlorine atom on the left-hand carbon atom or the right-hand one.

You would write it as:

$$n \; \underset{\underset{H}{|}}{\overset{\overset{Cl}{|}}{C}} = \underset{\underset{H}{|}}{\overset{\overset{H}{|}}{C}} \longrightarrow \left(\underset{\underset{H}{|}}{\overset{\overset{Cl}{|}}{C}} - \underset{\underset{H}{|}}{\overset{\overset{H}{|}}{C}} \right)_n$$

Uses of poly(chloroethene)

Poly(chloroethene) – PVC – has lots of uses. It is quite strong and rigid and so can be used for things like drainpipes or replacement windows. It can also be made flexible by adding "plasticisers". That makes it useful for sheet floor coverings, and even clothing. These polymers don't conduct electricity and PVC is used for electrical insulation.

PVC is used to insulate electric cables...

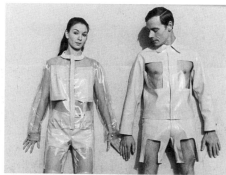

...and even for high fashion!

Thermosoftening and thermosetting plastics

Intermolecular forces are explained on pages 18–19.

Plastics made of polymer chains like poly(ethene), poly(propene) and PVC are described as **thermosoftening plastics**. Although the polymer chains are held together by strong covalent bonds, the intermolecular forces between the chains are much weaker. If you heat the plastic gently, there is enough energy to break the intermolecular forces, and the plastic melts. On cooling, the plastic becomes solid again.

In a **thermosetting plastic**, cross-links are set up between the individual chains during the polymerisation process. This means that the whole piece of plastic is essentially one huge molecule.

This won't melt on gentle heating, because you would have to break strong covalent bonds. Stronger heating causes the plastic to char, but it won't melt.

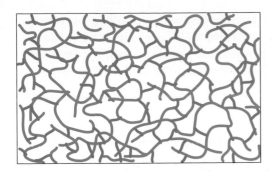

In a thermosetting plastic individual chains (shown in blue) are cross linked (shown in green).

Melamine (used as a plastic coating on some furniture) and certain glues are thermosetting plastics.

Dangers from burning plastics

When plastics burn they can produce a range of poisonous gases. Because of their high carbon content, they are likely to produce carbon monoxide as well as carbon dioxide, unless there is a very plentiful air supply. Carbon monoxide is poisonous. The thick black smoke produced by burning plastics is due to carbon – again produced by incomplete combustion.

Plastics containing chlorine (such as PVC) produce hydrogen chloride when they burn. Plastics containing nitrogen (such as polyurethane) produce hydrogen cyanide. Both hydrogen chloride and hydrogen cyanide are extremely poisonous.

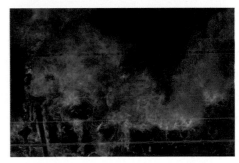

Burning plastics produce poisonous gases as well as heat and flames.

The Oil Industry

The oil industry is BIG business!

Benefits of the oil industry

It is hard to imagine what life would be like if the oil industry suddenly stopped producing oil and gas. Transport – land, sea and air – would collapse. No more plastics. No more artificial fibres like polyester or nylon. No heating oil. Major power cuts as oil and gas fired power stations stopped producing electricity. Modern detergents would disappear. The pharmaceutical industry couldn't get its essential raw materials, so medicines would run out. Disease would flourish.

Nitrogen fertilisers come from the Haber Process which uses hydrogen from natural gas or oil fractions. Modern agriculture also uses lots of pesticides, fungicides and weedkillers – most of which are based on organic compounds from the oil industry. It is hard to think of a single aspect of modern life that the oil industry doesn't touch.

The down-side of the oil industry

Simply because it is so big and important, the oil industry is also responsible for many of the world's environmental problems. Extraction and transport of oil results in obvious problems like oil spills, but there are all sorts of more important secondary effects from oil use.

Obvious pollution from an oil spill.

Global warming is the result of carbon dioxide produced from burning fuels. Acid rain is produced by sulphur dioxide caused by the presence of sulphur in fuels, or by nitrogen oxides produced when they burn in air. The holes in the ozone layer are caused by chlorinated solvents – again a product of the oil industry. All of these pollution problems are discussed in detail in Chapter 17.

The majority of plastics in use are **non-biodegradable**. This means that they aren't broken down by living organisms and so they last in the environment almost for ever. Some of this problem can be overcome by recycling, but there are so many different sorts of plastics with different characteristics that this isn't easy.

Employment of scientists in the oil industry

The oil industry is a major employer of scientists. For example:

geologists	in the search for new sources of oil
chemical engineers	in the design and running of chemical plants
analytical chemists	in checking the purity of products
research chemists	in the search for easier reactions, better catalysts and new products

End of Chapter Checklist

If you haven't got a copy of your specification, read the introduction on page vi.

You will need to be able to do some or all of the following. Check your Awarding Body's specification (syllabus) to find out exactly what you need to know.

- Understand that crude oil is a mixture of hydrocarbons, and know how the simple properties of those hydrocarbons vary with molecular size.
- Know how crude oil is separated into fractions, taking advantage of differences in boiling points.
- Know the uses of the main fractions from crude oil distillation.
- Understand what is meant by *cracking*, and why it is important.
- Understand what is meant by polymerisation as applied to poly(ethene), poly(propene) and poly(chloroethene).
- Relate some simple uses of the polymers to their properties.
- Understand what is meant by a *thermosoftening plastic* and a *thermosetting plastic*.
- Give examples of environmental or safety problems concerned with the use of plastics.
- Discuss benefits and problems relating to the oil industry.

Questions

More questions on the oil industry can be found at the end of Section E on page 259.

1   Six of the fractions which are obtained by the fractional distillation of crude oil are (in alphabetical order): bitumen, diesel oil (gas oil), fuel oil, gasoline, kerosine, petroleum gases. Draw up a simple table, listing these fractions in order of increasing boiling point. Give one use for each of the fractions.

2   Hydrocarbons burn in an excess of air or oxygen to give carbon dioxide and water.

   a)   What do you understand by the term *hydrocarbon*?

   b)   Write an equation for the complete combustion of the hydrocarbon heptane, $C_7H_{16(l)}$.

   c)   The more volatile a hydrocarbon is, the more flammable it is. In a liquid, reaction with oxygen can only take place at the surface. In a gas, the oxygen molecules can mix easily with the hydrocarbon molecules.

   i)   What do you understand by the word *volatile*?

   ii)   Which is the more volatile hydrocarbon, $C_{15}H_{32}$ or C_8H_{18}? Explain your answer.

   d)   Explain why the *incomplete combustion* of hydrocarbons causes safety problems.

3   Cracking is a process which splits larger hydrocarbons into smaller ones.

   a)   Give two reasons why an oil producer might want to crack a hydrocarbon.

   b)   Give the conditions under which cracking is carried out.

   c)   A molecule of the hydrocarbon $C_{11}H_{24}$ was cracked to give two molecules of ethene, C_2H_4, and one other molecule. Write an equation for the reaction which took place. (You can omit the state symbols from your equation.)

   d)   Write an equation for an alternative cracking reaction involving the same hydrocarbon, $C_{11}H_{24}$.

   e)   State one important use for ethene.

4   Propene, C_3H_6, can be polymerised to make poly(propene).

   a)   What do you understand by the term *polymerisation*?

   b)   Draw a displayed formula (showing all the bonds) for propene.

c) Draw a diagram to show the structure of a poly(propene) chain. Restrict yourself to showing 3 repeating units.

d) Give one use for poly(propene) and explain the properties which make it suitable for that use.

e) Styrene has the formula $C_6H_5CH\!\!=\!\!CH_2$. Write an equation to show what happens when styrene is polymerised to make polystyrene. Your equation should clearly show the structure of the polystyrene. (Show the C_6H_5 group as a whole without worrying about its structure.)

f) Both poly(propene) and polystyrene are thermosoftening plastics. Describe a simple experiment that you could do to show this.

5 Imagine a world in which fossil fuels like coal, natural gas and oil had never formed. This would have effects other than the immediately obvious ones. For example, the iron and steel industry depends on coke made from coal, although in the past it used (on a much smaller scale) charcoal made from wood. Choose any one aspect of the modern world and explain in no more than 300 words how it would be different in the absence of fossil fuels. You could choose from transport, use of materials, landscape, disease prevention, and power generation, for example, but you needn't restrict yourself to this list.

Try not to be too simplistic about this. Human beings are inventive! For example, it is possible to obtain organic chemicals from alcohol (from fermented sugar) and fuels from plant oils.

Chapter 21: Alkanes and Alkenes

There are literally millions of different organic compounds. They all contain carbon, and almost all contain hydrogen. Atoms like oxygen or nitrogen or chlorine crop up quite commonly as well. This chapter uses two simple families of hydrocarbons (compounds containing only carbon and hydrogen) to explore some basic ideas in organic chemistry.

The Alkanes

The alkanes are a family of simple hydrocarbons. They contain carbon–carbon single bonds. Alkanes are described as **saturated** hydrocarbons in the sense that they contain the maximum possible number of hydrogen atoms for a given number of carbons.

The three smallest alkanes are:

methane, CH_4 ethane, C_2H_6 propane, C_3H_8

Methane is the major component of natural gas. Ethane and propane are also present in small quantities in natural gas, and are important constituents of the petroleum gases from crude oil distillation.

Homologous series

A homologous series is a family of compounds with similar properties because they have similar bonding. The alkanes are the simplest homologous series.

Members of a homologous series have a general formula

In the case of the alkanes, if there are "n" carbons, there are "2n+2" hydrogens.

The general formula for the alkanes is C_nH_{2n+2}.

So, for example, if there are 3 carbons, there are $(2 \times 3) + 2 = 8$ hydrogens. The formula for propane is C_3H_8.

If you wanted the formula for an alkane with 15 carbons, you could easily work out that it was $C_{15}H_{32}$ – and so on.

Members of a homologous series have trends in physical properties

The molecules of the members of a homologous series increase in size in a regular way. There is always a difference of CH_2 between one member and the next.

Intermolecular forces are explained on pages 18–19.

As the molecules get bigger, the intermolecular forces between them increase. This means that more energy has to be put in to break the attractions between one molecule and its neighbours.

One effect of this is that the boiling points increase in a regular way.

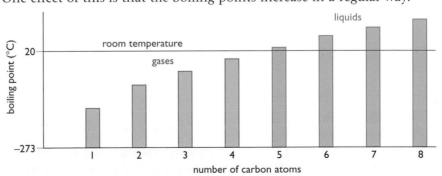

Boiling points of the alkanes.

Notice that the first four alkanes are gases at room temperature. All the other ones you are likely to come across at GCSE are liquids. Solids start to appear at about $C_{17}H_{36}$.

Members of a homologous series have similar chemical properties

Chemical properties are dependent on bonding. Because alkanes only contain carbon–carbon single bonds and carbon–hydrogen bonds, they are all going to behave in the same way. These are strong bonds and the alkanes are fairly unreactive. All you will need to know for GCSE is that they burn and that they can be cracked. These reactions have already been covered in Chapter 20.

If you want to be fussy, you might argue that methane doesn't have a carbon–carbon bond, and so might have different properties. It does – to the extent that it can't be cracked to produce smaller hydrocarbons!

The Alkenes

The alkenes are another family (homologous series) of hydrocarbons. They all contain a carbon–carbon double bond. Alkenes are **unsaturated** hydrocarbons. The presence of the double bond means that they don't contain as many hydrogen atoms as the corresponding alkane.

The general formula

Alkenes have a general formula C_nH_{2n}. There isn't an alkene with just one carbon atom. The two smallest alkenes are ethene and propene.

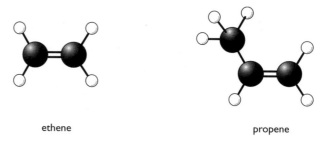

ethene   propene

Physical properties

These are very similar to the alkanes. Remember that the small alkanes with up to 4 carbon atoms are gases. The same thing is true for the alkenes. They are gases up to C_4H_8, and the next dozen or so are liquids.

Chemical reactions of the alkenes

In common with all hydrocarbons, alkenes burn in air or oxygen to give carbon dioxide and water. For example:

$$C_2H_{4(g)} + 3O_{2(g)} \rightarrow 2CO_{2(g)} + 2H_2O_{(l)}$$

More importantly, alkenes undergo **addition reactions**. Part of the double bond breaks and the electrons are used to join other atoms onto the two carbon atoms.

The addition of hydrogen (hydrogenation)

Hydrogenation is the addition of hydrogen to a compound. If a mixture of ethene and hydrogen is passed over a nickel catalyst at about 150°C, ethane is formed.

$$CH_2{=}CH_{2(g)} + H_{2(g)} \rightarrow CH_3CH_{3(g)}$$

This isn't an important reaction for ethene itself. This reaction happens with any molecule containing a carbon–carbon double bond, and ethene is taken as a simple example. Ethene is too useful to turn into ethane. Hydrogenation of a double bond is important industrially in the manufacture of margarine (see page 247).

Notice the way that ethene is written in the equation to show the double bond. The ethane formed is also written in a way that shows exactly how the hydrogen has been added.

The addition of bromine

Bromine adds to alkenes without any need for heat or a catalyst. The reaction is often carried out using bromine solution ("bromine water"). For example, with ethene:

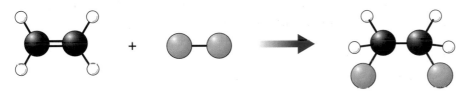

The product is called 1,2-dibromoethane and is a colourless liquid.

You can write this as an equation in two ways. The first is very close to the way the reaction is shown using the models:

You can find out how to name complicated organic molecules later in this chapter.

$$\begin{array}{c} H \quad H \\ | \quad | \\ C{=}C \\ | \quad | \\ H \quad H \end{array} + Br_2 \longrightarrow \begin{array}{c} H \quad H \\ | \quad | \\ H{-}C{-}C{-}H \\ | \quad | \\ Br \quad Br \end{array}$$

The other method takes up rather less space:

$$CH_2{=}CH_{2(g)} + Br_{2(aq)} \rightarrow CH_2BrCH_2Br_{(l)}$$

Any compound with a carbon–carbon double bond will react with bromine in a similar way. This is used to **test for a carbon–carbon double bond**.

If you shake an unknown organic compound with bromine water and the orange bromine water is decolorised, the compound contains a carbon–carbon double bond.

The left-hand tube in the photograph shows the effect of shaking a liquid alkene with bromine water. The organic layer is on top. You can see that the bromine water has been completely decolorised – showing the presence of the carbon–carbon double bond.

The right-hand tube in the photograph shows what happens if you use a liquid alkane – which doesn't have a carbon–carbon double bond. The

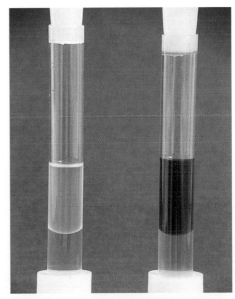

The results of shaking a liquid alkene (left-hand tube) or alkane (right-hand tube) with bromine water.

colour of the bromine is still there. Interestingly, most of the colour is now in the top organic layer. That is because the covalent bromine is more soluble in the organic compound than it is in water.

The polymerisation of alkenes

This has already been described on pages 213–216.

Types of Formula for Organic Molecules

Molecular formulae

A molecular formula simply counts the numbers of each sort of atom present in the molecule, but tells you nothing about the way they are joined together. For example, the molecular formula of propane is C_3H_8, and the molecular formula of ethene is C_2H_4.

Molecular formulae are very rarely used in organic chemistry, because they don't give any useful information about the bonding in the molecule. You might use them in equations for the combustion of simple hydrocarbons where the structure of the molecule doesn't matter. For example:

$$C_3H_{8(g)} + 5O_{2(g)} \rightarrow 3CO_{2(g)} + 4H_2O_{(l)}$$

Structural formulae

A structural formula shows how the atoms are joined up. There are two ways of representing structural formulae – they can be drawn as a displayed formula or they can be written out as, for example, $CH_3CH_2CH_3$. You need to be confident about doing it either way.

Displayed formulae

A displayed formula shows all the bonds in the molecule as individual lines. You need to remember that each line represents a pair of shared electrons.

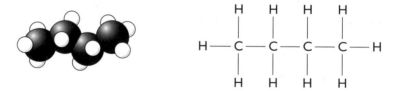

The diagram shows a model of butane, together with its displayed formula.

Notice that the way the displayed formula is drawn bears no resemblance to the shape of the actual molecule. Displayed formulae are always drawn with the molecule straightened out and flattened. They do, however, show exactly how all the atoms are joined together.

The normal way to draw a structural formula

For anything other than the smallest molecules, drawing a fully displayed formula is very time-consuming. You can simplify the formula by writing, for example, CH_3 or CH_2 instead of showing all the carbon–hydrogen bonds.

Butane could be written quite quickly as $CH_3CH_2CH_2CH_3$ – and this shows all the necessary detail. But you have to be very careful. For example, all of these structures represent butane, even though they look different:

$CH_3-CH_2-CH_2-CH_3$

CH_3-CH_2
$\quad\quad\;\; |$
$\quad\quad\; CH_2$
$\quad\quad\;\; |$
$\quad\quad\; CH_3$

CH_3-CH_2
$\quad\quad\;\; |$
CH_3-CH_2

Each structure shows four carbon atoms joined up in a chain, but the chain has simply twisted. This happens in real molecules as well.

Not one of the structural formulae accurately represents the shape of butane. The convention is to write it with all the carbon atoms in a straight line.

A molecule like propene, C_3H_6, has a carbon–carbon double bond. That is shown by drawing two lines between the carbon atoms to show the two pairs of shared electrons. You would normally write this in a simplified structural formula as $CH_3CH=CH_2$.

$$H-\overset{\displaystyle H}{\underset{\displaystyle H}{C}}-\overset{\displaystyle H}{C}=\overset{\displaystyle H}{\underset{\displaystyle H}{C}}$$

Naming Organic Compounds

Names for organic compounds can look quite complicated, but they are no more than a code which describes the molecule. Each part of a name tells you something specific about the molecule. One part of a name tells you how many carbon atoms there are in the longest chain, another part tells you whether there are any carbon–carbon double bonds, and so on.

Coding the chain length

Look for the following letters in the name:

Code letters	No. of carbons in chain
meth	1
eth	2
prop	3
but	4
pent	5
hex	6

For example, **but**ane has a chain of 4 carbon atoms; **prop**ane has a chain of 3 carbon atoms.

If you are asked to draw the structure for a molecule in a GCSE exam, always draw it in the fully displayed form. You can't lose any marks by doing this, whereas you might if you use the simplified form. If, on the other hand, you are just writing a structure in an equation, you can use whichever version you prefer.

The best way to understand this is to make some models. If you don't have access to atomic models, use blobs of Plasticine joined together with bits of match sticks to represent the bonds You will find that you can change the shape of the model by rotating the bonds. That's what happens in real molecules.

You have to learn these! The first 4 are the tricky ones because there isn't any pattern; "pent" means 5 (as in **pent**agon), and "hex" means 6 (as in **hex**agon).

Coding for the type of compound

Alkanes

Alkanes are hydrocarbons in which all the carbons are joined to each other with single covalent bonds. Compounds like this are coded with the ending "**ane**". For example, eth**ane** is a 2-carbon chain (because of "eth") with a carbon–carbon single bond, CH_3CH_3.

Alkenes

Alkenes contain a carbon–carbon double bond. This is shown in their name by the ending "**ene**". For example, eth**ene** is a 2-carbon chain containing a carbon–carbon double bond, $CH_2\!\!=\!\!CH_2$. With longer chains, the position of the double bond could vary in the chain. This is shown by numbering the chain and noting which carbon atom the double bond *starts* from.

$CH_2\!\!=\!\!CHCH_2CH_3$	but-1-ene	a 4-carbon chain with a double bond starting on the first carbon
$CH_3CH\!\!=\!\!CHCH_3$	but-2-ene	a 4-carbon chain with a double bond starting on the second carbon
$CH_3CH_2CH\!\!=\!\!CHCH_2CH_3$	hex-3-ene	a 6-carbon chain with a double bond starting on the third carbon

How do you know which end of the chain to number from? The rule is that you number from the end which produces the smaller numbers in the name.

The diagrams both show the same molecule, but one has been flipped over so that what was originally on the left is now on the right, and vice versa.

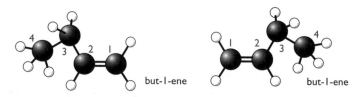

but-1-ene   but-1-ene

It would be silly to change the name every time the molecule moved! Both of them are called but-1-ene.

Coding for branched chains

Hydrocarbon chains can have side branches on them. You are only likely to come across two small side chains:

Side chain	Coded
CH_3-	methyl
CH_3CH_2-	ethyl

The name of a molecule is always based on the *longest* chain you can find in it. The position of the chain is shown by numbering exactly as before.

The longest chain in the molecule in the next diagram has 4 carbon atoms ("**but**") with no double bonds ("**ane**"). The name is based on butane. There is a **methyl** group branching off the **number 2** carbon. (Remember to number from the end which produces the smaller number.)

In more complicated molecules, the presence of the code "**an**" in the name again shows that the carbons are joined by single bonds. For example, you can tell that propan-1-ol contains 3 carbon atoms ("prop") joined together by carbon–carbon single bonds ("an"). The coding on the end gives you more information about the molecule. This is explained in the next chapter.

Don't worry too much about this. It isn't a big issue at GCSE, but it does arise from time to time later on in this chapter.

Notice that the count of the number of carbons in the side chain is coded exactly as before. "meth" shows a 1-carbon side chain; "eth" shows 2 carbons.

The compound is called 2-methylbutane.

$$\overset{4}{C}H_3-\overset{3}{C}H_2-\overset{2}{\underset{\underset{\displaystyle CH_3}{|}}{C}}H-\overset{1}{C}H_3$$

Where there is more than one side chain, you describe the position of each of them.

$$\overset{|}{C}H_3-\overset{\overset{\displaystyle CH_3}{\overset{\displaystyle 2|}{}}}{\underset{\underset{\displaystyle CH_3}{|}}{C}}-\overset{3}{C}H_3$$

The longest chain in the molecule in the diagram has 3 carbon atoms and no double bonds. Therefore the name is based on propane.

There are 2 methyl groups attached to the second carbon. The compound is 2,2-dimethylpropane. The "di" in the name shows the presence of the two methyl groups. "2,2-" shows that they are both on the second carbon atom.

You can reverse the process and draw a structural formula from a name. All you have to do is decode the name.

For example, what is the structural formula for **2,3-dimethylbut-2-ene**?

Start by looking for the code for the longest chain length. "**but**" shows a 4-carbon chain. "**ene**" shows that it contains a carbon–carbon double bond starting on the second carbon atom ("**-2-ene**").

There are 2 methyl groups ("**dimethyl**") attached to the second and third carbon atoms in the chain ("**2,3-**"). All you have to do now is to fit all this together into a structure.

> It is actually much more important to be able to decode names to give structures than the other way around. If you don't know what a teacher or an examiner is talking about you are completely lost!

$$\overset{|}{C}-\overset{\overset{\displaystyle CH_3}{\overset{\displaystyle 2|}{}}}{C}=\overset{3}{C}-\overset{4}{\underset{\underset{\displaystyle CH_3}{|}}{C}}$$

Start by drawing the structure without any hydrogen atoms on the main chain. It doesn't matter whether you draw the CH_3 groups pointing up or down. Then add enough hydrogens so that each carbon atom is forming four bonds.

The final structure is:

$$CH_3-\overset{\overset{\displaystyle CH_3}{|}}{C}=\overset{\underset{\displaystyle CH_3}{|}}{C}-CH_3$$

Structural Isomerism

Structural isomers are molecules with the same molecular formula, but with different structural formulae. Examples will make this clear.

Structural isomerism in the alkanes

Isomers of butane, C_4H_{10}

If you had some atomic models and picked out 4 carbon atoms and 10 hydrogen atoms, you would find that it was possible to fit them together in more than one way. The two different molecules formed are known as

> **Warning!** Don't confuse the word **isomer** with **isotope**. Isotopes are atoms with the same atomic number but different mass numbers.

isomers. Both have the molecular formula C_4H_{10}, but they have different structures. **Structural isomerism** is the existence of two or more different structures with the same molecular formula.

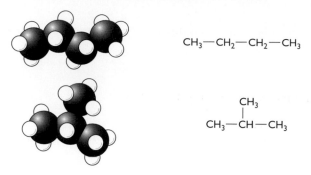

$$CH_3-CH_2-CH_2-CH_3$$

$$\begin{array}{c} CH_3 \\ | \\ CH_3-CH-CH_3 \end{array}$$

If you look carefully at the models in the diagram, you can see that you couldn't change one into the other just by bending or twisting the molecule. You would have to take the model to pieces and rebuild it. That's a simple way of telling that you have got isomers.

A "straight chain" is an unbranched chain.

The "straight chain" isomer is called butane. The branched chain has a 3-carbon chain with no carbon–carbon double bond ("propane") with a methyl group on the second carbon. The name is 2-methylpropane.

Isomers of pentane, C_5H_{12}

There are three isomers of pentane:

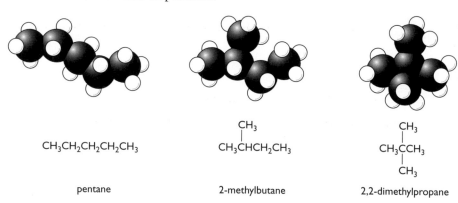

$CH_3CH_2CH_2CH_2CH_3$	$\begin{array}{c} CH_3 \\ \| \\ CH_3CHCH_2CH_3 \end{array}$	$\begin{array}{c} CH_3 \\ \| \\ CH_3CCH_3 \\ \| \\ CH_3 \end{array}$
pentane	2-methylbutane	2,2-dimethylpropane

$$\begin{array}{c} CH_3CHCH_3 \\ | \\ CH_2 \\ | \\ CH_3 \end{array}$$

Students frequently think that they can find another isomer as well. If you look closely at this "fourth" isomer, you will see that it is just 2-methylbutane rotated in space.

To avoid this sort of problem, always draw your isomers so that the longest carbon chain is drawn horizontally. Check each isomer after you have drawn it to be sure that you have done that.

Physical differences between the isomers

The various isomers will have slightly different physical properties because they will experience slightly different intermolecular forces.

Branched chains have weaker intermolecular attractions than straight ones. Intermolecular forces are only effective over very short distances. The more branching there is in a chain, the more difficult it is for the molecules to get close to each other.

You can see the effect of this on the boiling points of the isomers of pentane:

	Boiling point (°C)
pentane	36.3
2-methylbutane	27.9
2,2-dimethylpropane	9.5

On the graph showing the relationship between number of carbons and boiling point for the alkanes on page 221, the boiling points were for the straight chain alkanes.

As the amount of branching increases, boiling point falls.

Structural isomerism in the alkenes

Ethene and propene

Ethene, $CH_2{=}CH_2$, doesn't have any isomers. Propene, $CH_3CH{=}CH_2$, doesn't have a structural isomer which is still an alkene. (You can find a structural isomer of C_3H_6 which doesn't have a carbon–carbon double bond by joining the carbon atoms in a ring.)

Butene, C_4H_8

Butene has three structural isomers containing a carbon–carbon double bond.

$$CH_3CH_2CH{=}CH_2 \qquad CH_3CH{=}CHCH_3 \qquad \underset{\underset{CH_3}{|}}{CH_3C}{=}CH_3$$

but-1-ene   but-2-ene   2-methylpropene

If you aren't comfortable with the names, re-read the earlier part of this chapter (starting on page 225) on organic names.

Notice the way that you can vary the position of the double bond as well as branch the chain.

Pentene, C_5H_{10}

These are all structural isomers of pentene:

$$CH_3CH_2CH_2CH{=}CH_2 \qquad\qquad CH_3CH_2CH{=}CHCH_3$$

pent-1-ene   pent-2-ene

$$\underset{\underset{CH_3}{|}}{CH_3CH_2C}{=}CH_2 \qquad \underset{\underset{CH_3}{|}}{CH_3CHCH}{=}CH_2 \qquad \underset{\underset{CH_3}{|}}{CH_3CH{=}CCH_3}$$

2-methylbut-1-ene   3-methylbut-1-ene   2-methylbut-2-ene

At first sight this looks really worrying! There are two issues. First, do you understand the names? If not, get that sorted out before you do anything else.

Second, could you draw at least a few of these isomers? Use a scrap of paper and see how many isomers of C_5H_{10} you can find. Draw fully displayed formulae showing all the bonds. Remember that you must have a carbon–carbon double bond. Better still – use some models.

End of Chapter Checklist

If you haven't got a copy of your specification, read the introduction on page vi.

You will need to be able to do some or all of the following. Check your Awarding Body's specification (syllabus) to find out exactly what you need to know.

- Understand what is meant by a homologous series, and know the general formulae for alkanes and alkenes.

- Know and explain the trends in boiling point in the alkanes and alkenes as the number of carbon atoms increases.

- Understand that intermolecular attractions fall as the amount of branching increases.

- Know that alkenes undergo addition reactions with hydrogen and with bromine, and that the reaction with bromine water is used to test for carbon–carbon double bonds.

- Understand what is meant by molecular and structural (including displayed) formulae.

- Know how to name alkanes and alkenes (with up to 5 carbons), including the possibility of branching.

- Understand what is meant by structural isomerism, and draw the structures of isomers of both alkanes and alkenes with up to 5 carbon atoms.

Questions

More questions on alkanes and alkenes can be found at the end of Section E on page 259.

1 **a)** Alkanes are *saturated* hydrocarbons. What do you understand by the term *saturated*?

b) Undecane is an alkane with 11 carbon atoms.

 i)   Write down the molecular formula for undecane.

 ii) What physical state (solid, liquid or gas) would you expect undecane to be in at room temperature?

 iii) Write an equation for the complete combustion of undecane.

2 A gaseous hydrocarbon with 3 carbon atoms decolorised bromine water.

a) Write the displayed formula for the hydrocarbon.

b) Write the equation for the reaction between the hydrocarbon and bromine.

c) Write the equation for the complete combustion of the hydrocarbon in oxygen.

d) The hydrocarbon was mixed with hydrogen and passed over a hot nickel catalyst. Write an equation for the reaction, and name the product.

3 **a)** Write down the names of the following hydrocarbons:
 i) CH_4, ii) $CH_3CH_2CH_3$, iii) C_5H_{12}, iv) $CH_3CH=CH_2$, v) C_2H_4, vi) $CH_2=CHCH_2CH_3$.

b) Write fully displayed formulae (showing all the bonds) for: i) butane, ii) ethane, iii) but-2-ene, iv) 2-methylbut-2-ene.

4 **a)** What do you understand by the term *structural isomerism*?

b) There are two structural isomers of C_4H_{10}.

 i)   Draw their structures and name them.

 ii) Which of these isomers has the lowest boiling point? Explain your reasoning.

c) There are five structural isomers of C_6H_{14}. Draw their structures and name them.

d) How many structural isomers can you find with a molecular formula C_4H_8? You don't need to restrict yourself to alkenes. Draw their structures and name as many as you can.

Section E: Organic Chemistry

Chapter 22: Alcohols, Acids and Esters

This chapter deals with some quite complicated compounds. To make sense of it you must understand terms like *homologous series*, and know how to draw structures for organic molecules and understand how they are named. This is all covered in Chapter 21.

Alcohols

What everybody knows as "alcohol" is just one member of a large family (homologous series) of similar compounds. Alcohols all contain an –OH group covalently bonded onto a carbon chain. Their general formula is $C_nH_{2n+1}OH$.

The familiar alcohol in drinks is C_2H_5OH (or, better, CH_3CH_2OH), and should properly be called **ethanol**.

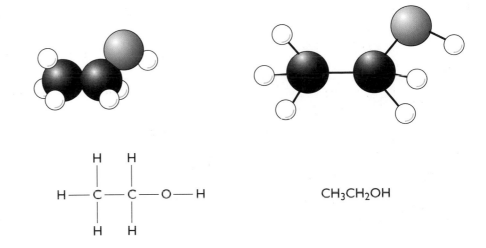

$$CH_3CH_2OH$$

Various ways of drawing ethanol.

Alcohol oils social occasions.

Alcohol in drinks

Alcohol depresses the activity of some of the higher parts of the brain. As a result, it reduces inhibitions. Even small quantities of alcohol reduce concentration.

The alcohol content of different drinks varies enormously. By law, bottles or cans have to tell you the percentage of alcohol by volume they contain. This means that you can work out the volume of pure alcohol that the drink contains.

Wine typically contains about 12% alcohol by volume. A reasonably sized wine glass holds about 125 cm³. 12% of 125 cm³ is 15 cm³. Drinking a glass of wine is therefore the equivalent of drinking 15 cm³ of pure alcohol.

In the UK, 10 cm³ of alcohol is defined as "1 unit of alcohol". The maximum recommended weekly consumption for adult men was 21–28 units, and for women was 14–21 units in 2001. Recommendations change from time to time. It is easy to find out the current recommendations by doing an internet search.

Drinking excessive amounts of alcohol is the cause of all sorts of social problems – both domestic and public violence, for example. More than half of road traffic accidents involving injury are alcohol related.

A human liver showing the effect of cirrhosis.

Sugar cane – the raw material for making ethanol.

The reason you can't get 100% pure alcohol by simple distillation is very complicated.

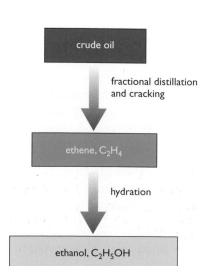

From crude oil to ethanol.

Long term consumption of alcohol beyond the recommended limits can cause serious liver damage. In the Western world, cirrhosis of the liver is the third most common cause of death in people aged 45 to 65. Almost half of all cirrhosis cases are caused by alcohol.

The production of alcohol (ethanol – C_2H_5OH)

Making ethanol by fermentation

Yeast is added to a sugar solution and left in the warm for several days in the absence of air ("**anaerobic**" conditions). **Enzymes** in the yeast convert the sugar into ethanol and carbon dioxide. The process is known as **fermentation**. (You will find more about enzymes in Chapter 24.)

The biochemistry is very complicated. First, the sugar (sucrose) is split into two smaller sugars, glucose and fructose. Glucose and fructose have the same molecular formulae, but different structures. They are isomers. (You will find more about sugars in Chapter 23.)

Enzymes in the yeast convert these sugars into ethanol and water in a multitude of small steps. All we normally write are the overall equations for the reactions.

$$C_{12}H_{22}O_{11(aq)} + H_2O_{(l)} \rightarrow C_6H_{12}O_{6(aq)} + C_6H_{12}O_{6(aq)}$$
$$\text{sucrose} \qquad\qquad\qquad \text{glucose} \qquad\quad \text{fructose}$$

$$C_6H_{12}O_{6(aq)} \rightarrow 2C_2H_5OH_{(aq)} + 2CO_{2(g)}$$
$$\text{glucose} \qquad\quad \text{ethanol} \qquad \text{carbon dioxide}$$

Yeast is killed by more than about 15% of alcohol in the mixture, and so it is impossible to make pure alcohol by fermentation. The alcohol is purified by fractional distillation. This takes advantage of the difference in boiling point between ethanol and water. Water boils at 100°C whereas ethanol boils at 78°C.

The liquid distilling over at 78°C is 96% pure ethanol. The rest is water. It is impossible to remove this last 4% of water by simple distillation.

Making ethanol by the hydration of ethene

Ethanol is also made by reacting ethene with water – a process known as hydration.

$$CH_2{=}CH_{2(g)} + H_2O_{(g)} \rightarrow CH_3CH_2OH_{(g)}$$

Starting materials: ethene and steam
Temperature: 300°C
Pressure: 60–70 atmospheres
Catalyst: phosphoric acid

Only a small proportion of the ethene reacts. The ethanol produced is condensed as a liquid and the unreacted ethene is recycled through the process.

Comparing the two methods of producing ethanol

At the moment, countries which have easy access to crude oil produce ethanol mainly from ethene, but one day the oil will start to run out. At that point, the production of ethanol from sugar will provide an alternative route to many of the organic chemicals we need.

	Fermentation	Hydration of ethene
Use of resources	Uses renewable resources – sugar beet or sugar cane.	Uses finite resources. Once all the oil has been used up there won't be any more.
Type of process	A batch process. Everything is mixed together in a reaction vessel and then left for several days. That batch is then removed and a new reaction is set up. This is inefficient.	A continuous flow process. A stream of reactants is constantly passed over the catalyst. This is more efficient than a batch process.
Rate of reaction	Slow, taking several days for each batch.	Rapid.
Quality of product	Produces very impure ethanol which needs further processing.	Produces much purer ethanol.
Reaction conditions	Uses gentle temperatures and ordinary pressure.	Uses high temperatures and pressures, needing a high input of energy.

Uses of ethanol

Ethanol is sold as "industrial methylated spirit". This is ethanol with a small amount of another alcohol, methanol, added to it. Methanol is poisonous, and makes the industrial methylated spirit unfit to drink, and so avoids the high taxes on alcoholic drinks.

Ethanol is widely used as a solvent – for example, for cosmetics and perfumes. It is relatively safe and is a good solvent for the complex organic molecules which don't dissolve in water.

Ethanol is also a useful fuel. It burns to form carbon dioxide and water, producing about two-thirds as much energy per litre as petrol.

$$C_2H_5OH_{(l)} + 3O_{2(g)} \rightarrow 2CO_{2(g)} + 3H_2O_{(l)}$$

Mixtures of petrol with 10–20% ethanol – known as **gasohol** – are increasingly used in some countries, such as Brazil. These are countries which have little or no oil industry to produce their own petrol. On the other hand, they often have a climate which is good for growing sugar cane. Ethanol can be produced by fermenting the sugar, and then mixed with imported petrol. This saves money on imports.

The other alcohols

All alcohols contain an –OH group covalently bonded to a carbon chain. This is coded for in the name by the ending "**ol**".

The two smallest alcohols are **methanol** and **ethanol**. Neither of these have structural isomers containing an –OH group. Ethanol has a 2-carbon chain ("**eth**"), only carbon–carbon single bonds ("**an**") and an –OH group ("**ol**").

If you aren't sure about names, you **must** read pages 225–227 before you go on. If you aren't sure about structural isomerism, you should read pages 227–229.

$$CH_3—OH \qquad CH_3CH_2—OH$$
methanol ethanol

Once you have a 3-carbon chain, it is possible to have two structural isomers, depending on where the –OH group is attached to the chain. The two isomers are **propan-1-ol** and **propan-2-ol**.

$$CH_3CH_2CH_2—OH \qquad\qquad CH_3\underset{\underset{OH}{|}}{C}HCH_3$$

propan-1-ol   propan-2-ol

Notice that the chain is numbered from the end which produces the smaller number in the name: **propan-1-ol** and not **propan-3-ol**.

With 4 carbons, you can branch the chain as well as move the –OH group around.

$$CH_3CH_2CH_2CH_2—OH \qquad CH_3CH_2\underset{\underset{OH}{|}}{C}HCH_3 \qquad CH_3\underset{\underset{OH}{|}}{\overset{\overset{CH_3}{|}}{C}}CH_3$$

butan-1-ol   butan-2-ol   2-methylpropan-2-ol

And with 5 carbons, things get really complicated! These are all structural isomers of $C_5H_{11}OH$:

$$CH_3CH_2CH_2CH_2CH_2—OH \qquad CH_3CH_2CH_2\underset{\underset{OH}{|}}{C}HCH_3 \qquad CH_3CH_2\underset{\underset{OH}{|}}{C}HCH_2CH_3$$

$$CH_3CH_2\underset{\underset{CH_3}{|}}{C}HCH_2—OH \qquad\qquad CH_3\underset{\underset{CH_3}{|}}{C}HCH_2CH_2—OH$$

$$CH_3CH_2\underset{\underset{OH}{|}}{\overset{\overset{CH_3}{|}}{C}}CH_3 \qquad CH_3\underset{\underset{OH}{|}}{\overset{\overset{CH_3}{|}}{C}}HCHCH_3 \qquad CH_3\underset{\underset{CH_3}{|}}{\overset{\overset{CH_3}{|}}{C}}CH_2—OH$$

> One of the Awarding Bodies (OCR – Extension Block A) expects you to be able to draw structural isomers of alcohols with up to 5 carbon atoms. Don't try to *remember* all of these. Use a bit of paper and see how many you can find. Don't forget to draw the longest carbon chain horizontally. Compare your structures with these – and don't worry if you don't find all of them. It is inconceivable that you could be asked to draw all of them in a GCSE exam.

Burning the ethanol in brandy.

Some reactions of the alcohols

Alcohols burn

All alcohols burn to form carbon dioxide and water. For example, with methanol:

$$2CH_3OH_{(l)} + 3O_{2(g)} \rightarrow 2CO_{2(g)} + 4H_2O_{(l)}$$

Methanol is being tried as an alternative fuel for cars. To find out the current state of research into methanol as an alternative fuel, try an internet search on **methanol fuel**.

The use of ethanol in fuel has already been mentioned on page 233.

Dehydrating alcohols

Dehydration refers to the removal of water from a compound. The dehydration of ethanol produces ethene. Ethanol vapour is passed over hot aluminium oxide acting as a catalyst.

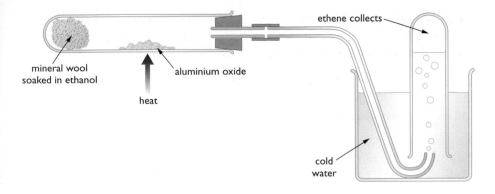

$$CH_3CH_2OH_{(g)} \rightarrow CH_2\!=\!CH_{2(g)} + H_2O_{(l)}$$

Notice that the –OH group is removed, together with a hydrogen from the neighbouring carbon atom.

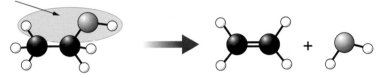

H and OH lost from neighbouring carbon atoms

Other alcohols would dehydrate in a similar way. For example:

$$CH_3CH_2CH_2OH_{(g)} \rightarrow CH_3CH\!=\!CH_{2(g)} + H_2O_{(l)}$$

Alcohols react with sodium

Alcohols react gently with sodium to produce hydrogen. For example, with ethanol:

$$2CH_3CH_2OH_{(l)} + 2Na_{(s)} \rightarrow 2CH_3CH_2ONa_{(solution\ in\ ethanol)} + H_{2(g)}$$

The organic product is called sodium ethoxide. This is a useful way of disposing of small amounts of unwanted sodium, or of treating small sodium spills on the bench in the lab.

> This is similar to the reaction between sodium and water to produce sodium hydroxide and hydrogen, but is much less vigorous.

Alcohols can be oxidised

A bottle of wine left open to the air turns sour. The French for "sour wine" is *vin aigre*, which has been distorted into **vinegar**. The ethanol in the wine is oxidised by air with the help of bacteria to form ethanoic acid, CH_3COOH. The old name for ethanoic acid was acetic acid.

ethanol   ethanoic acid

Or, in symbols:

$$CH_3CH_2OH \xrightarrow{\text{oxidation}} CH_3COOH$$

You will find more about acids like ethanoic acid in the next few pages – including their reactions with alcohols to make **esters**.

Carboxylic Acids

Acids such as ethanoic acid are known as **carboxylic acids**. The term "carboxylic" refers to the presence of the –COOH group. These acids form another homologous series.

Names and structures

The carboxylic acid group in a molecule is coded by the ending "**oic acid**". The code for the number of carbon atoms ("meth", "eth", etc) includes the one in the –COOH group. For example, ethanoic acid has a total of 2 carbon atoms and so is CH_3COOH.

Methanoic acid, ethanoic acid and propanoic acid don't have any structural isomers which are also acids.

If you aren't sure about structural isomerism, then you really ought to read pages 227–229 before you go on.

methanoic acid   ethanoic acid   propanoic acid

With 4 carbon atoms, you can find two structural isomers which are acids:

butanoic acid   2-methylpropanoic acid

When you are naming branches in a chain with a –COOH group, the carbon in the –COOH group automatically counts as the number 1 carbon.

Citric acid (found in citrus fruits like oranges and lemons, and used to flavour soft drinks), is a more complicated carboxylic acid.

Uses of ethanoic acid

Vinegar is a dilute solution of ethanoic acid. The familiar smell of vinegar is due to the acid. Naturally produced vinegars will have their origin described on the label – for example, "wine vinegar" or "cider vinegar". Cheap vinegar may well be a product of the chemical industry and now has to be called "non-brewed condiment".

Vinegar is used as a flavouring and a preservative.

Ethanoic acid is also used in the production of **acetate rayon** – an artificial fibre made from cellulose from wood pulp.

Using ethanoic acid as a preservative in pickled onions.

Reactions of the carboxylic acids

Ethanoic acid is taken as typical of the carboxylic acids. It is the one most commonly used in the lab because it is the cheapest.

Acid properties

You might want to revise properties of acids from Chapter 7. Weak acids are described on pages 72–73.

Ethanoic acid is a weak acid with a pH about 2 to 3, depending on the concentration of the solution. It will turn blue litmus paper red, and reacts with all the things you expect acids to react with.

Reactions with alkalis

Dilute ethanoic acid reacts with an alkali like sodium hydroxide solution to give a salt and water. A colourless solution of sodium ethanoate is formed.

$$CH_3COOH_{(aq)} + NaOH_{(aq)} \rightarrow CH_3COONa_{(aq)} + H_2O_{(l)}$$

Students often find the formula of the sodium ethanoate confusing, because it is written with the metal at the end instead of at the beginning. This is to reflect its structure. Notice that the sodium ethanoate is ionic.

sodium ethanoate

Reaction with metals

Dilute ethanoic acid reacts with metals like magnesium to produce a salt and hydrogen. The salt formed is magnesium ethanoate – as a colourless solution.

$$2CH_3COOH_{(aq)} + Mg_{(s)} \rightarrow (CH_3COO)_2Mg_{(aq)} + H_{2(g)}$$

Reactions with carbonates and hydrogencarbonates

Carbonates react with acids to give a salt, carbon dioxide and water. Ethanoic acid behaves like any other acid. For example, with sodium carbonate, you would get lots of fizzing and a colourless solution of sodium ethanoate.

$$2CH_3COOH_{(aq)} + Na_2CO_{3(s)} \rightarrow 2CH_3COONa_{(aq)} + CO_{2(g)} + H_2O_{(l)}$$

The ionic equation for this is the same as for any other acid/carbonate reaction:

$$2H^+_{(aq)} + CO_3^{2-}_{(s)} \rightarrow H_2O_{(l)} + CO_{2(g)}$$

Hydrogencarbonates, like sodium hydrogencarbonate, behave exactly like carbonates. You wouldn't see any obvious difference between this reaction and the last.

$$CH_3COOH_{(aq)} + NaHCO_{3(s)} \rightarrow CH_3COONa_{(aq)} + CO_{2(g)} + H_2O_{(l)}$$

> This is a good example of a case where writing an ionic equation is far easier than writing a full equation.

Reactions between carboxylic acids and alcohols

Heating a mixture of ethanoic acid and ethanol with a few drops of concentrated sulphuric acid produces a sweet smelling liquid called **ethyl ethanoate**. This is one member of a family (homologous series) of compounds called **esters**.

$$\underset{\text{ethanoic acid}}{CH_3COOH_{(l)}} + \underset{\text{ethanol}}{CH_3CH_2OH_{(l)}} \rightleftharpoons \underset{\text{ethyl ethanoate}}{CH_3COOCH_2CH_{3(l)}} + H_2O_{(l)}$$

The concentrated sulphuric acid isn't written into the equation because it is a catalyst and isn't used up in the reaction.

The ethyl ethanoate has the lowest boiling point of any of the substances present, and is distilled off as soon as it is formed.

> This is a reversible reaction. If you aren't sure about reversible reactions, they are discussed in detail in Chapter 16.

Esters

Names and structures

GCSE (and A-level) students often have more trouble with the names and structures of esters than with any other organic compounds. Look again at the formation of ethyl ethanoate:

To be absolutely accurate, the oxygen attached to the ethyl group in the ethyl ethanoate comes from the ethanol and ought really to be coloured red.
You won't need to worry about this unless you do a Chemistry degree!

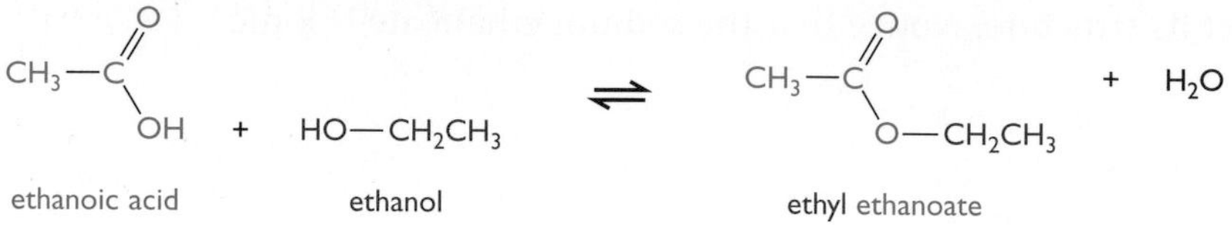

ethanoic acid   ethanol   ethyl ethanoate

The ethanol has been written back-to-front so that you can more easily see how everything fits together. The confusing thing about the name of the ethyl ethanoate is that the two halves are written the other way around from the way they appear in the formula.

Here is another example:

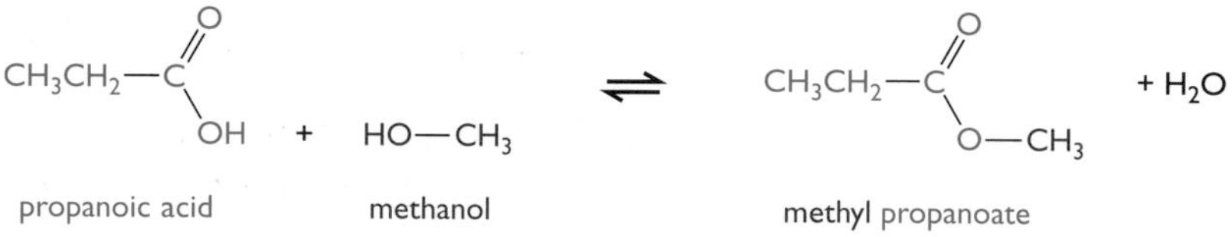

propanoic acid   methanol   methyl propanoate

Uses for esters

The small esters like ethyl ethanoate are commonly used as solvents, but most esters are used in flavourings and perfumes. The typical smell of bananas, raspberries, pears or any other fruit is due in part to naturally occurring esters. Food chemists create artificial food flavourings using mixtures of esters and other organic compounds.

You have to read labels *very* carefully. For example, "strawberry **flavoured**" means that by law the flavouring has to come from real strawberries. "Strawberry **flavour**" means that the flavouring can be entirely artificial. ICI's chemists have produced as many as 200 different versions of strawberry flavour designed for different products.

End of Chapter Checklist

If you haven't got a copy of your specification, read the introduction on page vi.

You will need to be able to do some or all of the following. Check your Awarding Body's specification (syllabus) to find out exactly what you need to know.

- Describe some of the social and health consequences of drinking alcohol.

- Know how ethanol is manufactured by fermentation or by hydration of ethene, and give the advantages and disadvantages of each method.

- Know that ethanol is used as a solvent and a fuel.

- Draw structures for alcohols with up to 5 carbon atoms.

- Know that alcohols burn, can be dehydrated, react with sodium, can be oxidised to carboxylic acids, and react with those acids to make esters.

- Draw structures for carboxylic acids with up to 4 carbon atoms.

- Know that ethanoic acid is used for flavouring and as a preservative (vinegar), and is used to make acetate rayon.

- Know how carboxylic acids react with metals, bases, carbonates and hydrogencarbonates, and alcohols.

- Know that esters are made from the reaction between alcohols and acids, and be able to draw structures for esters involving the acids and alcohols mentioned in your specification.

- Know that esters are used in flavourings and perfumes.

Questions

More questions on these compounds can be found at the end of Section E on page 259.

1   a) Given some sugar and some yeast, describe briefly how you would produce an impure solution of ethanol (alcohol).

    b) How would you produce a reasonably pure sample of ethanol from the impure mixture?

    c) Industrially, much alcohol is made by the *hydration* of ethene. Explain, with the help of an equation, what you understand by the term *hydration*.

    d) Give the conditions for the manufacture of ethanol from ethene.

2   a) Write the equation for the combustion of ethanol, C_2H_5OH.

    b) Carbon dioxide is a "greenhouse gas", and increased amounts of it in the atmosphere are a cause of global warming. It could be argued that burning ethanol made

by fermentation doesn't add to global warming, but burning ethanol made from the hydration of ethene does. How could you justify this strange statement?

    c) What other environmental advantages are there in producing ethanol by fermentation?

3   Some studies have suggested that 1–2 glasses of red wine a day may have beneficial effects on health. Other studies have found that people who drink 1 unit of alcohol a day may have better long term health than people who don't drink alcohol at all. By doing an internet (or other) search, find out the most recent scientific view on this. Write a short article (maximum 300 words) suitable for a lifestyle magazine, summarising what you find. Add diagrams or pictures if you think they help.

4   a) When a small lump of calcium carbonate was dropped into some dilute ethanoic acid, a gas G was given off and the calcium carbonate reacted to give a solution S. Name G and S.

b) Write the full and ionic equations for the reaction between calcium carbonate and ethanoic acid.

c) How could you show that dilute ethanoic acid was acidic, using *i)* a piece of universal indicator paper, *ii)* a length of magnesium ribbon?

5 Draw structures for:

a) propan-1-ol

b) the compound formed when propan-1-ol is dehydrated by passing its vapour over hot aluminium oxide

c) the carboxylic acid which would be obtained if propan-1-ol was oxidised

d) the ester formed when propan-1-ol reacts with ethanoic acid.

6 a) A unit of alcohol is 10 cm^3 of pure alcohol. How many units are there in:

i) half of a 750 cm^3 bottle of white wine containing 12% alcohol by volume

ii) 3 pints of beer containing 4% alcohol by volume (1 pint = 568 cm^3)

iii) a small (350 cm^3) bottle of brandy containing 36% alcohol by volume?

b) The maximum daily recommended alcohol consumption by women is 2–3 units (less for younger or lighter women). Work out the volume of each of the following that you would need to drink in order to consume 2 units of alcohol.

i) red wine containing 12.5% alcohol by volume

ii) vodka containing 40% alcohol by volume

iii) beer containing 4% alcohol by volume. How many pints of beer is that? (1 pint = 568 cm^3)

7 This question relates to the following structures:

$$CH_3CH_2CH_2CHCH_3 \quad \overset{\textstyle CH_3CHCH_2CH_2-OH}{\underset{\textstyle CH_3}{|}} \quad CH_3CH_2CH_2-C\overset{O}{\underset{OH}{}}$$

A   OH   **B**   **C**

a) **A** and **B** are isomers of $C_5H_{11}OH$. Write the names of these two alcohols.

b) Draw the structures of any two other alcohols which are also isomers of $C_5H_{11}OH$.

c) **C** is butanoic acid. 2-Methylpropanoic acid is an isomer of this. Draw its structure.

d) The molecular formula of butanoic acid is $C_4H_8O_2$. Draw the structural formulae of any two esters which have this molecular formula. Name the esters you have drawn.

e) Draw the structure of the product formed when **A** is heated with **C** in the presence of concentrated sulphuric acid.

Chapter 23: Food and Drugs

This chapter looks at the range of chemical compounds found in foods – either naturally or as additives. It ends by looking at some simple drugs.

Glucose and fructose have the same molecular formula, but different structures. They are isomers. If you aren't sure about isomerism, read pages 227–229.

Food

Carbohydrates

Carbohydrates are compounds of carbon ("**carbo**") together with hydrogen and oxygen in the same 2:1 ratio that they are found in water ("**hydrate**"). Simple carbohydrates (often referred to as **sugars**) include:

glucose	$C_6H_{12}O_6$
fructose	$C_6H_{12}O_6$
sucrose	$C_{12}H_{22}O_{11}$

More complicated carbohydrates include **starch** and **cellulose**.

Glucose

Glucose isn't as sweet as sucrose (ordinary sugar) and is absorbed very rapidly by the body. Cells in your body use the breakdown of glucose as a major source of energy. High energy drinks contain glucose to give an almost instant release of energy into the bloodstream. They also contain more complex carbohydrates which are absorbed more slowly.

Glucose is one of the building blocks which make up larger carbohydrates. It is described as a **monosaccharide**.

Its structure is quite difficult to draw in a simple way. Take your time over this, and be sure that you understand what is going on.

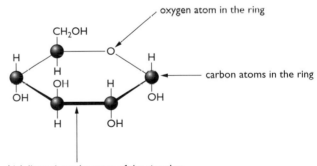

The carbons in the ring have been shown as atoms to make the essential structure more obvious.

The diagram is cluttered and not easy to follow. It can be simplified still more by missing out the carbons in the ring entirely.

When you draw a structure like this, remember that there is a carbon atom at each corner of the ring apart from where the oxygen is.

In future diagrams the single hydrogen atoms attached to the ring have been drawn very small so that it is easier to see the patterns of "ups and downs" of the other groups.

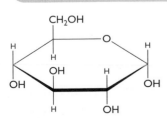

A simplified structure for glucose.

Fructose

Fructose is an isomer of glucose – it has the same molecular formula, but a different structure. It is another monosaccharide.

It is sweeter than ordinary sugar (sucrose). The sweetness of honey is due to significant amounts of fructose.

Unlike glucose, fructose is based on a ring of 5 atoms – 4 carbons and an oxygen.

Confusingly, you may find drawings of two apparently different versions of fructose. It is often drawn as in the left-hand diagram. However, when you need the structure as a part of a sucrose molecule (see below), it is usually drawn as in the right-hand diagram.

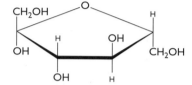

Two versions of fructose.

These are exactly the same molecule. If you made models of them, you would find that one of them is just a "turned over" version of the other one.

Sucrose

Sucrose is familiar everyday sugar. It consists of a glucose unit and a fructose unit joined together to make a **disaccharide**. A disaccharide is a carbohydrate made from two simple sugars.

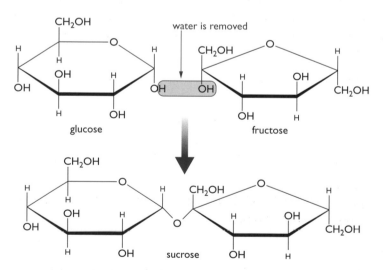

A reaction in which two molecules join together with the loss of a small molecule (in this case, water) is known as a **condensation reaction**.

You can reverse this reaction, and split sucrose up into glucose and fructose by reacting it with water. This is known as **hydrolysis**. The sucrose can be boiled with a dilute acid like dilute sulphuric acid. The acid acts as a catalyst – it is the water in the acid which actually reacts. Alternatively, it can be split

Fructose makes honey especially sweet.

If you are going to learn just one of these structures, it would make sense to learn the right-hand one. You will need that version for the sucrose structure.

242

into the simpler sugars by the enzyme, **invertase**. Invertase is also known as "sucrase".

You will find more about enzymes in Chapter 24.

Starch

Starch makes up the reserve energy supply for plants, and is present in common foods like potatoes, wheat, maize and rice.

Starch is a **polysaccharide**. It consists of long chains of glucose units joined together in the same way that glucose and fructose are joined in sucrose.

There are two forms of starch. Soluble starch consists of chains of about 200 glucose units. Insoluble starch is made of heavily branched chains containing more than a thousand units.

Maize (sweetcorn) is a rich source of starch.

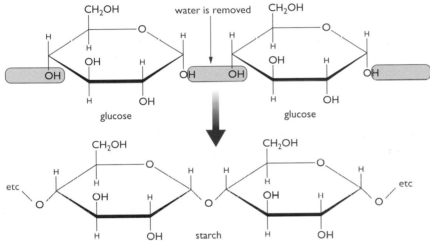

This is an example of polymerisation – joining up lots of small molecules to make a big one. Because of the loss of the water molecule every time a link is made between two glucose units, this is known as **condensation polymerisation**.

You could simplify this diagram to show the basic pattern in the starch chain:

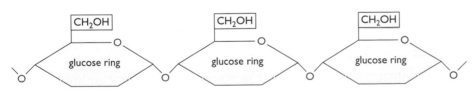

Starch can be hydrolysed back into smaller chains or individual glucose molecules by various enzymes.

Cellulose

Cellulose is also a polymer of glucose and is probably the most widely occurring organic material. It makes up the structural parts of plants. Cotton, for example, is almost pure cellulose, and dry wood is about 70% cellulose.

Cellulose is another polymer of glucose, but based on a slightly different glucose molecule called **beta glucose**. In this molecule, the –OH group on the right-hand carbon atom in the ring (as we have been drawing it) is pointing up rather than down.

General Sherman, the world's biggest tree in the Sequoia National Park, California. Most of the world's largest living thing is made of cellulose.

Chapter 23: Food and Drugs

243

In beta glucose, this –OH group points upwards.

In order for this –OH group to be lined up correctly...

...this glucose molecule has had to be turned upside-down

This –OH group is lined up correctly to react with another glucose molecule like the left-hand one.

This causes the polymer to have a different shape, which essentially involves every second glucose unit being turned upside-down.

cellulose

The basic pattern is therefore:

CH_2OH glucose ring glucose ring glucose ring CH_2OH CH_2OH

This small change stops cellulose being processed by enzymes in animals. This means that, without help, animals can't digest cellulose.

Cows and similar animals have bacteria in their digestive tracts which break down the cellulose in the grass. The smaller molecules produced can be used by the animals' own enzymes.

Proteins

After carbohydrates, proteins make up the second important group of compounds in the diet. Meat, fish and dairy products are important sources. Vegetarians or vegans can get their protein perfectly well from plant sources such as soya, nuts, peas, beans and cereals.

Amino acids

Proteins are built out of various arrangements of 20 or so **amino acids**. An amino acid is an organic compound with a carboxylic acid group, –COOH, and an amino group, $-NH_2$.

We can write a general structural formula for any 2-amino acid. In this structure, "R" can represent a whole range of different groups of various complexity. In glycine, for example, "R" represents another hydrogen atom; in alanine, it represents a CH_3 group.

Bacteria in their digestive system break down cellulose into smaller molecules that the cows can absorb.

$$R-CH-COOH$$
$$\quad\quad |$$
$$\quad NH_2$$

A general 2-amino acid.

All the amino acids in proteins are 2-amino acids. This means that the amino group is attached to the carbon next to the –COOH group. (You may also find them described by their old name which is alpha-amino acids.)

The simplest amino acid is **glycine**. Glycine is an old name – the modern chemical name for it is 2-aminoethanoic acid.

the amino group ⟶ CH_2—COOH ⟵ the acid group

NH_2

glycine: 2-aminoethanoic acid

The next most complicated amino acid is **alanine** or 2-aminopropanoic acid. Notice that the amino group is again on the carbon atom next to the COOH group – the number 2 carbon.

CH_3CH—COOH

NH_2

alanine: 2-aminopropanoic acid

The human body can build all but eight of the amino acids it needs to make its own proteins. These eight have to be obtained from the diet, and are called **essential amino acids.**

For meat-eaters this isn't a problem – those essential amino acids will be present in meat because the animal will already have made them. Some plant proteins are short of some of the essential amino acids but, provided you eat a variety of plant products, that won't matter.

Polypeptide chains

Proteins are formed when amino acids join up into long chains called **polypeptide chains.** The amino group on one amino acid reacts with the acid group on another one.

H—N—CH—C—OH (with H, R, O above)

To see this happening it is helpful to draw the amino acids slightly differently, with the two important groups both on the same line, and with the "R" group sticking up. Notice that the structures of the amino group and the acid group have been shown in more detail.

H—N—CH—C—OH   H—N—CH—C—OH   H—N—CH—C—OH etc

water is removed

R^1 and R^2 and R^3 represent the rest of each amino acid. They may be the same or different, depending on what protein you are making. This is another example of **condensation polymerisation.** Water is being lost as the molecules combine together.

The product molecule (as far as we've drawn it) looks like this:

H—N—CH—C—N—CH—C—N—CH—C— etc

peptide links

Amino acids are usually called by their old names which sound easier than the modern chemical name. Unfortunately, the simple name tells you nothing about the structure.

If you have come straight to this page via the index, you should read page 243 to understand what is meant by condensation polymerisation.

Chapter 23: Food and Drugs

The –CONH– link is called a **peptide** link. Because the protein molecule contains lots of these links, it is a **polypeptide**.

The peptide link can be broken, and the original amino acids re-formed, by reacting with water. This is another example of hydrolysis. Like the hydrolysis of disaccharides or polysaccharides, this can be done by boiling with dilute acids. Again, the acid acts as a catalyst. Alternatively, the hydrolysis can be carried out by enzymes.

When you eat proteins, they are broken down into separate amino acids, and then rebuilt into the new proteins that your body needs.

Fats

Fats are the third group of major nutrients. One gram of fat provides more than twice the energy of 1 gram of carbohydrate.

The structure of fats

Fats and oils from animals or plants are **triglycerides**. These are very large esters. You may remember that esters are formed when carboxylic acids and alcohols react together by a condensation reaction – losing a molecule of water in the process.

For example, the small ester, ethyl ethanoate, comes from ethanoic acid and ethanol.

> Notice that the ethanol has been written backwards so that you can easily see where the water is removed from.

$$CH_3COOH + HOCH_2CH_3 \rightleftharpoons CH_3COOCH_2CH_3 + H_2O$$

Triglycerides are all made from an alcohol with 3 –OH groups called **glycerol**.

$$
\begin{array}{l}
HOCH_2 \\
|\\
HOCH \quad \text{glycerol}\\
|\\
HOCH_2
\end{array}
$$

This forms an ester with long-chain carboxylic acids at each of its –OH groups. Some important acids include:

> The old names are given for these acids (and for glycerol). Nobody is going to expect you to remember the formulae of the complicated ones, although you might perhaps remember stearic acid as a simple example.

palmitic acid	$CH_3(CH_2)_{14}COOH$
stearic acid	$CH_3(CH_2)_{16}COOH$
oleic acid	$CH_3(CH_2)_7CH=CH(CH_2)_7COOH$
linoleic acid	$CH_3(CH_2)_4CH=CHCH_2CH=CH(CH_2)_7COOH$
linolenic acid	$CH_3CH_2CH=CHCH_2CH=CHCH_2CH=CH(CH_2)_7COOH$

The fat which is an ester of stearic acid and glycerol is called **glyceryl tristearate** and has the structure on the left:

> You will often find the structure of glyceryl tristearate reversed, with the hydrocarbon groups drawn on the right-hand side. Drawing it the way we have here is better, because you can see how it relates to simple esters like ethyl ethanoate.

$$
\begin{array}{l}
CH_3(CH_2)_{16}COOCH_2 \\
|\\
CH_3(CH_2)_{16}COOCH\\
|\\
CH_3(CH_2)_{16}COOCH_2
\end{array}
\qquad
\begin{array}{l}
RCOOCH_2 \\
|\\
RCOOCH\\
|\\
RCOOCH_2
\end{array}
$$

glyceryl tristearate   a general triglyceride

Hydrocarbon groups are frequently represented by the symbol "R". A triglyceride would then have the general formula on the right.

Saturated and unsaturated fats

A **saturated fat** is one which has no carbon–carbon double bonds in the hydrocarbon chains. If the hydrocarbon chain contains one carbon–carbon double bond (as in oleic acid), it is called **mono-unsaturated**. If it has more than one carbon–carbon double bond (like linoleic and linolenic acids), it is said to be **polyunsaturated**.

(1) The relationship between degree of saturation and melting point

As a general guide, the more carbon–carbon double bonds there are in the fat, the lower its melting point. Unsaturated fats tend to have melting points below room temperature. These are therefore liquids, and tend to be called **oils** – as in "sunflower oil", "olive oil" or "fish oil", depending on where they come from.

For some purposes, you need a solid fat. It is a bit messy to spread olive oil on your piece of bread! You can raise the melting point of an oil to above room temperature by removing some or all of the carbon–carbon double bonds.

(2) The manufacture of margarine – hydrogenated fats

You may remember that ethene and hydrogen react if they are heated together in the presence of a nickel catalyst. Ethane is formed.

$$CH_2{=}CH_{2(g)} + H_{2(g)} \rightarrow CH_3CH_{3(g)}$$

This is an example of **hydrogenation**. This works perfectly well for any carbon–carbon double bond, however complicated the rest of the molecule might be.

$$CH_3(CH_2)_7CH{=}CH(CH_2)_7COOCH_2$$
$$CH_3(CH_2)_7CH{=}CH(CH_2)_7COOCH \qquad \text{unsaturated fat}$$
$$CH_3(CH_2)_7CH{=}CH(CH_2)_7COOCH_2$$

↓ hydrogen, heat, nickel catalyst

$$CH_3(CH_2)_7CH_2CH_2(CH_2)_7COOCH_2$$
$$CH_3(CH_2)_7CH_2CH_2(CH_2)_7COOCH \qquad \text{saturated fat}$$
$$CH_3(CH_2)_7CH_2CH_2(CH_2)_7COOCH_2$$

Margarine is manufactured from vegetable oils like sunflower oil. The oil is reacted with hydrogen and a nickel catalyst under controlled conditions, so that some, but not all, of the double bonds are hydrogenated. The more double bonds are removed, the harder the margarine becomes.

Health issues

(1) Fat consumption

If you take in more energy than you need, you gain weight. Fats in your diet are a very rich source of energy. If your body doesn't use that much energy, the excess food energy is stored as body fat. There is no great mystery about the increase in obese or overweight people in the developed world – they eat too much and don't take enough exercise.

Olive oil contains a very high proportion of a mono-unsaturated compound.

Fats are a very rich source of energy in the diet.

Obesity is linked with serious diseases like heart disease, diabetes and strokes, and there is thought to be a link to some cancers, although the evidence is less strong.

As well as the need to reduce overall fat consumption, the type of fat you eat also matters. To see why that is, you have to know about cholesterol.

(2) Cholesterol

Cholesterol is a very complicated alcohol involved in digesting fats and making cell membranes and hormones. It is produced by the body and is also present in animal-based foods.

Cholesterol is insoluble in water and is carried around the blood by molecules known as **lipoproteins**. There are two sorts of lipoprotein:

- Low density lipoprotein (LDL) is the major carrier of cholesterol. It contributes to the furring up of arteries leading to blood clots, heart attacks and strokes. Cholesterol carried by LDL is often described as "bad" cholesterol.

- High density lipoprotein (HDL) carries cholesterol to the liver where it is broken down. Cholesterol carried like this is described as "good" cholesterol.

Saturated fats – from dairy and other animal products – are associated with raised cholesterol levels. Unsaturated fats tend to decrease the total cholesterol level – particularly the harmful LDL cholesterol. Unsaturated fats are present in fish and plant oils (except coconut oil which is mainly saturated).

There is controversy about the relative benefits of mono-unsaturated versus polyunsaturated fats and oils. There is a worry that a high intake of polyunsaturated fats may carry an increased cancer risk, although the evidence isn't very strong. There is at present no evidence against mono-unsaturated fats. Olive oil is a particularly good source of these.

Ideas about the wisdom of eating different forms of fats change all the time as more studies are done. Find out the current state of play by doing an internet search.

Hydrolysing fats

Fats can be split into the original carboxylic acids and glycerol by enzymes. This involves a reaction with water.

$$CH_3(CH_2)_{16}COOCH_2$$
$$CH_3(CH_2)_{16}COOCH \quad + \quad 3H_2O \quad \longrightarrow \quad 3CH_3(CH_2)_{16}COOH \quad + \quad HOCH$$
$$CH_3(CH_2)_{16}COOCH_2 \qquad\qquad\qquad\qquad\qquad\qquad\qquad HOCH_2$$

glyceryl tristearate   stearic acid   glycerol

They can also be split by heating with sodium hydroxide solution – although in this case you get sodium salts rather than acids. Soap is made in this way.

$$CH_3(CH_2)_{16}COOCH_2 \qquad\qquad\qquad\qquad\qquad\qquad\qquad HOCH_2$$
$$CH_3(CH_2)_{16}COOCH \quad + \quad 3NaOH \quad \longrightarrow \quad 3CH_3(CH_2)_{16}COONa \quad + \quad HOCH$$
$$CH_3(CH_2)_{16}COOCH_2 \qquad\qquad\qquad\qquad\qquad\qquad\qquad HOCH_2$$

glyceryl tristearate   sodium stearate (soap)

NUTRITION	
Typical Values per 100 g	
Energy	724 kcals / 2975 kJ
Protein	0.4 g
Carbohydrate	0.4 g
of which sugars	0.4 g
Fat	80.5 g
of which saturates	36.0 g
monounsaturates	18.1 g
polyunsaturates	21.7 g
Sodium	0.4 g

INGREDIENTS: Organic Butter (60%), Organic Sunflower Oil (31%), Water, Salt (1%), Organic Carrot Juice (1%)

Nutrition list from a spread made from butter and sunflower oil.

Other food chemistry

Vitamin C

Vitamin C (ascorbic acid) is present in fresh fruit and vegetables. Prolonged storage or over-cooking reduces the amount of vitamin C. A deficiency leads to a disease called scurvy.

An internet search on "**vitamin c**" will throw up all sorts of claims about the benefits of taking vitamin C supplements. Some of these have little or no scientific basis, but there may be a long term advantage in taking small additional amounts. Vitamin C is an **anti-oxidant** and can destroy harmful free radicals (atoms or groups of atoms with single unpaired electrons) produced in the breakdown of some foods. This helps to prevent damage to DNA which could lead to cancer.

Raising agents

Raising agents are used to produce the light open structure of bread and some cakes. In bread, the flour and water is mixed with **yeast** and left in the warm to "rise". Fermentation produces bubbles of carbon dioxide which are trapped in the dough. When the bread is baked, the yeast is killed, but the bubble structure remains.

You can also use **raising agents**, which are mixtures of sodium hydrogen-carbonate and an acid such as tartaric acid – known as **baking powder**. When it is moistened, the mixture reacts to form bubbles of carbon dioxide.

Food additives

Lots of different substances may be added to food to improve its look, its texture or flavour, or to help to preserve it. Although a few of these additives are natural, the majority are the product of the chemical industry.

In Europe, many permitted additives are given **E-numbers**. These have to be shown among the list of ingredients. Some manufacturers list them by name rather than E-number.

Some of these additives are colours. For example, E102 (tartrazine) is an orange dye which used to be used for things as varied as orange drinks and the coatings for fish fingers. Tartrazine is believed to cause hyperactivity in children. Other colours like E162 (beetroot red) come from a natural source, and are harmless.

Other additives are used to preserve food. Cold meats like ham may contain additives with E-numbers between E249 and E252. These are sodium and potassium nitrites and nitrates which prevent the growth of bacteria. There is a *theoretical* risk that these could produce stomach cancer. On the other hand, if they weren't used there is a *certainty* that food would quickly go bad, and result in food poisoning which could cause death.

Some preservatives are entirely safe. For example, E300 is vitamin C (ascorbic acid). This is used as an anti-oxidant and can be added to processed fruit to stop it discolouring.

E-numbers in the low 400s include many emulsifiers and stabilisers. These are used to produce stable mixtures of, for example, fats and water to make creamy sauces.

The only way of finding out the current scientific view on vitamin C is to do an internet search. Make sure that you don't get a one-sided view. Many sites are simply out to promote the use of vitamin C.

Bread is full of gas bubbles!

ideas
evidence

If you are interested, it is easy to find complete lists of E-numbers on the internet. Search for "**e-numbers**".

Willow trees.

The hexagon with the circle inside it is a way of drawing a **benzene ring**. There is a carbon atom at each corner of the hexagon, and a hydrogen atom attached to each of the carbons with nothing else already attached. In this case, there would be a hydrogen atom on each carbon apart from the ones attached to –OH and –COOH. The circle in the ring shows some bonding which is impossible to describe at GCSE. You may also find this drawn as a hexagon with alternating double and single carbon–carbon bonds. Either form is acceptable for exam purposes.

The diagrams show two versions of the aspirin structure. You may also find it drawn with the molecule flipped over with the CH$_3$COO– group drawn on the right-hand side.

In the example, if E-numbers were used instead of names, you would find that the emulsifiers and stabilisers were E410, E412, E415 and E472.

Additives are also used to add or enhance flavours. Artificial flavours are added because they are stronger and cheaper than natural flavours. It is, for example, cheaper to use a compound which tastes of raspberries than to use real raspberries. They can also be used to replace flavours which are lost during food processing. Artificial sweeteners (such as aspartame) are used instead of sugar in "diet" drinks.

Although individual additives have been approved as safe, many people worry about their use. It would be impossible to check every *combination* of additives to see whether they produced any new harmful effect, and it is impossible to be sure about long term effects.

Drugs

A drug can be any substance that you eat, drink, inhale or inject and which affects chemical reactions in your body. Drugs range from commonly used substances like alcohol and nicotine to sophisticated and expensive compounds designed to fight cancer or AIDS.

Here, we are only concerned with **analgesics** – drugs designed to reduce pain. Common analgesics include **aspirin**, **paracetamol** and **ibuprofen**.

Aspirin

Since ancient times the leaves and bark of willow trees and the sap of poplar trees have been used as remedies for all sorts of illnesses. In the eighteenth century, they were discovered to be effective against pain and fever.

In the 1820s, several chemists isolated the active ingredient, called **salicin**. A decade later, **salicylic acid** was produced from salicin.

salicylic acid

Both salicin and salicylic acid were used to treat rheumatic fever, but salicylic acid had the advantage that it could be made cheaply by other chemical methods. Unfortunately, salicylic acid caused irritation to the mouth and wasn't so safe to use as salicin.

Chemists then looked for ways of modifying salicylic acid to make it safer, without losing its essential effects on pain and fever. Eventually, Felix Hoffmann came up with a commercial process for manufacturing aspirin in 1897.

aspirin

Soluble aspirin

Aspirin is insoluble in water. If it could be made to dissolve, it would be absorbed by the body more quickly. That would make it more effective in rapidly reducing pain. It was also hoped that faster absorption would cut down one of the major side effects of aspirin – it causes internal bleeding and ulceration. Unfortunately, that turned out not to be the case.

Aspirin is an acid and so it reacts with a carbonate to produce a salt. Soluble aspirin tablets contain aspirin mixed with calcium carbonate and starch. The mixture doesn't react as long as it is dry, but in contact with water a calcium salt of aspirin is formed. This is ionic and water-soluble. The starch helps the water to penetrate the tablet effectively. The diagram shows the important negative ion produced in the reaction.

the ion present when
soluble aspirin dissolves

Aspirin today

Since aspirin was invented, it has been estimated that people have consumed a trillion (1,000,000,000,000) tablets. Current consumption is about 50 billion tablets a year worldwide.

For simple aches and pains, alternative drugs such as paracetamol or ibuprofen are preferred. Paracetamol doesn't cause internal bleeding, and the risk is much less with ibuprofen than with aspirin. An important modern use for aspirin is in helping to prevent heart disease. Very low doses of aspirin taken daily help to prevent the blood clots which can lead to heart attacks.

Issues relating to the testing of new drugs

The testing process

Getting a drug from its original design in the lab to its final use in patients is a long and painstaking process. There are a series of stages:

- The drug is tested on animals to check for any toxic effects. Animals don't always show the same reactions as humans, so if there aren't any harmful effects in animals, the drug is next tested on humans.

- In the next stage, a small number of healthy people use the drug over a period of a few months to find out how big a dose can be taken without too many side effects. If that is completed without problems, the drug is then tried in a larger sample of people who actually have the disease which the drug will be used to treat. This is to find out whether the drug does what it is supposed to in the short term – perhaps up to two years.

- Finally, the long term safety and effectiveness of the drug is tested. This involves thousands of people taking the drug for several years.

At each stage of the testing, **controlled trials** are done. Some people will get the new drug, others may get a **placebo** – for example, a pill made of sugar

About 50 billion aspirin tablets are consumed every year.

ideas
evidence

To find a good example of a testing process which was inadequate and led to devastating results, do an internet search on **thalidomide**.

which is designed to look like the real thing. Another group might get an existing drug. The various groups have to be matched so that they are as similar as possible.

To avoid bias, double-blind studies are done. In these everything is coded in such a way that neither the researchers nor the people being tested know who is getting the new drug, a placebo, or an old drug. Only when the study is complete do the researchers find out which people got which treatment.

Some ethical issues

- Is it morally right to test drugs on animals? What if there is currently no alternative way of doing the initial safety testing? What if waiting for alternative testing methods to be developed leads to delays in the production and use of an effective anti-cancer drug?

- Is it morally right to give placebos to some people suffering from a serious illness when others in their trial may be getting an effective drug which will cure them?

Questions of this sort don't have simple answers that everybody can agree on. Question 8 at the end of this chapter will help you to decide your own views.

End of Chapter Checklist

If you haven't got a copy of your specification, read the introduction on page vi.

You will need to be able to do some or all of the following. Check your Awarding Body's specification (syllabus) to find out exactly what you need to know.

- Understand the terms *carbohydrate*, *monosaccharide*, *disaccharide* and *polysaccharide*, and understand that polysaccharides are made by *condensation polymerisation*.

- Draw structural formulae for glucose, fructose and sucrose, and simplified structural formulae for starch and cellulose.

- Explain the hydrolysis of polysaccharides.

- Know that proteins are polymers of amino acids and are formed by condensation polymerisation.

- Draw the structural formula for glycine, and recognise other amino acids from their structures.

- Explain the hydrolysis of proteins.

- Know that fats are triglycerides, and draw a structural formula for a triglyceride.

- Understand what is meant by saturated, unsaturated and polyunsaturated fats, and discuss the health implications of eating the various types of fat.

- Describe the manufacture of margarine.

- Explain the hydrolysis of triglycerides.

- Know that vitamin C is an anti-oxidant and is believed to have a protective effect on tissues. Know that vitamin C content is reduced by cooking or storage.

- Explain the use of raising agents in bread and cake making.

- Understand the purpose and the advantages and disadvantages of various food additives (including use of the E-number system).

- Understand what is meant by the term *analgesic*, and name some common analgesics.

- Describe the history and use of aspirin, including soluble aspirin.

- Draw the structures for aspirin and the ion present in soluble aspirin.

- Outline the way that new drugs are tested, and discuss some of the moral issues involved.

Questions

1 Choose from the following compounds to answer the questions:

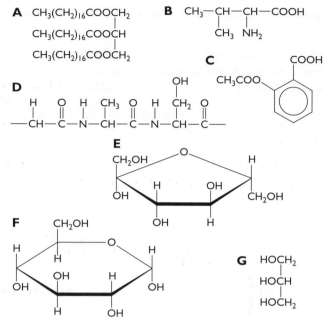

A $CH_3(CH_2)_{16}COOCH_2$
$CH_3(CH_2)_{16}COOCH$
$CH_3(CH_2)_{16}COOCH_2$

B CH_3—CH—CH—$COOH$
CH_3 NH_2

a) Which of these compounds is aspirin?

b) Which of these compounds is the amino acid, valine?

c) Which of these compounds is glycerol?

d) Which of these compounds are monosaccharides?

e) Which of these compounds is glucose?

f) Which of these compounds is a triglyceride?

g) Which of these structures represents a part of a protein chain?

h) Which of these compounds is an ester of stearic acid and glycerol?

i) Which of these compounds is a polypeptide?

j) Which of these compounds combine to make sucrose?

2 a) Draw the structure of sucrose, and then show what happens to that structure when sucrose is hydrolysed by boiling it with dilute acid.

b) Starch is a *condensation polymer* of glucose. Explain what this means with the aid of a diagram.

c) Cellulose is also a polymer of glucose, but with a different structural arrangement. Animals (including humans) cannot digest cellulose. So how do cows and sheep survive by eating grass which is largely cellulose?

d) Maltose is a *disaccharide* consisting of two glucose units joined together. Suggest a structural formula for maltose.

3 a) Draw the structural formula for the amino acid, glycine.

b) Draw the structural formula of a small polypeptide made of 3 glycine units joined by peptide links.

4 a) What do you understand by the terms *unsaturated*, *mono-unsaturated* and *polyunsaturated* as applied to fats and oils?

b) Write a brief paragraph outlining the health effects of eating the various types of fats and oils.

c) Describe the manufacture of margarine from plant oils, and explain what is happening in the process.

d) Using the symbol "R" for the hydrocarbon chain in the carboxylic acid, draw a structural formula for a triglyceride and write an equation to show what happens when it is hydrolysed.

5 Look at the labels on any 3 different tins, bottles or packets of food or drink, and in each case list the food additives and explain what their purpose is. If you eat mainly organic or fresh food, you may find this difficult. In that case, investigate packaged food or drink in your local shop or supermarket. Even if the additives aren't listed by E-number, you may find it useful to have a description of the various E-numbers. You can find this quickly by doing an internet search.

6 a) What do you understand by the term *analgesic*? Name one other analgesic apart from aspirin.

b) Give a disadvantage of aspirin which means that many people now prefer to use an alternative drug for pain control.

c) Apart from pain control, give another modern use for aspirin.

d) Write the structural formula for aspirin. Explain how this structure is modified when soluble aspirin is dissolved in water.

7 People who eat a Mediterranean diet tend to have a lower incidence of heart disease than people in the UK or America. Two of the several explanations that have been put forward to account for this have been the possible protective effect of red wine, and the relatively high consumption of olive oil, which is rich in mono-unsaturated fats.

Suppose you were a member of a team of scientists who were going to do a study to see which of these two effects was the more important. How would you go about it? You can assume that you would have access to willing groups of people and their health records over as long a period of time as you need.

8 Suppose your school was organising a debate on the topic: "Testing drugs on animals is morally wrong". Write two short speeches (to last about 2 minutes each) – one in favour of animal tests, and the other one against them. Try to make your "for" and "against" speeches equally strong – whatever your feelings about the matter. Be careful not to get side-tracked into the argument about testing substances like cosmetics on animals.

Chapter 24: Enzymes

This chapter looks briefly at enzymes as catalysts and gives some examples of their everyday uses.

Enzymes as Biological Catalysts

Enzymes are protein molecules which catalyse the reactions happening in living cells.

How enzymes work

Enzymes are very **specific**. An enzyme catalysing one reaction will have no effect whatsoever on a different reaction. Even small changes in the molecules taking part in a reaction can mean that a particular enzyme will no longer work with those modified molecules.

Each enzyme molecule has an **active site**. This has a specific shape into which only one type of molecule will fit, just as a particular lock will only take a key of exactly the right shape.

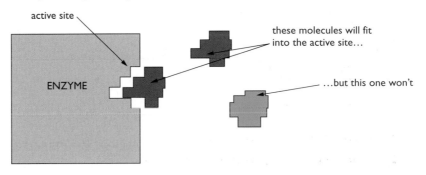

The **substrate** molecule bonds to the active site. The bonds in the molecule may then be weakened, which helps to make it react more easily.

In Chemistry, the substances taking part in a reaction are known as **reactants**. In Biochemistry, they are often known as **substrates**.

Obviously, anything which affects the active site will affect the working of the enzyme.

The effect of temperature on enzyme catalysis

With most chemical reactions, the rate increases as you increase the temperature. The molecules are moving faster and hit each other more often. More importantly, lots more of them hit each other hard enough to exceed the activation energy for the reaction.

It would be helpful if you read pages 43 and 46 on the effect of temperature on ordinary chemical reactions before you go on.

A similar effect happens with enzyme catalysis, but only up to a point. After a certain temperature, the effectiveness of the enzyme falls very quickly.

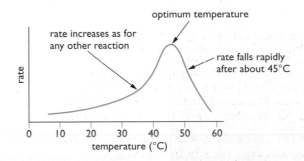

The temperature at which the rate is highest is described as the **optimum temperature**.

With most enzymes, at temperatures above about 45°C, the structure of the protein starts to break down. The protein becomes **denatured**. This changes the shape of the active site so that the substrate molecules won't fit it properly any more.

The effect of pH on enzymes

Enzymes also work best at a particular pH. This is described as the **optimum pH**. Many enzymes have an optimum pH around 7, but some work best under more extreme conditions. The enzyme pepsin, for example, works in the acidic environment of the stomach and has an optimum pH of about 2.

Proteins are polypeptide chains with side groups attached. These side groups may contain acidic or basic groups which may well be involved in the active site. If you change the pH, these acidic or basic groups may react. For example, the acidic –COOH group would turn into a $-COO^-$ ion in the presence of an alkali. If that –COOH group was an essential part of the active site, then the substrate molecule would no longer attach itself there.

At extremes of pH, covalent bonds in the enzyme may start to break, leading to the protein becoming denatured, or even to the complete break-up of the polypeptide chains.

Uses of enzymes

Enzymes in the food and drink industries

(1) Processes involving yeast

Wine is made by fermenting sugars present in grapes to form alcohol (ethanol) and carbon dioxide. The reaction is controlled by enzymes in the yeast.

Beer is made by fermenting carbohydrates present in germinating barley. The reaction is controlled by enzymes in the barley and in the yeast which is later added to it.

Bread rises because of the bubbles of carbon dioxide produced as enzymes in the yeast process carbohydrates in the flour.

(2) Making yoghurt

Yoghurt is made by the action of enzymes in special bacteria (for example, *Lactobacillus bulgaricus* and *Streptococcus thermophilus*) on milk. A sugar in the milk called **lactose** is converted into **lactic acid**. This is responsible for the typical yoghurt taste.

(3) Making chocolates with soft centres

Covering soft centres with chocolate would be a messy job! Instead, you start with a hard centre which includes the enzyme **invertase**. This splits sucrose into two smaller sugars, glucose and fructose, which form a soft mixture. You will find the structures of sucrose, glucose and fructose on pages 241–242.

(4) Some other processes

- Enzymes called **proteases** are used to break up proteins into smaller peptide chains or individual amino acids. This makes them more digestible. This is used in some baby foods. Peptide chains and amino acids are explained on pages 244–246.

If you are unsure about what a polypeptide chain is, read pages 245–246.

You will find the essential chemistry of this on page 232.

Yoghurt making.

Soft-centred chocolates start off with hard centres.

- Enzymes called **carbohydrases** break carbohydrates into smaller sugars. They are used to convert starch syrup into sugar syrup.

- Fructose is sweeter than glucose. **Isomerase** converts glucose into fructose. Because fructose is sweeter you need less of it. This would be an advantage in some slimming foods, avoiding the need for artificial sweeteners.

Biological detergents

Biological detergents contain enzymes which attack protein ("proteases") and fat ("lipases") and convert them into smaller molecules which are more easily removed from the fabric.

Some factors affecting the industrial use of enzymes

Enzymes have the advantage that they work at normal temperatures and pressures. If an industrial process can be controlled by enzymes, it avoids the expense associated with high temperatures and pressures.

Unfortunately, it may not be possible to find a micro-organism or a specific enzyme which produces the chemical you want. Or it may produce a complex mixture from which it would be expensive to separate what you want.

Enzyme reactions often happen over fairly long periods of time, needing a "batch process". This is less convenient and more expensive than using a continuous process.

For the advantages and disadvantages of producing ethanol using enzymes, or from the oil industry, see page 233.

Whether you are using a batch or a continuous process, you need to choose conditions (for example, temperature and pH) that keep the enzyme active as long as possible. It is also helpful if the enzyme stays where you put it! Some processes immobilise the enzyme by trapping it on some sort of inert material. For example, concentrated sugar syrups are made from whey (a by-product from cheese-making) by passing it over an enzyme trapped on alginate beads. Alginates are derived from seaweed.

End of Chapter Checklist

If you haven't got a copy of your specification, read the introduction on page vi.

You will need to be able to do some or all of the following. Check your Awarding Body's specification (syllabus) to find out exactly what you need to know.

- Describe enzymes as biological catalysts, and know that they are protein molecules.

- Know that enzymes are specific, and have an optimum temperature and pH. Offer simple explanations for this.

- Describe some applications of enzymes in food and drink manufacture and in biological detergents.

- Discuss some of the factors affecting the use of enzymes in industrial processes.

Questions

More questions on enzymes can be found at the end of Section E on page 259.

1   a)   What is an enzyme?

b)   Enzymes are described as being *specific*. Explain what this means.

c)   Food can be preserved by keeping it in a fridge at about 5°C. This slows down the chemical reactions in the food. Explain why keeping the food in a fridge slows down the chemical reactions.

d)   Before freezing vegetables for long-term storage, they are often "blanched" by dipping them into boiling water for a minute or so. This has the effect of stopping all further reactions controlled by enzymes present in the food. Explain why heating the food to 100°C has this effect.

e)   Enzymes have an *optimum pH*. Explain what this means.

2   Write a short, illustrated article on the production of beer starting from its raw materials: barley, hops, yeast and water. Explain what effect enzymes are having at each stage of the process. Your article should be produced on a computer, and should not exceed 300 words.

3   In the years shortly before the First World War, scientists were trying to find a way of producing butanol using bacteria. It was intended that the butanol would be converted into buta-1,3-diene, which is used in making artificial rubber. In 1914 Chaim Weizmann found a bacillus (a sort of bacteria) which converted maize starch into butanol, propanone, ethanol, carbon dioxide and water. Propanone is an important solvent used in the manufacture of the explosive cordite. By 1917, thousands of tonnes of propanone were being produced every year by this process. Nowadays, propanone is made from propene – produced when petroleum fractions are cracked. Buta-1,3-diene is also formed during cracking.

List the advantages and disadvantages of producing molecules like propanone, *a)* using bacteria, *b)* from the oil industry.

End of Section Questions

1 Crude oil (petroleum) is a complex mixture of *hydrocarbons*. The crude oil can be separated into simpler mixtures (called *fractions*) by taking advantage of differences in boiling points between the various hydrocarbons.

a) What do you understand by the term *hydrocarbon*?
(2 marks)

b) What is the relationship between the number of carbon atoms in a hydrocarbon and its boiling point?
(1 mark)

c) What name is given to the process of producing the simpler mixtures from the crude oil?
(1 mark)

d) One of the fractions produced from crude oil is called *kerosine*. Give one use for kerosine.
(1 mark)

e) One of the hydrocarbons present in kerosine is an alkane containing 10 carbon atoms, called decane. Write the molecular formula for decane.
(1 mark)

f) The hydrocarbon $C_{15}H_{32}$ (also present in kerosine) burns to form carbon dioxide and water. Write the equation for the reaction.
(2 marks)

g) How would you test the products when $C_{15}H_{32}$ burns to show that carbon dioxide has been formed?
(2 marks)

h) Name the environmental problem that is caused by the formation of carbon dioxide during the combustion of hydrocarbons.
(1 mark)

i) Liquefied petroleum gas (LPG), used for domestic heating and cooking, is propane, C_3H_8. Burning propane in badly maintained appliances or in poorly ventilated rooms can cause death. Explain why that is.
(4 marks)

Total 15 marks

2 Some of the gas oil fraction from crude oil is broken into smaller molecules by heating it in the presence of a catalyst. A mixture of *saturated* and *unsaturated* hydrocarbons is formed.

a) Explain the difference between a saturated and an unsaturated hydrocarbon.
(2 marks)

b) What name is given to the process of breaking up the gas oil fraction in this way?
(1 mark)

c) When a molecule $C_{17}H_{36}$ was heated in the presence of a catalyst, it broke up to give two molecules of ethene, C_2H_4, one molecule of propene, C_3H_6, and another molecule, X.

i) Write a balanced equation for the reaction.
(2 marks)

ii) Is molecule X an alkane or an alkene?
(1 mark)

d) Some propene is converted into propenenitrile which is polymerised to make a fibre used for textiles. "Orlon", "Acrilan" and "Courtelle" are all poly(propenenitrile). The structure of poly(propenenitrile) is:

$$\begin{array}{c}
\quad H \quad CN \; H \quad CN \; H \quad CN \; H \quad CN \\
\quad | \quad | \quad | \quad | \quad | \quad | \quad | \quad | \\
-C-C-C-C-C-C-C-C- \\
\quad | \quad | \quad | \quad | \quad | \quad | \quad | \quad | \\
\quad H \quad H \quad H \quad H \quad H \quad H \quad H \quad H
\end{array}$$

i) What do you understand by the term *polymerisation*?
(2 marks)

ii) By looking at the structure of the polymer, suggest a structural formula for the monomer, propenenitrile.
(2 marks)

e) Poly(chloroethene) (also called PVC) is a polymer produced from chloroethene.

i) State one use for poly(chloroethene). *(1 mark)*

ii) Poly(chloroethene) is non-biodegradable and produces hydrogen chloride gas when it burns. What are the implications of this for disposal of waste PVC?
(4 marks)

Total 15 marks

3 a) Draw fully displayed structural formulae for
i) propane, ii) propene. *(2 marks)*

b) Propane and propene are both gases. Describe a simple test which would enable you to distinguish between them. You should describe what you would do and what you would see in each case. Write an equation for any reaction(s) you describe.
(4 marks)

c) Propene can be *hydrogenated* by passing it, together with hydrogen, over a catalyst.

i) Name the catalyst used. *(1 mark)*

ii) Write an equation for the reaction. *(1 mark)*

iii) Give an example of a large-scale process which involves the hydrogenation of a carbon–carbon double bond. *(1 mark)*

Total 9 marks

4 This question is about *structural isomerism*.

a) What are *structural isomers*? *(2 marks)*

b) The diagram shows the structure of one of the isomers of C_5H_{12}.

$$H-\underset{\underset{H}{|}}{\overset{\overset{H}{|}}{C}}-\underset{\underset{H}{|}}{\overset{\overset{H}{|}}{C}}-\underset{\underset{H}{|}}{\overset{\overset{H}{|}}{C}}-\underset{\underset{H}{|}}{\overset{\overset{H}{|}}{C}}-\underset{\underset{H}{|}}{\overset{\overset{H}{|}}{C}}-H$$

    i)   Draw the structures of the other two isomers of C_5H_{12}.   *(2 marks)*

    ii) Which of the three isomers has the highest boiling point? Explain your answer.   *(3 marks)*

c) Draw the structures of the two alcohols which have the molecular formula C_3H_8O.   *(2 marks)*

d) The molecular formula $C_3H_6O_2$ is shared by one carboxylic acid and two esters. Draw the structures of the acid and one of the esters. Name the acid and the ester that you have drawn.   *(4 marks)*

                               Total 13 marks

5 Ethanol, CH_3CH_2OH, can be made by *fermentation* followed by *fractional distillation*, or by the *hydration of ethene*.

a) Describe briefly how an impure dilute solution of ethanol is made by fermentation.   *(5 marks)*

b) Fermentation is controlled by *enzymes*. What are enzymes?   *(2 marks)*

c) State the different boiling points of ethanol and water which enable them to be separated by fractional distillation.   *(2 marks)*

d) Write an equation to show the hydration of ethene.   *(1 mark)*

e) State any two conditions used during the hydration reaction.   *(2 marks)*

f) Explain which of the two processes has the advantage in terms of:

    i)   the use of resources   *(2 marks)*

    ii) the conditions used   *(2 marks)*

    iii) the speed of production.   *(2 marks)*

                               Total 18 marks

6 This question is about ethanol, CH_3CH_2OH, and its reactions.

a) Ethanol can be used as a fuel. Write the equation for the complete combustion of ethanol.   *(2 marks)*

b) Ethanol is likely to be found among the ingredients of some perfumes and cosmetics. Explain why it is used in these products.   *(2 marks)*

c) If ethanol vapour is passed over hot aluminium oxide, it dehydrates and a gas is produced.

    i)   Name the gas produced and write an equation for its formation.   *(2 marks)*

    ii) Draw a simple diagram of the apparatus you could use to produce and collect a few test tubes of the gas using this reaction.   *(4 marks)*

d) If ethanol is warmed with ethanoic acid in the presence of some concentrated sulphuric acid, ethyl ethanoate is formed.

    i)   What type of compound is ethyl ethanoate?   *(1 mark)*

    ii) What is the function of the concentrated sulphuric acid?   *(1 mark)*

    iii) Write an equation for the reaction, clearly showing the structure of the ethyl ethanoate.   *(2 marks)*

                               Total 14 marks

7 a) When wine is exposed to bacteria in the air, it turns sour as the ethanol is oxidised to ethanoic acid, CH_3COOH.

    i)   What common name is given to the solution of ethanoic acid formed in this way?   *(1 mark)*

    ii) Give two different uses for the solution.   *(2 marks)*

b) Ethanoic acid is a *weak acid*.

    i)   What do you understand by the term *weak* as applied to acids?   *(1 mark)*

    ii) Describe a simple experiment by which you could show that dilute ethanoic acid was acidic, without using either indicators or a pH meter.   *(2 marks)*

c) This part of the question concerns the compound:

$$CH_3-\overset{\overset{\textstyle O}{\|}}{C}\diagdown_{\textstyle O-CH_3}$$

    i)   Name this compound.   *(1 mark)*

    ii) Suggest a possible commercial use for the compound.   *(1 mark)*

    iii) What reagents would you need to make this compound from ethanoic acid?   *(2 marks)*

                               Total 10 marks

Chapter 25: RAMs and Moles

> This chapter introduces the ideas that you need in order to do some simple chemistry calculations. Don't worry – the level of maths involved is very low.

Each gold bar contains almost 4×10^{25} gold atoms. That's 4 followed by 25 noughts!

> Remember that isotopes are atoms of the same element but with different masses. Isotopes are explained on page 3.

> This is a slight approximation. To be accurate, each of these hydrogen atoms has a mass of 1.008 on the carbon-12 scale. For GCSE purposes, we take it as being exactly 1.

You can make iron(II) sulphide by heating a mixture of iron and sulphur.

$$Fe_{(s)} + S_{(l)} \rightarrow FeS_{(s)}$$

How do you know what proportions to mix them up in? You can't just mix equal masses of them because iron and sulphur atoms don't weigh the same. Iron atoms contain more protons and neutrons than sulphur atoms so that an iron atom is one and three-quarters times heavier than a sulphur atom. In this, or any other reaction, you can only get the proportions right if you know about the masses of the individual atoms taking part.

Relative Atomic Mass (RAM)

Defining relative atomic mass

Atoms are amazingly small. In order to get a gram of hydrogen, you would need to count out 602,204,500,000,000,000,000,000 atoms (to the nearest 100,000,000,000,000,000).

It would be silly to measure the masses of atoms in conventional mass units like grams. Instead, their masses are compared with the mass of an atom of the carbon-12 isotope, taken as a standard. We call this the "**carbon-12 scale**".

On this scale one atom of the carbon-12 isotope weighs *exactly* 12 units.

An atom of the commonest isotope of magnesium weighs twice as much as this and is therefore said to have a **relative isotopic mass** of 24.

An atom of the commonest isotope of hydrogen weighs only one twelfth as much as a carbon-12 atom, and so has a relative isotopic mass of 1.

12 ^1H atoms have the same mass as...

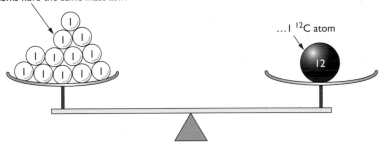

...1 ^{12}C atom

The basic unit on the carbon-12 scale is one twelfth of the mass of a ^{12}C atom – approximately the mass of the commonest hydrogen atom. For example, a fluorine-19 atom has a relative isotopic mass of 19 because its atoms have a mass 19 times that basic unit.

The **relative atomic mass (RAM)** of an element (as opposed to one of its isotopes) is given the symbol A_r and is defined like this:

> The relative atomic mass of an element is the weighted average mass of the isotopes of the element. It is measured on a scale on which a carbon-12 atom has a mass of exactly 12 units.

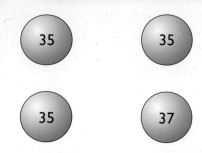

The masses of four typical chlorine atoms. The average of these isn't 36.

Explaining the term "weighted average"

In any sample of chlorine, some atoms have a relative mass of 35; others a relative mass of 37. A simple average of 35 and 37 is, of course, 36 – but that isn't the relative atomic mass of chlorine. The problem is that there aren't equal numbers of ^{35}Cl and ^{37}Cl atoms.

A typical sample of chlorine has:
^{35}Cl 75%
^{37}Cl 25%

If you had 100 typical atoms of chlorine, 75 would be ^{35}Cl, and 25 would be ^{37}Cl.

The total mass of the 100 atoms would be $(75 \times 35) + (25 \times 37)$
=   3550

The average mass of 1 atom would be $\dfrac{3550}{100}$
=   35.5

The weighted average is closer to 35 than to 37, because there are more ^{35}Cl atoms than ^{37}Cl atoms. A weighted average allows for the unequal proportions.

35.5 is the relative atomic mass (RAM) of chlorine.

More examples of calculating relative atomic masses

Magnesium

The isotopes of magnesium and their percentage abundances are:

^{24}Mg 78.6%
^{25}Mg 10.1%
^{26}Mg 11.3%

Again, assume that you have 100 typical atoms.

The total mass would be $(78.6 \times 24) + (10.1 \times 25) + (11.3 \times 26)$
=   2432.7

The RAM would be $\dfrac{2432.7}{100}$
=   24.3 (to 3 significant figures)

The relative atomic mass of magnesium is 24.3.

The original percentages are only quoted to 3 significant figures. You mustn't quote your answer any more accurately. As a general guide at GCSE, if in doubt quote answers to 3 significant figures. You aren't likely to be penalised for doing that.

Lithium

The abundance data might be given in a different form. You might get a graph with the most common isotope being given a relative abundance of 100, and the others quoted relative to that.

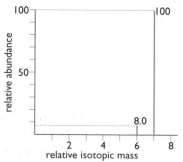

Relative abundance of lithium isotopes.

In this case, there would be 8.0 atoms of ^{6}Li for every 100 of ^{7}Li.

The total mass of 108 atoms would be		$(8.0 \times 6) + (100 \times 7)$
	=	748
The average mass of 1 atom (the RAM)	=	$\dfrac{748}{108}$
	=	6.9

This time you can only quote your answer to 2 significant figures because one of the relative abundances is only given to 2 figures.

The relative atomic mass of lithium is 6.9.

Relative Formula Mass (RFM)

You can measure the masses of compounds on the same carbon-12 scale. For example, it turns out that a water molecule, H_2O, has a mass of 18 on the ^{12}C scale. Where you are talking about compounds, you use the term **relative formula mass (RFM)**. Relative formula mass is sometimes called relative molecular mass (RMM).

Relative formula mass is given the symbol M_r.

Avoid the term "relative molecular mass", because it can only properly be applied to substances which are actually molecules – in other words, to covalent substances. You shouldn't use it for things like magnesium oxide or sodium chloride which are ionic. RFM covers everything.

Working out some relative formula masses

To find the RFM of magnesium carbonate, MgCO$_3$

Relative atomic masses: C = 12; O = 16; Mg = 24

All you have to do is to add up the relative atomic masses to give you the relative formula mass of the whole compound. In this case, you need to add up the masses of 1 × Mg, 1 × C and 3 × O.

You will never have to remember RAMs. They will always be given to you in an exam – either in the question, or on the Periodic Table. *If you use a Periodic Table, be sure to use the right number!* It will always be the larger of the two numbers given.

RFM	= 24 + 12 + (3 × 16)
	= 84

To find the RFM of calcium hydroxide, Ca(OH)$_2$

Relative atomic masses: H = 1; O = 16; Ca = 40

RFM	= 40 + (16 + 1) × 2
	= 74

To find the RFM of copper(II) sulphate crystals, CuSO$_4$·5H$_2$O

Relative atomic masses: H = 1; O = 16; S = 32; Cu = 64

You will find the RAM of copper quoted variously as 64 or 63.5. In an exam, just use the value you are given.

RFM	= 64 + 32 + (4 × 16) + 5 × [(2 × 1) + 16]
	= 250

Most people have no difficulty with RFM sums until they get to an example involving water of crystallisation (the 5H$_2$O in this example). 5H$_2$O means 5 molecules of water – so to get the total mass of this, work out the RFM of water (18) and then multiply it by 5. It is dangerous to do the hydrogens and oxygens separately. The common mistake is to work out 10 hydrogens (quite correctly!), but then only count 1 oxygen rather than 5.

Using relative formula mass to find percentage composition

Having found the relative formula mass of a compound, it is then easy to work out the percentage by mass of any part of it. Examples make this clear.

To find the percentage by mass of copper in copper(II) oxide, CuO

Relative atomic masses: O = 16; Cu = 64

RFM of CuO $= 64 + 16$
$= 80$

Of this, 64 is copper.

Percentage of copper $= \dfrac{64}{80} \times 100$
$= 80\%$

To find the percentage by mass of copper in malachite, $CuCO_3 \cdot Cu(OH)_2$

Relative atomic masses: H = 1; C = 12; O = 16; Cu = 64

RFM of $CuCO_3 \cdot Cu(OH)_2 = 64 + 12 + (3 \times 16) + 64 + 2 \times (16 + 1)$
$= 222$

Of this, (2×64) is copper.

Percentage of copper $= \dfrac{(2 \times 64)}{222} \times 100$
$= 57.7\%$

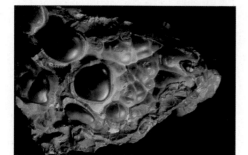

Malachite is a copper ore. You can think of it as behaving like a mixture of copper(II) carbonate and copper(II) hydroxide.

To find the percentage of water in alabaster, $CaSO_4 \cdot 2H_2O$

Relative atomic masses: H = 1; O = 16; S = 32; Ca = 40

RFM of alabaster $= 40 + 32 + (4 \times 16) + 2 \times [(2 \times 1) + 16]$
$= 172$

Of this, $2 \times [(2 \times 1) + 16]$ g is water. That's 36 g.

Percentage of water $= \dfrac{36}{172} \times 100$
$= 20.9\%$

The Mole

In Chemistry, the mole is a measure of **amount of substance**. A mole is a particular mass of that substance. You can use such expressions as:

- a mole of copper(II) sulphate crystals, $CuSO_4 \cdot 5H_2O$

- a mole of oxygen gas, O_2

- 0.1 mole of zinc oxide, ZnO

- 3 moles of magnesium, Mg

Alabaster, $CaSO_4 \cdot 2H_2O$, is a soft mineral which is easily carved.

The abbreviation for mole is **mol**.

You find the mass of 1 mole of a substance in the following way:

Work out the relative formula mass, and attach the units "grams".

Working out the masses of a mole of substance

1 mole of oxygen gas, O_2

Relative atomic mass: O = 16

RFM of oxygen, O_2 = 2 × 16
= 32

1 mole of oxygen, O_2, weighs 32 g.

1 mole of iron(II) sulphate crystals, $FeSO_4 \cdot 7H_2O$

Relative atomic masses: H = 1; O = 16; S = 32; Fe = 56

RFM of crystals = 56 + 32 + (4 × 16) + 7 × [(2 × 1) + 16]
= 278

1 mole of iron(II) sulphate crystals weighs 278 g.

Moles in chemistry may not be as soft as their furry namesakes, but that doesn't mean they are hard!

The importance of quoting the formula

Whenever you talk about a mole of something, you **must** quote its formula, otherwise there is the risk of confusion.

For example, if you talk about 1 mole of oxygen, this could mean:

- 1 mole of oxygen atoms, O, weighing 16 g
- 1 mole of oxygen molecules, O_2, weighing 32 g.

Or if you were talking about 1 mole of copper(II) sulphate, this could mean:

- 1 mole of anhydrous copper(II) sulphate, $CuSO_4$ (160 g)
- 1 mole of copper(II) sulphate crystals, $CuSO_4 \cdot 5H_2O$ (250 g)

Simple calculations with moles

You need to be able to interconvert between a mass in grams and a number of moles for a given substance. There is a simple formula that you can learn:

$$\text{number of moles} = \frac{\text{mass (g)}}{\text{mass of 1 mole (g)}}$$

You can rearrange that to find whatever you want. If rearranging this expression causes you problems, you can learn a simple triangular arrangement which does the whole thing for you.

1 mole of copper(II) sulphate crystals, $CuSO_4 \cdot 5H_2O$

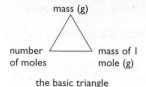

the basic triangle

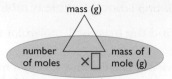

mass (g) = number of moles × mass of 1 mole (g)

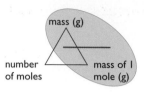

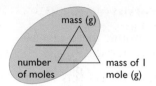

$$\text{number of moles} = \frac{\text{mass (g)}}{\text{mass of 1 mole (g)}}$$

$$\text{mass of 1 mole (g)} = \frac{\text{mass (g)}}{\text{number of moles}}$$

Look at this carefully and make sure that you understand how you can use it to work out the three equations that you might need.

Finding the mass of 0.2 moles of calcium carbonate, CaCO$_3$

Relative atomic masses: C = 12; O = 16; Ca = 40

First find the relative formula mass of calcium carbonate.

RFM of CaCO$_3$ = 40 + 12 + (3 × 16)
= 100

1 mol of CaCO$_3$ weighs 100 g.

mass (g) = no. of moles × mass of 1 mole (g)
= 0.2 × 100 g
= 20 g

0.2 moles of CaCO$_3$ has a mass of 20 g.

Finding the number of moles in 54 g of water, H$_2$O

Relative atomic masses: H = 1; O = 16

1 mol of H$_2$O weighs 18 g.

$$\text{number of moles} = \frac{\text{mass (g)}}{\text{mass of 1 mole (g)}}$$

$$= \frac{54}{18}$$

$$= 3 \text{ mol}$$

54 g of water is 3 moles.

Moles and the Avogadro Constant

Suppose you had 1 mole of ^{12}C. It would have a mass of 12 g, and contain a huge number of carbon atoms – in fact, about 6×10^{23} carbon atoms – that's 6 followed by 23 noughts. This number of atoms in 12 g of ^{12}C is called the Avogadro Constant.

1 mole of anything else contains this same number of particles. For example:

- 1 mole of magnesium contains 6×10^{23} magnesium atoms, Mg, and has a mass of 24 g.
- 1 mole of water contains 6×10^{23} water molecules, H_2O, and has a mass of 18 g.
- 1 mole of sodium chloride contains 6×10^{23} formula units, NaCl, and has a mass of 58.5 g. (You can't say "molecules of NaCl" because sodium chloride is ionic.)
- 1 mole of oxide ions contains 6×10^{23} O^{2-} ions and has a mass of 16 g.
- 1 mole of electrons contains 6×10^{23} electrons.

1×10^{32} water molecules go over Niagara Falls every second during the summer. That's 100 million million million million million water molecules per second.

Using Moles to Find Formulae

Interpreting symbols in terms of moles

Assume that you know the formula for something like copper(II) oxide, for example. The formula is CuO.

When you are doing sums, it is often useful to interpret a symbol as meaning more than just "an atom of copper" or "an atom of oxygen". For calculation purposes we take the symbol Cu to mean **1 mole of copper atoms**.

In a formula, "Cu" means "64 g of copper". "O" means "16 g of oxygen". (RAMs: O = 16; Cu = 64.) So in copper(II) oxide, the copper and oxygen are combined in the ratio of 64 g of copper to 16 g of oxygen.

You can read a formula like H_2O as meaning that 2 moles of hydrogen atoms are combined with 1 mole of oxygen atoms. In other words, 2 g of hydrogen are combined with 16 g of oxygen. (RAMs: H = 1; O = 16)

Working out formulae

The formula for magnesium oxide

Suppose you did an experiment to find out how much magnesium and oxygen reacted together to form magnesium oxide. Suppose 2.4 g of magnesium combined with 1.6 g of oxygen. You can use these figures to find the formula of magnesium oxide. (RAMs: O = 16; Mg = 24)

	Mg		**O**
Combining masses:	2.4 g		1.6 g
No. of moles of atoms:	$\dfrac{2.4}{24}$		$\dfrac{1.6}{16}$
	= 0.10		= 0.10
Ratio of moles:	1	:	1
Simplest formula:		MgO	

> **Remember:** Number of moles is mass in grams divided by the mass of 1 mole in grams.

This simplest formula is called the **empirical formula**. The empirical formula just tells you the **ratio** of the various atoms. Without more information it isn't possible to work out the "true" or "molecular" formula which could be Mg_2O_2, Mg_3O_3, etc. For ionic substances, the formula quoted is always the empirical formula.

Crystals of the copper-containing mineral cuprite, Cu_2O.

The formula for red copper oxide

You might get the data in a more complicated form. This example is about a less common form of copper oxide. If hydrogen is passed over the hot oxide, it is reduced to metallic copper. These figures could be obtained using the apparatus shown on page 59.

Mass of empty tube	= 52.2 g
Mass of tube + copper oxide (before experiment)	= 66.6 g
Mass of tube + copper (after experiment)	= 65.0 g

The tube loses mass because oxygen has been removed from the copper oxide, leaving metallic copper.

Mass of oxygen	= 66.6 – 65.0 g	= 1.6 g
Mass of copper	= 65.0 – 52.2 g	= 12.8 g

Now you have all the information to find the empirical formula. (RAMs: O = 16; Cu = 64)

	Cu	**O**
Combining masses:	12.8 g	1.6 g
No. of moles of atoms:	$\dfrac{12.8}{64}$	$\dfrac{1.6}{16}$
	= 0.20	= 0.10
Ratio of moles:	2 :	1
Simplest (empirical) formula:	Cu_2O	

Working out formulae using percentage composition figures

Often, figures for the compound are given as percentages by mass. For example: Find the empirical formula of a compound containing 85.7% C, 14.3% H by mass. (RAMs: H = 1; C = 12)

This isn't a problem! Those percentage figures apply to any amount of substance you choose – so choose 100 g. In which case, the percentages convert simply into masses. 85.7% of 100 g is 85.7 g.

	C	**H**
Given percentages	85.7%	14.3%
Combining masses in 100 g	85.7 g	14.3 g
No. of moles of atoms:	$\dfrac{85.7}{12}$	$\dfrac{14.3}{1}$
	= 7.14	= 14.3
Ratio of moles:	1 :	2
Simplest (empirical) formula:	CH_2	

> Usually the ratio will be fairly obvious, but if you can't spot it at once, divide everything by the smallest number and see if that helps. Sometimes, you may find that a ratio comes out as, for example, 1:1.5. In that case, all you have to do is double both numbers to give the true ratio of 2:3.

Converting empirical formulae into molecular formulae

In the example we have just looked at, CH_2 can't possibly be the real formula of the hydrocarbon – the carbon would have spare unbonded electrons. The **molecular formula** (the true formula) would have to be some multiple of CH_2, like C_2H_4 or C_3H_6 or whatever – as long as the ratio is still 1 carbon to 2 hydrogens.

You could find the molecular formula if you knew the relative formula mass of the compound (or the mass of 1 mole – which is just the RFM expressed in grams).

In the previous example, suppose you knew that the relative formula mass of the compound was 56.

CH_2 has a relative formula mass of only 14. (RAMs: H = 1; C = 12)

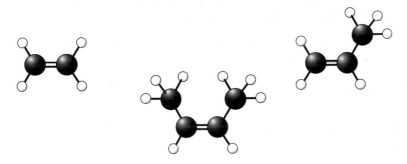

All of these have an empirical formula CH_2.

All you need to find out is how many times 14 goes into 56.

$\frac{56}{14}$ = 4 and so you need 4 lots of CH_2 – in other words, C_4H_8.

The molecular formula is C_4H_8.

> Even having got the molecular formula, you still don't know the structure. There are several isomers of C_4H_8. Isomerism is discussed on pages 227–229.

Empirical formula calculations involving water of crystallisation

When some substances crystallise from solution, water becomes chemically bound up with the salt. This is called **water of crystallisation**. The salt is said to be **hydrated**. Examples include $CuSO_4 \cdot 5H_2O$ and $MgCl_2 \cdot 6H_2O$.

Finding the "n" in $BaCl_2 \cdot nH_2O$

When you heat a salt which contains water of crystallisation, the water is driven off, leaving the anhydrous salt behind. Hydrated barium chloride is a commonly used example because the barium chloride itself doesn't decompose even on quite strong heating.

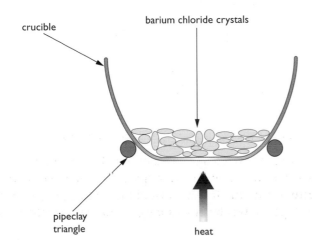

crucible

barium chloride crystals

pipeclay
triangle

heat

If you heated barium chloride crystals in a crucible you might end up with these results:

Mass of crucible $= 30.00$ g

Mass of crucible + barium chloride crystals, $BaCl_2 \cdot nH_2O$ $= 32.44$ g

Mass of crucible + anhydrous barium chloride, $BaCl_2$ $= 32.08$ g

To find "n" you need to find the ratio of the number of moles of $BaCl_2$ to the number of moles of water. It's just another empirical formula sum. (RAMs: H = 1; O = 16; Cl = 35.5; Ba = 137)

Mass of $BaCl_2$ $= 32.08 - 30.00$ g $= 2.08$ g

Mass of water $= 32.44 - 32.08$ g $= 0.36$ g

	$BaCl_2$	H_2O
Combining masses:	2.08 g	0.36 g
No. of moles:	$\dfrac{2.08}{208}$	$\dfrac{0.36}{18}$
	$= 0.01$	$= 0.02$
Ratio of moles:	1 :	2
Empirical formula:	$BaCl_2 \cdot 2H_2O$	

208 is the RFM of $BaCl_2$. 18 is the RFM of water. Check them if you aren't sure.

End of Chapter Checklist

If you haven't got a copy of your specification, read the introduction on page vi.

You will need to be able to do some or all of the following. Check your Awarding Body's specification (syllabus) to find out exactly what you need to know.

● Define relative atomic mass.

● Calculate the relative atomic mass of an element from information about the relative abundances of its isotopes.

● Calculate relative formula mass and use it to find percentage composition.

● Know how to work out the mass of a mole of a substance.

● Understand what is meant by the Avogadro Constant.

● Convert from moles to mass and vice versa.

● Calculate empirical and molecular formulae.

● Calculate the "n" in a formula of the type "salt·nH_2O".

Questions

More calculation questions can be found at the end of Section F on page 294.

1   Calculate the relative atomic mass of gallium given the percentage abundances:

   ^{69}Ga – 60.2%, ^{71}Ga – 39.8%.

2   Calculate the relative atomic mass of silicon given:

Relative isotopic mass	Relative abundance
28	100
29	5.10
30	3.36

3   a)   Define relative atomic mass.

   b)   Calculate the relative atomic masses of copper and sulphur from the percentage abundances of their isotopes. Use your answers to find the relative formula mass of copper(II) sulphide, CuS.

   ^{63}Cu – 69.1%; ^{65}Cu – 30.9%.
   ^{32}S – 95.0%; ^{33}S – 0.76%; ^{34}S – 4.22%; ^{36}S – 0.020%.

4   Calculate the relative formula masses of the following compounds.

   (RAMs: H = 1; C = 12; N = 14; O = 16; Na = 23; S = 32; Ca = 40; Cr = 52; Fe = 56)

   a) CO_2

   b) $(NH_4)_2SO_4$

   c) $Na_2CO_3 \cdot 10H_2O$

   d) $Cr_2(SO_4)_3$

   e) $(NH_4)_2SO_4 \cdot FeSO_4 \cdot 6H_2O$

5   Find the percentage of the named substance in each of the following examples.

   (RAMs: H = 1; C = 12; O = 16; Mg = 24; S = 32)

   a)   Carbon in propane, C_3H_8.

   b)   Water in magnesium sulphate crystals, $MgSO_4 \cdot 7H_2O$.

6   Work out the percentage of nitrogen in each of the following substances (all used as nitrogen fertilisers).

   (RAMs: H = 1; C = 12; N = 14; O = 16; S = 32; K = 39)

   a)   urea, $CO(NH_2)_2$

   b)   potassium nitrate, KNO_3

   c)   ammonium nitrate, NH_4NO_3

   d)   ammonium sulphate, $(NH_4)_2SO_4$.

7   Work out the mass of the following:

   (RAMs: H = 1; C = 12; N = 14; O = 16; Na = 23; Pb = 207)

   a)   1 mole of lead(II) nitrate, $Pb(NO_3)_2$

   b)   4.30 moles of methane, CH_4

   c)   0.24 moles of sodium carbonate crystals, $Na_2CO_3 \cdot 10H_2O$.

8 How many moles are represented by each of the following:

(RAMs: H = 1; O = 16; S = 32; Fe = 56; Cu = 64)

a) 50 g of copper(II) sulphate crystals, $CuSO_4 \cdot 5H_2O$

b) 1 tonne of iron, Fe (1 tonne is 1000 kg)

c) 0.032 g of sulphur dioxide, SO_2.

9 Some more questions about converting between moles and grams of a substance:

(RAMs: H = 1; O = 16; Na = 23; Cl = 35.5; Ca = 40; Cu = 64)

a) What is the mass of 4 mol of sodium chloride, NaCl?

b) How many moles is 37 g of calcium hydroxide, $Ca(OH)_2$?

c) How many moles is 1 kg (1000 g) of calcium, Ca?

d) What is the mass of 0.125 mol of copper(II) oxide, CuO?

e) 0.1 mol of a substance weighed 4 g. What is the weight of 1 mole?

f) 0.004 mol of a substance weighed 1 g. What is the relative formula mass of the compound?

10 (RAMs: H = 1; C = 12; N = 14; O = 16; Na = 23; S = 32; K = 39; Br = 80)

Find the empirical formulae of the following compounds which contained:

a) 5.85 g K; 2.10 g N; 4.80 g O

b) 3.22 g Na; 4.48 g S; 3.36 g O

c) 22.0% C; 4.6% H; 73.4% Br (by mass).

11 1.24 g of phosphorus was burnt completely in oxygen to give 2.84 g of phosphorus oxide. Find a) the empirical formula of the oxide, b) the molecular formula of the oxide, given that 1 mole of the oxide weighs 284 g. (RAMs: O = 16; P = 31)

12 An organic compound contained C – 66.7%, H – 11.1%, O – 22.2% by mass. Its relative formula mass was 72. Find a) the empirical formula of the compound, b) the molecular formula of the compound. (RAMs: H = 1; C = 12; O = 16)

13 In an experiment to find the number of molecules of water of crystallisation in sodium sulphate crystals, $Na_2SO_4 \cdot nH_2O$, 3.22 g of sodium sulphate crystals were heated gently. When all the water of crystallisation had been driven off, 1.42 g of anhydrous sodium sulphate was left. Find the value for "n" in the formula. (RAMs: H = 1; O = 16; Na = 23; S = 32)

14 Gypsum is hydrated calcium sulphate, $CaSO_4 \cdot nH_2O$. A sample of gypsum was heated in a crucible until all the water of crystallisation had been driven off. The following results were obtained:

Mass of crucible = 37.34 g
Mass of crucible + gypsum, $CaSO_4 \cdot nH_2O$ = 45.94 g
Mass of crucible + anhydrous calcium sulphate, $CaSO_4$ = 44.14 g

Calculate the value of "n" in the formula $CaSO_4 \cdot nH_2O$.

(RAMs: H = 1; O = 16; S = 32; Ca = 40)

15 How many water molecules, H_2O, are there in 1 drop of water? Assume 1 drop of water is 0.05 cm^3, and that the density of water is 1 g cm^{-3}. (RAMs: H = 1; O = 16. The Avogadro Constant = 6×10^{23} mol^{-1})

Chapter 26: Calculations from Equations

This chapter shows how you can use moles to work out the amounts of substances taking part in chemical reactions. Before you start you must be confident about relative formula mass (RFM) and how to do simple sums with moles. If you aren't, read pages 263–266 first.

The RFM of $CaCO_3$ is 100, and the RFM of CaO is 56. Work them out!

Your maths may be good enough that you don't need to take all these steps to get to the answer. If you can do it more quickly, that's fine. You must, however, still show all your working.

When you have finished a chemistry calculation, the impression should nearly always be that there a lot of words with a few numbers scattered between them – *not* vice versa.

Galena, PbS.

Calculations Involving Only Masses

Typical calculations will give you a mass of starting material and ask you to calculate how much product you are likely to get. You will also meet examples done the other way around, where you are told the mass of the product and are asked to find out how much of the starting material you would need. In almost all the cases you will meet at GCSE, you will be given the equation for the reaction.

A problem involving heating limestone

When limestone, $CaCO_3$, is heated, calcium oxide is formed. Suppose you wanted to calculate the mass of calcium oxide produced by heating 25 g of limestone. (Relative atomic masses: C = 12; O = 16; Ca = 40)

The calculation

First write the equation:

$$CaCO_{3(s)} \rightarrow CaO_{(s)} + CO_{2(g)}$$

Interpret the equation in terms of moles:

Remember that each formula represents 1 mole of that substance.

1 mol $CaCO_3$ produces 1 mol CaO (and 1 mol CO_2)

Substitute masses where relevant:

100 g (1 mol) $CaCO_3$ produces 56 g (1 mol) CaO

(Notice that we haven't calculated the mass of carbon dioxide. In this question you aren't interested in it, and so working it out is just a waste of time – and potentially confusing.)

Do the simple proportion sum:

If   100 g of calcium carbonate gives 56 g of calcium oxide

1 g of calcium carbonate gives $\frac{56}{100}$ g of calcium oxide = 0.56 g

25 g of calcium carbonate gives 25 × 0.56 g of calcium oxide
= 14 g of calcium oxide

A problem involving the manufacture of lead

Lead is extracted from galena, PbS. The ore is roasted in air to produce lead(II) oxide, PbO.

$$2PbS_{(s)} + 3O_{2(g)} \rightarrow 2PbO_{(s)} + 3SO_{2(g)}$$

The lead(II) oxide is reduced to lead by heating it with carbon in a blast furnace.

$$PbO_{(s)} + C_{(s)} \rightarrow Pb_{(l)} + CO_{(g)}$$

The molten lead is tapped from the bottom of the furnace.

Calculate:

(a) The mass of sulphur dioxide produced when 1 tonne of galena is roasted.

(b) The mass of lead that would eventually be produced from that 1 tonne of galena.

(Relative atomic masses: O = 16; S = 32; Pb = 207)

Calculation part (a)

First write the equation:

$$2PbS_{(s)} + 3O_{2(g)} \rightarrow 2PbO_{(s)} + 3SO_{2(g)}$$

Interpret the equation in terms of moles:

2 mol PbS produces 3 mol SO_2 (the others aren't important for this calculation)

Substitute masses where relevant:

2 × 239 g PbS produces 3 × 64 g SO_2
478 g PbS produces 192 g SO_2

Now there seems to be a problem. The question is asking about tonnes and not grams. You could work out how many grams there are in a tonne and then do hard sums with large numbers. However, it's much easier to think a bit, and realise that the ratio is always going to be the same whatever the units – so that...

if   478 g PbS produces 192 g SO_2
then   478 tonnes PbS produces 192 tonnes SO_2

Do the simple proportion sum:

If   478 tonnes PbS produces 192 tonnes SO_2

1 tonne PbS gives $\dfrac{192}{478}$ tonnes SO_2 = 0.402 tonnes SO_2

Calculation part (b)

First write the equation:

This time there are two equations to think about:

$$2PbS_{(s)} + 3O_{2(g)} \rightarrow 2PbO_{(s)} + 3SO_{2(g)}$$
$$PbO_{(s)} + C_{(s)} \rightarrow Pb_{(l)} + CO_{(g)}$$

Interpret the equation in terms of moles:

Trace the lead through the equations:

2 mol PbS produces 2 mol PbO
2 mol PbO produces 2 mol Pb

We've doubled the second equation so that we can trace what happens to all the 2PbO from the first one.

In other words, every 2 moles of PbS produces 2 moles of lead.

Substitute masses where relevant:

In this case, the relevant masses are only the PbS and the final lead.

2 × 239 g PbS produces 2 × 207 g Pb
478 g PbS produces 414 g Pb
So: 478 tonnes PbS produces 414 tonnes Pb

Lead ingots.

If you need the practice, work out the RFMs for yourself.

You could well save a bit of time here by realising that if 2 mol PbS produces 2 mol Pb, that's exactly the same as 1 mol PbS producing 1 mol Pb. That would save you having to multiply two numbers by 2. On the other hand, you would have to think!

Do the simple proportion sum:

If   478 tonnes PbS produces 414 tonnes Pb

      1 tonne PbS gives $\frac{414}{478}$ tonnes Pb = 0.866 tonnes

0.866 tonnes of lead are produced from 1 tonne of galena.

Calculations Involving Gas Volumes

Units of volume

Volumes (of gases or liquids) are measured in: cubic centimetres (cm^3)

                                               or   cubic decimetres (dm^3)

                                               or   litres (l)

I litre = I dm^3 = I000 cm^3

Avogadro's Law

> If you want to talk about 1000 cm^3, the cubic decimetre is the preferred unit rather than the litre, but it doesn't actually matter.

Equal volumes of gases at the same temperature and pressure contain equal numbers of molecules.

This means that if you have 100 cm^3 of hydrogen at some temperature and pressure, it contains exactly the same number of molecules as there are in 100 cm^3 of carbon dioxide or any other gas under those conditions – irrespective of the size of the molecules.

Avogadro's Law can be used for some very simple calculations involving gases.

Calculating the volume of oxygen needed to burn 100 cm^3 of methane

Methane (natural gas) burns in air according to the equation:

$$CH_{4(g)} + 2O_{2(g)} \rightarrow CO_{2(g)} + 2H_2O_{(l)}$$

The equation says that you need twice as many molecules of oxygen as you do of methane. According to Avogadro's Law, this means that you will need twice the volume of oxygen as you have of methane.

So, if you have to burn 100 cm^3 of methane, you will need 200 cm^3 of oxygen.

Similarly, for every molecule of methane you burn, you will get a molecule of carbon dioxide formed. Therefore every 100 cm^3 of methane will give 100 cm^3 of carbon dioxide, because 100 cm^3 of methane and 100 cm^3 of CO_2 contain the same number of molecules.

Avogadro's Law only applies to gases. If the water is formed as a liquid, you can't say anything about the volume of water produced. However, if it was formed as steam, then you could say that every 100 cm^3 of methane will produce 200 cm^3 of steam.

Calculating the volume of air needed to burn 100 cm^3 of methane

When you do a sum of this sort, be careful to notice whether you are being asked about air or oxygen. Air is only approximately one fifth oxygen – and so you need five times more air than you would need oxygen.

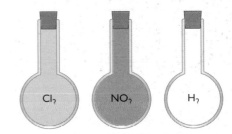

Three identical flasks containing different gases at the same temperature and pressure all contain equal numbers of molecules.

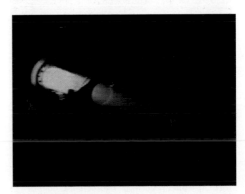

Burning methane in air.

In this example, if you need 200 cm³ of oxygen, you will need five times as much air – in other words, 1000 cm³.

The molar volume of a gas

1 mole of any gas contains the same number of molecules and so occupies the same volume as 1 mole of any other gas at the same temperature and pressure.

The volume occupied by 1 mole of a gas is called the **molar volume**. At room temperature and pressure, the molar volume is approximately 24 dm³ (24 000 cm³).

1 mole of any gas occupies 24 dm³ (24 000 cm³) at rtp.

Simple calculations with the molar volume

Calculating the volume of a given mass of gas

Calculate the volume of 0.01 g of hydrogen at rtp.

(Relative atomic mass: H = 1)

> 1 mol H_2 has a mass of 2 g
>
> 0.01 g of hydrogen is $\dfrac{0.01}{2}$ moles = 0.005 moles
>
> 1 mole of hydrogen occupies 24 000 cm³
> 0.005 mol of hydrogen occupies 0.005 × 24 000 cm³
> = 120 cm³

Calculating the mass of a given volume of gas

The Hindenburg disaster. The skin of the airship and the hydrogen inside caught fire.

On 6 May 1937, the airship Hindenburg caught fire and 36 people died. Suppose an airship contained 180 000 cubic metres of hydrogen (a little bit less than the Hindenburg). We can calculate the mass of this hydrogen.

180 000 cubic metres is 180 000 000 dm³. Assuming the gas was at room temperature and pressure, each 24 dm³ represents one mole of hydrogen.

$$\text{Number of moles of } H_2 = \frac{180\,000\,000}{24} = 7\,500\,000 \text{ mol}$$

Each mole of hydrogen, H_2, weighs 2 g. (RAM: H = 1)

> Mass of hydrogen = 7 500 000 × 2 g
> = 15 000 000 g

If you were to convert that into more reasonable units (you wouldn't do that in an exam!), the mass of hydrogen is 15 tonnes. That's a lot of hydrogen!

Using the molar volume in calculations from equations

Working out the volume of gas produced during a reaction

The calculation shows that 1 g of limpet shells, $CaCO_3$, would react with hydrochloric acid to give 0.24 dm³ of carbon dioxide.

Calculate the volume of carbon dioxide produced at room temperature and pressure when an excess of dilute hydrochloric acid is added to 1.00 g of calcium carbonate. (You use an excess of acid to make sure that all the calcium carbonate reacts.)

(RAMs: C = 12; O = 16; Ca = 40. Molar volume = 24 dm³ at rtp)

First write the equation:

$$CaCO_{3(s)} + 2HCl_{(aq)} \rightarrow CaCl_{2(aq)} + CO_{2(g)} + H_2O_{(l)}$$

Interpret the equation in terms of moles:

1 mol $CaCO_3$ gives 1 mol CO_2

Substitute masses and volumes where appropriate:

100 g $CaCO_3$ gives 24 dm^3 CO_2 at rtp

Do the simple proportion sum:

If   100 g $CaCO_3$ gives 24 dm^3 CO_2 at rtp

then   1 g $CaCO_3$ gives $\frac{1}{100} \times 24$ dm^3

         = 0.24 dm^3

So 1 g of calcium carbonate gives 0.24 dm^3 of carbon dioxide.

A problem involving making hydrogen

What is the maximum mass of aluminium which you could add to an excess of dilute hydrochloric acid so that you produced no more than 100 cm^3 of hydrogen at room temperature and pressure?

(RAM: Al = 27. Molar volume = 24 000 cm^3 at rtp)

What you are being asked is what mass of aluminium will give 100 cm^3 of hydrogen at rtp.

First write the equation:

$$2Al_{(s)} + 6HCl_{(aq)} \rightarrow 2AlCl_{3(aq)} + 3H_{2(g)}$$

Interpret the equation in terms of moles:

2 mol Al gives 3 mol H_2

Substitute masses and volumes where appropriate:

2 × 27 g Al gives 3 × 24 000 cm^3 H_2
54 g Al gives 72 000 cm^3 H_2

Do the simple proportion sum:

If   72 000 cm^3 H_2 comes from 54 g Al

then   1 cm^3 H_2 comes from $\dfrac{54}{72\,000}$ g Al   = 0.000 75 g

and   100 cm^3 H_2 comes from 100 × 0.000 75 g Al   = 0.075 g Al

To get 100 cm^3 of hydrogen, you would need 0.075 g of aluminium.

An industrial example

Coal contains sulphur compounds. When these burn, sulphur dioxide is produced. To remove it from the waste gases from power stations, the sulphur dioxide is reacted with limestone (calcium carbonate) and air.

$$2CaCO_{3(s)} + 2SO_{2(g)} + O_{2(g)} \rightarrow 2CaSO_{4(s)} + 2CO_{2(g)}$$

The calcium sulphate produced can be used to make plaster board for building.

Important: The commonest mistake in a sum of this kind is to work out the **mass** of 1 mole of CO_2. Once you have that figure of 44 g, you feel you have to do something with it, and will probably work out the mass of CO_2 produced instead of the volume.

You would almost certainly be given this equation in an exam.

The flue gas desulphurisation plant at Drax power station uses limestone to remove sulphur dioxide from the waste gases.

If you don't like this, there is no reason why you can't do the sum with 2 moles of everything!

Unfortunately, this time you can't work directly in tonnes – you have to work in grams. 1 tonne is a million grams. 10 000 tonnes is 10 000 000 000 grams.

It would be much more sensible to do this in scientific notation! 10,000 tonnes is 1×10^{10} g.

The volume of SO_2 then turns out to be 2.4×10^9 dm^3.

The Drax power station uses 10 000 tonnes of crushed limestone every week. We are going to calculate what mass of calcium sulphate is produced and what volume of sulphur dioxide the limestone removes from the flue gases (assuming the sulphur dioxide is at rtp).

(RAMs: C = 12; O = 16; S = 32; Ca = 40. Molar volume = 24 dm^3 at rtp)

First write the equation:

$$2CaCO_{3(s)} + 2SO_{2(g)} + O_{2(g)} \rightarrow 2CaSO_{4(s)} + 2CO_{2(g)}$$

Interpret the equation in terms of moles:

2 mol $CaCO_3$ reacts with 2 mol SO_2 and produces 2 mol $CaSO_4$

That's exactly the same as saying that:

1 mol $CaCO_3$ reacts with 1 mol SO_2 and produces 1 mol $CaSO_4$

Substitute masses and volumes where appropriate:

100 g $CaCO_3$ reacts with 24 dm^3 SO_2 and produces 136 g $CaSO_4$

Do the simple proportion sum for the calcium sulphate:

If   100 g $CaCO_3$ produces 136 g $CaSO_4$

then   100 tonnes $CaCO_3$ produces 136 tonnes $CaSO_4$

and   1 tonne $CaCO_3$ produces 1.36 tonnes $CaSO_4$

So   10 000 tonnes $CaCO_3$ produces 10 000 × 1.36 tonnes $CaSO_4$
          = 13 600 tonnes

10 000 tonnes of calcium carbonate produces 13 600 tonnes of $CaSO_4$.

Do the simple proportion sum for the SO_2:

If   100 g $CaCO_3$ reacts with 24 dm^3 SO_2

then   1 g $CaCO_3$ reacts with 0.24 dm^3 SO_2

and   10 000 000 000 g $CaCO_3$ reacts with 10 000 000 000 × 0.24 dm^3 SO_2
          = 2 400 000 000 dm^3 SO_2

10 000 tonnes of calcium carbonate reacts with 2 400 000 000 dm^3 SO_2.

End of Chapter Checklist

If you haven't got a copy of your specification, read the introduction on page vi.

You will need to be able to do some or all of the following. Check your Awarding Body's specification (syllabus) to find out exactly what you need to know.

- Calculate masses of reactants or products from given equations.
- State Avogadro's Law and use it to calculate volumes of reacting gases.
- Understand what is meant by the molar volume.
- Use the molar volume to convert between volume and mass for a particular gas.
- Carry out calculations from equations involving masses of solids and volumes of gas.

Questions

More calculation questions can be found at the end of Section F on page 294.

1 Titanium is manufactured by heating titanium(IV) chloride with sodium.

$$TiCl_{4(g)} + 4Na_{(l)} \rightarrow Ti_{(s)} + 4NaCl_{(s)}$$

What mass of sodium is required to produce 1 tonne of titanium? (RAMs: Na = 23; Ti = 48)

2 2.67 g of aluminium chloride was dissolved in water and an excess of silver nitrate solution was added to give a precipitate of silver chloride.

$$AlCl_{3(aq)} + 3AgNO_{3(aq)} \rightarrow Al(NO_3)_{3(aq)} + 3AgCl_{(s)}$$

What mass of silver chloride precipitate would be formed? (RAMs: Al = 27; Cl = 35.5; Ag = 108)

3 Calcium hydroxide is manufactured by heating calcium carbonate strongly to produce calcium oxide, and then adding a controlled amount of water to produce calcium hydroxide.

$$CaCO_{3(s)} \rightarrow CaO_{(s)} + CO_{2(g)}$$
$$CaO_{(s)} + H_2O_{(l)} \rightarrow Ca(OH)_{2(s)}$$

a) What mass of calcium oxide would you produce from 1 tonne of calcium carbonate?

b) What mass of water would you need to add to that calcium oxide?

c) What mass of calcium hydroxide would you eventually produce? (RAMs: H = 1; C = 12; O = 16; Ca = 40)

4 Copper(II) sulphate crystals, $CuSO_4 \cdot 5H_2O$, can be made by heating copper(II) oxide with dilute sulphuric acid and then crystallising the solution formed. Calculate the maximum mass of crystals that could be made from 4.00 g of copper(II) oxide using an excess of sulphuric acid.

$$CuO_{(s)} + H_2SO_{4(aq)} \rightarrow CuSO_{4(aq)} + H_2O_{(l)}$$
$$CuSO_{4(aq)} + 5H_2O_{(l)} \rightarrow CuSO_4 \cdot 5H_2O_{(s)}$$

(RAMs: H = 1; O = 16; S = 32; Cu = 64)

5 Chromium is manufactured by heating a mixture of chromium(III) oxide with aluminium powder.

$$Cr_2O_{3(s)} + 2Al_{(s)} \rightarrow 2Cr_{(s)} + Al_2O_{3(s)}$$

a) Calculate the mass of aluminium needed to react with 1 tonne of chromium(III) oxide. (RAMs: O = 16; Al = 27; Cr = 52)

b) Calculate the mass of chromium produced from 1 tonne of chromium(III) oxide.

6 If the mineral pyrite, FeS_2, is heated strongly in air, iron(III) oxide and sulphur dioxide are produced. What mass of **a)** iron(III) oxide, and **b)** sulphur dioxide could be made by heating 1 tonne of an ore which contained 50% by mass of pyrite? (RAMs: O = 16; S = 32; Fe = 56)

$$4FeS_{2(s)} + 11O_{2(g)} \rightarrow 2Fe_2O_{3(s)} + 8SO_{2(g)}$$

7 Carbon monoxide burns according to the equation:

$$2CO_{(g)} + O_{2(g)} \rightarrow 2CO_{2(g)}$$

a) Calculate the volume of oxygen needed for the complete combustion of 100 cm³ of carbon monoxide?

b) What volume of carbon dioxide will be formed?

8 What volume of air is needed for the complete combustion of 1 dm³ of propane?

$$C_3H_{8(g)} + 5O_{2(g)} \rightarrow 3CO_{2(g)} + 4H_2O_{(l)}$$

9 Take the molar volume to be 24 dm³ (24 000 cm³) at rtp.

a) Calculate the mass of 200 cm³ of chlorine gas (Cl_2) at rtp. (RAM: Cl = 35.5)

b) Calculate the volume occupied by 0.16 g of oxygen (O_2) at rtp. (RAM: O = 16)

c) If 1 dm³ of a gas at rtp weighs 1.42 g, calculate the mass of 1 mole of the gas.

279

10 Calculate the volume of hydrogen (measured at room temperature and pressure) obtainable by reacting 0.240 g of magnesium with an excess of dilute sulphuric acid. (RAM: Mg = 24. Molar volume = 24 000 cm^3 at rtp)

$$Mg_{(s)} + H_2SO_{4(aq)} \rightarrow MgSO_{4(aq)} + H_{2(g)}$$

11 What mass of potassium nitrate would you have to heat in order to produce 1.00 dm^3 of oxygen at rtp? (RAMs: N = 14; O = 16; K = 39. Molar volume = 24 dm^3 at rtp)

$$2KNO_{3(s)} \rightarrow 2KNO_{2(s)} + O_{2(g)}$$

12 Chlorine can be prepared by heating manganese(IV) oxide with an excess of concentrated hydrochloric acid. What is the maximum volume of chlorine (measured at room temperature and pressure) that could be obtained from 2.00 g of manganese(IV) oxide? (RAMs: O = 16; Mn = 55. Molar volume = 24 000 cm^3 at rtp)

$$MnO_{2(s)} + 4HCl_{(aq)} \rightarrow MnCl_{2(aq)} + Cl_{2(g)} + 2H_2O_{(l)}$$

13 Sodium sulphite, Na_2SO_3, oxidises very slowly in the air to form sodium sulphate, Na_2SO_4. 1.000 g of an old sample of sodium sulphite was analysed to find out how much had been oxidised to sodium sulphate.

The entire sample was dissolved in water and acidified with dilute hydrochloric acid. The remaining sodium sulphite reacts to form sodium chloride and takes no further part in the reaction.

$$Na_2SO_{3(s)} + 2HCl_{(aq)} \rightarrow 2NaCl_{(aq)} + SO_{2(g)} + H_2O_{(l)}$$

An excess of barium chloride solution was added to the resulting solution. A white precipitate of barium sulphate was produced.

$$Na_2SO_{4(aq)} + BaCl_{2(aq)} \rightarrow BaSO_{4(s)} + 2NaCl_{(aq)}$$

This was separated, dried and weighed. It was found to have a mass of 0.328 g.

(RAMs: O = 16; Na = 23; S = 32; Ba = 137)

a) Calculate the number of moles of barium sulphate formed.

b) How many moles of sodium sulphate were there in the mixture of sodium sulphate and sodium sulphite?

c) Calculate the mass of sodium sulphate present in the mixture.

d) What percentage by mass of sodium sulphite was left in the mixture?

Chapter 27: Electrolysis Calculations

> The calculations in this chapter are similar to those in Chapter 26, but with an added twist because of the electrons involved in electrolysis equations. It is important that you are confident about ordinary calculations from equations before you go on.

How to Interpret Electrode Equations

Moles of electrons

Magnesium is manufactured by electrolysing molten magnesium chloride. Magnesium is produced at the cathode (the negative electrode) and chlorine at the anode (the positive electrode). The electrode equations are:

$$Mg^{2+}_{(l)} + 2e^- \rightarrow Mg_{(l)}$$

$$2Cl^-_{(l)} \rightarrow Cl_{2(g)} + 2e^-$$

In terms of moles, you can say:

- 1 mole of Mg^{2+} ions gains 2 moles of electrons and produces 1 mole of magnesium, Mg.

- 2 moles of Cl^- ions form 1 mole of chlorine, Cl_2, and release 2 moles of electrons.

When you are doing calculations, you just read e^- as "**1 mole of electrons**".

Quantities of electricity

Coulombs

The coulomb is a measure of quantity of electricity. 1 coulomb is the quantity of electricity which passes if 1 ampere (amp) flows for 1 second.

> **number of coulombs = current in amps × time in sec**

So, if 2 amps flows for 20 minutes, you can calculate the quantity of electricity (not forgetting to convert the time into seconds) as:

Quantity of electricity $= 2 \times 20 \times 60$ coulombs
$= 2400$ coulombs

> You may find this written in symbols using **Q** for the quantity of electricity, **I** for the current in amps, and **t** for the time in seconds. **Q = I × t.**

The Faraday Constant

A flow of electricity is a flow of electrons. **1 faraday** is the quantity of electricity which represents 1 mole of electrons passing a particular point in the circuit – in other words, approximately 6×10^{23} electrons.

> 1 faraday is more accurately quoted as 96 500 coulombs. The value given is that required by OCR – the only examiners to want calculations involving the faraday.

> **1 faraday = 96 000 coulombs**

Interpreting electrode equations

In electrolysis calculations you are usually only interested in the quantity of electricity and the mass or volume of the product. For example:

$$Na^+_{(l)} + e^- \rightarrow Na_{(l)}$$

1 mole of sodium, Na, is produced by the flow of 1 mole of electrons (= 1 faraday).

$$Cu^{2+}_{(aq)} + 2e^- \rightarrow Cu_{(s)}$$

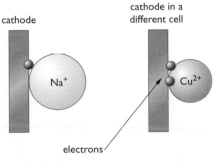

Each sodium ion needs 1 electron from the cathode to neutralise its charge. Each copper ion needs twice that number.

1 mole of copper, Cu, is produced by the flow of 2 moles of electrons (= 2 faradays). It takes twice as much electricity to produce a mole of copper as it does a mole of sodium. That's because the Cu^{2+} ion carries twice the charge, and needs twice as many electrons to neutralise it.

$$2Cl^-_{(l)} \rightarrow Cl_{2(g)} + 2e^-$$

1 mole of chlorine, Cl_2, is produced when 2 moles of electrons (= 2 faradays) flow around the circuit.

When 1 mole of chlorine, Cl_2, is produced at the anode, 2 moles of electrons are released. This means that the production of 1 mole of chlorine takes place when 2 moles of electrons flow in the circuit.

Some Sample Calculations

Electrolysing copper(II) sulphate solution

What mass of copper is deposited on the cathode during the electrolysis of copper(II) sulphate solution if 0.15 amps flows for 10 minutes?

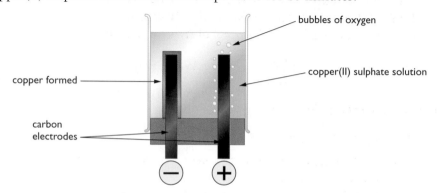

The electrode equation is:

$$Cu^{2+}_{(aq)} + 2e^- \rightarrow Cu_{(s)}$$

(RAM: Cu = 64. 1 faraday = 96 000 coulombs)

Start by working out the number of coulombs:

number of coulombs = amps × time in seconds
$$= 0.15 \times 10 \times 60$$
$$= 90$$

The 60 converts minutes into seconds.

Now work from the equation:

$$Cu^{2+}_{(aq)} + 2e^- \rightarrow Cu_{(s)}$$

2 moles of electrons give 1 mole of copper, Cu

2 × 96 000 coulombs give 64 g of copper

192 000 coulombs give 64 g of copper

90 coulombs give $\dfrac{90}{192\,000} \times 64$ g = 0.03 g

If you aren't happy with the last line, work out what 1 coulomb produces by dividing 64 by 192 000, and then multiply by 90.

A calculation involving gases

During the electrolysis of dilute sulphuric acid using unreactive platinum electrodes, hydrogen is released at the cathode and oxygen at the anode. Calculate the volumes of hydrogen and oxygen produced (measured at room temperature and pressure) if 1.0 amp flows for 20 minutes. The electrode equations are:

$$2H^+_{(aq)} + 2e^- \rightarrow H_{2(g)}$$

$$4OH^-_{(aq)} \rightarrow 2H_2O_{(l)} + O_{2(g)} + 4e^-$$

(The molar volume of a gas = 24 000 cm^3 at rtp. 1 faraday = 96 000 coulombs)

Again start by working out the number of coulombs:

$$\text{number of coulombs} = \text{amps} \times \text{time in seconds}$$
$$= 1.0 \times 20 \times 60$$
$$= 1200$$

Calculating the volume of hydrogen

$$2H^+_{(aq)} + 2e^- \rightarrow H_{2(g)}$$

2 moles of electrons give 1 mole of hydrogen, H$_2$
2 × 96 000 coulombs give 24 000 cm^3 of hydrogen at rtp
192 000 coulombs give 24 000 cm^3 of hydrogen at rtp

1200 coulombs give $\dfrac{1200}{192\,000} \times 24\,000$ cm^3 = 150 cm^3

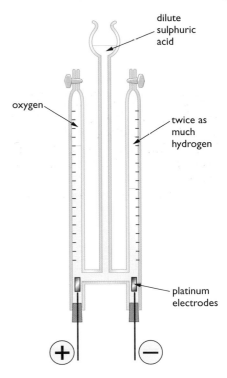

Apparatus for electrolysing dilute sulphuric acid and measuring the volume of gases produced.

Calculating the volume of oxygen

$$4OH^-_{(aq)} \rightarrow 2H_2O_{(l)} + O_{2(g)} + 4e^-$$

A flow of 4 moles of electrons produces 1 mole of oxygen, O$_2$
4 × 96 000 coulombs produces 24 000 cm^3 of oxygen
384 000 coulombs produces 24 000 cm^3 of oxygen

1200 coulombs produces $\dfrac{1200}{384\,000} \times 24\,000$ cm^3 = 75 cm^3

Therefore, 150 cm^3 of hydrogen and 75 cm^3 of oxygen are produced.

A reversed calculation

How long would it take to deposit 0.500 g of silver on the cathode during the electrolysis of silver(I) nitrate solution using a current of 0.250 amp? The cathode equation is:

$$Ag^+_{(aq)} + e^- \rightarrow Ag_{(s)}$$

(RAM: Ag = 108. 1 faraday = 96 000 coulombs)

1 mole of electrons gives 1 mole of silver, Ag
96 000 coulombs give 108 g of silver

To produce 0.500 g of silver you would need $\dfrac{0.500}{108} \times 96\,000$ = 444.4 coulombs.

$$\text{number of coulombs} = \text{amps} \times \text{time in seconds}$$
$$444.4 = 0.250 \times t$$
$$t = \frac{444.4}{0.250}$$
$$= 1780 \text{ sec}$$

The time needed to deposit 0.500 g of silver is 1780 seconds.

If you need help: Divide 96 000 by 108 to find out how many coulombs you need for 1 g of silver. Then multiply by 0.5 to find how many you need for 0.5 g.

Electrolysing more than one solution

Suppose you have two solutions connected together in series, so that the same quantity of electricity flows through both.

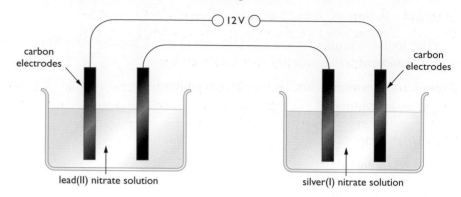

At the end of the electrolysis, it was found that 2.07 g of lead had been deposited on the cathode in the left-hand beaker.

(a) Calculate the quantity of electricity that passed during the experiment.

(b) Calculate the mass of silver that was deposited on the cathode in the right-hand beaker. (RAMs: Pb = 207; Ag = 108. 1 faraday = 96 000 coulombs)

Calculation part (a)

$$Pb^{2+}_{(aq)} + 2e^- \rightarrow Pb_{(s)}$$

2 moles of electrons give 1 mole of lead, Pb

2 × 96 000 coulombs give 207 g of lead

If   192 000 coulombs give 207 g of lead

then   1920 coulombs give 2.07 g of lead

The quantity of electricity which passed = 1920 coulombs.

Calculation part (b)

If 1920 coulombs passed through the beaker containing the lead(II) nitrate solution, then exactly the same amount passed through the rest of the circuit.

$$Ag^+_{(aq)} + e^- \rightarrow Ag_{(s)}$$

1 mole of electrons give 1 mole of silver, Ag
96 000 coulombs give 108 g of silver

1920 coulombs give $\dfrac{1920}{96\,000} \times 108$ g = 2.16 g

The mass of silver deposited is 2.16 g.

An alternative way of solving part (b)

If you weren't asked to find the quantity of electricity in part **(a)**, you could do part **(b)** much more easily without knowing anything at all about the Faraday Constant or even about coulombs.

Look again at the equations:

$$Pb^{2+}_{(aq)} + 2e^- \rightarrow Pb_{(s)}$$
$$Ag^+_{(aq)} + e^- \rightarrow Ag_{(s)}$$

How do you know which method to use? In an exam, look at the information you are given. If you are given a value for the Faraday Constant (96 000 coulombs), then you would be expected to use it. If you aren't given it, there must be another way of solving the problem.

If 2 moles of electrons flow, you will get 1 mole of lead and 2 moles of silver. However many electrons flow, you will always get twice as many moles of silver as of lead.

In this calculation, 2.07 g of lead were formed, which is 0.01 mol. You will therefore get 0.02 mol of silver = 0.02×108 g = 2.16 g.

A similar example involving just one solution

During the electrolysis of concentrated copper(II) chloride solution, 3.2 g of copper was deposited at the cathode. What volume of chlorine (measured at rtp) would be formed at the anode? (RAM: Cu = 64. Molar volume = $24\,000$ cm^3 at rtp)

> This time, you aren't given the Faraday Constant. There must be another way!

The electrode equations are:

$$Cu^{2+}_{(aq)} + 2e^- \rightarrow Cu_{(s)}$$
$$2Cl^-_{(aq)} \rightarrow Cl_{2(g)} + 2e^-$$

Notice that for every 2 moles of electrons that flow, you will get 1 mole of Cu and 1 mole of chlorine, Cl_2. You are bound to get the same number of moles of each.

In this case, 3.2 g of copper is $\dfrac{3.2}{64}$ mol = 0.05 mol.

So you will also get 0.05 mol of chlorine, which is $0.05 \times 24\,000$ cm^3 at rtp = 1200 cm^3.

End of Chapter Checklist

If you haven't got a copy of your specification, read the introduction on page vi.

You will need to be able to do some or all of the following. Check your Awarding Body's specification (syllabus) to find out exactly what you need to know.

- Calculate quantity of electricity in coulombs given the current and time.

- Understand that 1 mole of electrons is represented by 96 000 coulombs which is called the Faraday Constant.

- Perform calculations for electrolysis reactions involving the masses (or volumes) of products, the current and the time.

- Calculate the mass (or volume) of one product of electrolysis given the mass (or volume) of another one.

Questions

More calculation questions can be found at the end of Section F on page 294.

1 During the electrolysis of copper(II) sulphate solution using copper electrodes, copper from the anode goes into solution according to the equation:

$$Cu_{(s)} \rightarrow Cu^{2+}_{(aq)} + 2e^-$$

Calculate the loss in mass from the anode if a current of 0.50 amps flows for 1 hour.

(RAM: Cu = 64. 1 faraday = 96 000 coulombs)

2 During the electrolysis of lead(II) nitrate solution, lead is deposited at the cathode, and oxygen is released from the anode. If a current of 0.350 amp flows for 1000 sec, calculate **a)** the mass of lead deposited, **b)** the volume of oxygen (measured at room temperature and pressure) produced. (RAM: Pb = 207. The molar volume of a gas is 24 000 cm^3 at rtp. 1 faraday = 96 000 coulombs)

$$Pb^{2+}_{(aq)} + 2e^- \rightarrow Pb_{(s)}$$
$$4OH^-_{(aq)} \rightarrow 2H_2O_{(l)} + O_{2(g)} + 4e^-$$

3 Some copper(II) sulphate solution was electrolysed using a pure copper cathode and an impure copper anode. Copper is lost from the anode and deposited on the cathode. Insoluble impurities in the anode form a sludge underneath the anode.

Cathode equation: $Cu^{2+}_{(aq)} + 2e^- \rightarrow Cu_{(s)}$

Anode equation: $Cu_{(s)} \rightarrow Cu^{2+}_{(aq)} + 2e^-$

a) What mass of copper will be deposited on the cathode if 0.40 amps flows for 75 minutes? (RAM: Cu = 64. 1 faraday = 96 000 coulombs)

b) If the anode was found to have lost 0.80 g during the experiment, calculate the percentage purity of the impure copper anode, assuming that only insoluble impurities were present.

4 Aluminium is manufactured by electrolysing a solution of aluminium oxide, Al_2O_3, in molten cryolite. The electrode equation is:

$$Al^{3+}_{(l)} + 3e^- \rightarrow Al_{(l)}$$

A typical cell produces 1 tonne (1000 kg) of aluminium every 24 hours. What current (in amps) is needed to produce this amount of aluminium? (RAM: Al = 27. 1 faraday = 96 000 coulombs)

5 Two solutions were electrolysed in series using the apparatus on page 284. One beaker contained chromium(III) sulphate solution, and the other cobalt(II) sulphate solution. 0.295 g of cobalt was deposited on the cathode in the beaker containing cobalt(II) sulphate. The electrode equations are:

$$Cr^{3+}_{(aq)} + 3e^- \rightarrow Cr_{(s)}$$
$$Co^{2+}_{(aq)} + 2e^- \rightarrow Co_{(s)}$$

Calculate **a)** the quantity of electricity which flowed during the experiment, and **b)** the mass of chromium deposited on the cathode in the other beaker. (RAMs: Cr = 52; Co = 59. 1 faraday = 96 000 coulombs)

6 Copper(II) sulphate solution and lead(II) nitrate solution were electrolysed in two beakers connected in series. If 0.64 g of copper was deposited at the cathode in one beaker, calculate the mass of lead deposited in the other one. (RAMs: Cu = 64; Pb = 207)

$$Cu^{2+}_{(aq)} + 2e^- \rightarrow Cu_{(s)}$$
$$Pb^{2+}_{(aq)} + 2e^- \rightarrow Pb_{(s)}$$

Section F: Sums

Chapter 28: Titration Calculations

In previous chapters we have looked at calculations from equations involving masses of solids and volumes of gases. Many reactions are done in solution, and this chapter looks at how you handle problems involving concentrations of solutions.

You may also find the symbol "M" used. For example, dilute hydrochloric acid might have a concentration quoted as 2M. "M" means "mol dm^{-3}" and is described as the **molarity** of the solution. You can also read "2M" as "2 molar". This is all unnecessarily confusing, and won't be used in this book.

Sea water contains about 0.6 moles of NaCl per cubic decimetre.

Working with Solution Concentrations

Concentrations of solutions

Concentrations can be measured in either

- g dm^{-3}
- mol dm^{-3}

These are exactly the same as writing g/dm^3 and mol/dm^3. You read them as "grams per cubic decimetre" and "moles per cubic decimetre". 1 cubic decimetre is the same as 1 litre.

You have to be able to convert between g dm^{-3} and mol dm^{-3}. This is no different from converting moles into grams and vice versa. When you are doing the conversions in concentration sums, the amount of substance you are talking about happens to be dissolved in 1 dm^3 of solution. That doesn't affect the sum in any way.

Remember:

$$\text{number of moles} = \frac{\text{mass (g)}}{\text{mass of 1 mole (g)}}$$

Converting from g dm^{-3} to mol dm^{-3}

A sample of sea water had a concentration of sodium chloride of 35.1 g dm^{-3}. Find its concentration in mol dm^{-3}. (RAMs: Na = 23; Cl = 35.5)

> 1 mol NaCl weighs 58.5 g
>
> 35.1 g is $\frac{35.1}{58.5}$ mol = 0.6 mol

The concentration of the NaCl is 0.6 mol dm^{-3}.

Converting from mol dm^{-3} to g dm^{-3}

What is the concentration of a 0.050 mol dm^{-3} solution of sodium carbonate, Na_2CO_3, in g dm^{-3}? (RAMs: C = 12; O = 16; Na = 23)

> 1 mol Na_2CO_3 weighs 106 g
> 0.050 mol weighs 0.050 × 106 g = 5.3 g
> 0.050 mol dm^{-3} is therefore 5.3 g dm^{-3}

Making it as tricky as possible!

What is the concentration in mol dm^{-3} of a solution containing 2.1 g of sodium hydrogencarbonate, $NaHCO_3$, in 250 cm^3 of solution? (RAMs: H = 1; C = 12; O = 16; Na = 23)

The problem here is that the volume is wrong. The solid is dissolved in 250 cm^3 instead of 1000 cm^3 (1 dm^3).

> 250 cm^3 is $\frac{1}{4}$ of 1000 cm^3 (1 dm^3)

Therefore a solution containing 2.1 g in 250 cm^3 has the same concentration as one containing 4 × 2.1 g in 1000 cm^3.

Chapter 28: Titration Calculations

287

4×2.1 g is 8.4 g

1 mol $NaHCO_3$ weighs 84 g

8.4 g is $\frac{8.4}{84}$ mol $= 0.10$ mol

The concentration is therefore 0.10 mol dm^{-3}.

Calculations from equations involving solutions

A calculation involving hard water

A sample of hard water contained $0.002 \text{ mol dm}^{-3}$ of calcium hydrogencarbonate, $Ca(HCO_3)_2$. When this is heated, it decomposes to make calcium carbonate. This forms as a white precipitate known as "limescale". Calculate the mass of calcium carbonate which could be formed when 100 dm^3 (100 litres) of the hard water is heated. (RAMs: C = 12; O = 16; Ca = 40)

$$Ca(HCO_3)_{2(aq)} \rightarrow CaCO_{3(s)} + H_2O_{(l)} + CO_{2(g)}$$

1 dm^3 of hard water contains 0.002 mol of $Ca(HCO_3)_2$

100 dm^3 contains 100×0.002 mol = 0.2 mol

The equation says that 1 mol of $Ca(HCO_3)_2$ gives 1 mol of $CaCO_3$

That means that 0.2 mol of $Ca(HCO_3)_2$ gives 0.2 mol of $CaCO_3$

1 mol $CaCO_3$ weighs 100 g

0.2 mol weighs 0.2×100 g = 20 g

20 g of calcium carbonate would be formed.

Another calculation involving hard water

Limescale can be removed from, for example, electric kettles by reacting it with a dilute acid such as the ethanoic acid present in vinegar.

$$CaCO_{3(s)} + 2CH_3COOH_{(aq)} \rightarrow (CH_3COO)_2Ca_{(aq)} + CO_{2(g)} + H_2O_{(l)}$$

What mass of calcium carbonate can be removed by 50 cm^3 of a solution containing 2 mol dm^{-3} of ethanoic acid? (RAMs: C = 12; O = 16; Ca = 40)

In any question of this sort, it is always a good policy to start by working out the number of moles of any substance where you know both the volume and the concentration. In this case, we know both of these for the ethanoic acid.

Number of moles of ethanoic acid $= \frac{50}{1000} \times 2 = 0.1$ mol

Now look at the equation. (Don't be scared by this equation! As long as you know that CH_3COOH is ethanoic acid, that's all you need worry about for this calculation.)

The equation says that 1 mol $CaCO_3$ reacts with 2 mol ethanoic acid

That means that however many moles of ethanoic acid there are in the reaction, there will only be half as many moles of calcium carbonate.

Number of moles of $CaCO_3 = \frac{0.1}{2} = 0.05$ mol

1 mol $CaCO_3$ weighs 100 g

0.05 mol weighs 0.05×100 g = 5 g

The ethanoic acid would react with 5 g of calcium carbonate.

Water becomes hard when carbon dioxide in rain water reacts with limestone.

If you aren't happy with this, put in an extra step. If there are 2 moles in 1000 cm^3 (because it is 2 mol dm^{-3}), work out how many there are in 1 cm^3 by dividing by 1000. That gives 0.002 mol. Multiply that by 50 to find out how many moles there are in 50 cm^3.

Calculations from Titrations

A reminder about acid–alkali titrations

A solution of the alkali is measured into a conical flask using a pipette. The acid is run in from the burette – swirling the flask constantly. Towards the end, the acid is run in a drop at a time until the indicator just changes colour.

The photograph shows the end point of a titration using methyl orange as indicator. If the indicator changes to red (its acidic colour), you have added too much acid.

The end point colour for methyl orange. The flask on the left shows the colour before the end point.

The standard calculation

A simple titration problem will look like this:

25.0 cm^3 of 0.100 $mol\,dm^{-3}$ sodium hydroxide solution required 23.5 cm^3 of dilute hydrochloric acid for neutralisation. Calculate the concentration of the hydrochloric acid.

$$NaOH_{(aq)} + HCl_{(aq)} \rightarrow NaCl_{(aq)} + H_2O_{(l)}$$

You do a titration to find the concentration of one solution, knowing the concentration of the other one.

Planning a route through the calculation

- Start with what you know most about. In this case, you know both the volume and the concentration of the sodium hydroxide solution. Work out how many moles of this you have got.

- Look at the equation to work out how many moles of hydrochloric acid that amount of sodium hydroxide reacts with.

- Work out the concentration of the hydrochloric acid.

Doing the calculation

The experiment used 25.0 cm^3 of 0.100 $mol\,dm^{-3}$ NaOH solution.

$$\text{Number of moles of NaOH} = \frac{25.0}{1000} \times 0.100 \text{ mol} = 0.002\,50 \text{ mol}$$

The equation says that 1 mol NaOH reacts with 1 mol HCl
Therefore 0.002 50 mol NaOH reacts with 0.002 50 mol HCl

That 0.002 50 mol HCl must have been in the 23.5 cm^3 of hydrochloric acid that was added during the titration – otherwise neutralisation wouldn't have occurred.

All you need to do now is to find out how many moles there would be in 1000 cm^3 (1 dm^3) of this solution.

If 23.5 cm^3 contain 0.002 50 mol HCl

1000 cm^3 contain $\frac{1000}{23.5} \times 0.002\,50$ mol HCl = 0.106 mol

The concentration is therefore 0.106 $mol\,dm^{-3}$.

> Put in an extra step if you need to. Work out how many moles there are in 1 cm^3 by dividing 0.100 by 1000. (The concentration is 0.100 mol in 1000 cm^3.) Then multiply by 25 to find out how many there are in 25 cm^3.

> Again, insert an extra step if you need to. Work out the number of moles in 1 cm^3 by dividing by 23.5, and then multiply by 1000 to find out how many moles there are in 1000 cm^3.

A very slightly harder calculation

25 cm^3 of sodium hydroxide solution of unknown concentration was titrated with dilute sulphuric acid of concentration 0.050 $mol\,dm^{-3}$. 20.0 cm^3

of the acid was required to neutralise the alkali. Find the concentration of the sodium hydroxide solution in mol dm^{-3}.

$$2NaOH_{(aq)} + H_2SO_{4(aq)} \rightarrow Na_2SO_{4(aq)} + 2H_2O_{(l)}$$

This time you know everything about the sulphuric acid.

The experiment used 20.0 cm^3 of 0.050 mol dm^{-3} H$_2$SO$_4$.

$$\text{Number of moles of sulphuric acid used} = \frac{20.0}{1000} \times 0.050 \text{ mol}$$
$$= 0.0010 \text{ mol}$$

Use more steps for this, and similar future problems, if you need to.

The equation proportions aren't 1:1 this time. That's what makes the calculation slightly different from the last one. The equation says that each mole of sulphuric acid reacts with 2 moles of sodium hydroxide.

$$\text{Number of moles of sodium hydroxide} = 2 \times 0.0010 \text{ mol}$$
$$= 0.0020 \text{ mol}$$

That 0.0020 mol must have been in the 25 cm^3 of sodium hydroxide solution.

$$\text{Concentration} = \frac{1000}{25} \times 0.0020 \text{ mol dm}^{-3} = 0.080 \text{ mol dm}^{-3}$$

A straightforward titration sum with a sting in the tail

Washing soda crystals have the formula Na$_2$CO$_3$·nH$_2$O. The object of this calculation is to find the number of molecules of water of crystallisation, "n".

28.6 g of washing soda crystals were dissolved in pure water. More pure water was added to make the total volume of the solution up to 1000 cm^3.

A 25.0 cm^3 sample of this solution was neutralised by 40.0 cm^3 of 0.125 mol dm^{-3} hydrochloric acid using methyl orange as indicator.

$$Na_2CO_{3(aq)} + 2HCl_{(aq)} \rightarrow 2NaCl_{(aq)} + CO_{2(g)} + H_2O_{(l)}$$

Washing soda crystals.

(a) Calculate the concentration of the sodium carbonate solution in moles of sodium carbonate (Na$_2$CO$_3$) per cubic decimetre.

(b) Calculate the mass of Na$_2$CO$_3$ and mass of water in the washing soda crystals, and use these results to find a value for "n" in the formula Na$_2$CO$_3$·nH$_2$O. (RAMs: H = 1; C = 12; O = 16; Na = 23)

Part (a) is the straightforward titration calculation. Part (b) is an extra bit.

Part (a) – the titration calculation

You know the volume and concentration of the hydrochloric acid.

$$\text{Number of moles of HCl} = \frac{40.0}{1000} \times 0.125 \text{ mol} = 0.005\,00 \text{ mol}$$

From the equation you can see that you only need half that number of moles of sodium carbonate.

$$\text{Number of moles of Na}_2\text{CO}_3 = \frac{0.005\,00}{2} \text{ mol} = 0.002\,50 \text{ mol}$$

The sodium carbonate solution contained 0.002 50 mol in 25.0 cm^3.

$$\text{Concentration of Na}_2\text{CO}_3 = \frac{1000}{25.0} \times 0.002\,50 \text{ mol dm}^{-3}$$
$$= 0.100 \text{ mol dm}^{-3}$$

Part (b)

First calculate the mass of Na_2CO_3 in the total 1000 cm^3 (1 dm^3) of solution. Remember that you have just worked out that the solution is 0.100 mol dm^{-3}.

1 mol Na_2CO_3 weighs 106 g

0.100 mol Na_2CO_3 weighs 0.100 × 106 g = 10.6 g

Now for the mass of water in the crystals:

The original mass of the crystals dissolved in the water was 28.6 g. Of this, we have worked out that 10.6 g is Na_2CO_3.

Mass of water = 28.6 – 10.6 g = 18.0 g
But 1 mol H_2O weighs 18 g

There is therefore 1 mol of H_2O in the crystals together with 0.100 mol of Na_2CO_3.

Since there are ten times as many moles of H_2O as of Na_2CO_3, the formula is $Na_2CO_3 \cdot 10H_2O$.

Notice that we've only just got around to using the figure of 28.6 g, despite the fact that it was the very first number in the question. That is quite common in titration sums. You always start from the volume and concentration of the substance you know most about.

End of Chapter Checklist

If you haven't got a copy of your specification, read the introduction on page vi.

You will need to be able to do some or all of the following. Check your Awarding Body's specification (syllabus) to find out exactly what you need to know.

- Convert concentrations from $g\,dm^{-3}$ to $mol\,dm^{-3}$ and vice versa.
- Calculate the number of moles of substance given a volume of solution and a concentration in $mol\,dm^{-3}$.
- Work out unknown concentrations from titration results.
- Use results from titrations to calculate other things, given guidance on method.

Questions

More calculation questions can be found at the end of Section F on page 294.

1 Some dilute sulphuric acid, H_2SO_4, had a concentration of $4.90\,g\,dm^{-3}$. What is its concentration in $mol\,dm^{-3}$? (RAMs: H = 1; O = 16; S = 32)

2 What is the concentration in $g\,dm^{-3}$ of some potassium hydroxide, KOH, solution with a concentration of $0.200\,mol\,dm^{-3}$? (RAMs: H = 1; O = 16; K = 39)

3 What mass of sodium carbonate, Na_2CO_3, would be dissolved in $100\,cm^3$ of solution in order to get a concentration of $0.100\,mol\,dm^{-3}$? (RAMs: C = 12; O = 16; Na = 23)

4 What mass of barium sulphate would be produced by adding excess barium chloride solution to $20.0\,cm^3$ of copper(II) sulphate solution of concentration $0.100\,mol\,dm^{-3}$? (RAMs: O = 16; S = 32; Ba = 137)

$$BaCl_{2(aq)} + CuSO_{4(aq)} \rightarrow BaSO_{4(s)} + CuCl_{2(aq)}$$

5 What is the maximum mass of calcium carbonate which will react with $25.0\,cm^3$ of $2.00\,mol\,dm^{-3}$ hydrochloric acid? (RAMs: C = 12; O = 16; Ca = 40)

$$CaCO_{3(s)} + 2HCl_{(aq)} \rightarrow CaCl_{2(aq)} + H_2O_{(l)} + CO_{2(g)}$$

6 Copper(II) sulphate crystals, $CuSO_4 \cdot 5H_2O$, are made by adding an excess of copper(II) oxide to hot sulphuric acid, filtering the mixture and then crystallising the solution. What is the maximum mass of crystals that could be obtained by adding an excess of copper(II) oxide to $25\,cm^3$ of $1.0\,mol\,dm^{-3}$ sulphuric acid? (RAMs: H = 1; O = 16; S = 32; Cu = 64)

$$CuO_{(s)} + H_2SO_{4(aq)} \rightarrow CuSO_{4(aq)} + H_2O_{(l)}$$
$$CuSO_{4(aq)} + 5H_2O_{(l)} \rightarrow CuSO_4 \cdot 5H_2O_{(s)}$$

7 In each of these questions concerning simple titrations, calculate the unknown concentration in $mol\,dm^{-3}$.

a) $25.0\,cm^3$ of $0.100\,mol\,dm^{-3}$ sodium hydroxide solution was neutralised by $20.0\,cm^3$ of dilute nitric acid of unknown concentration.

$$NaOH_{(aq)} + HNO_{3(aq)} \rightarrow NaNO_{3(aq)} + H_2O_{(l)}$$

b) $25.0\,cm^3$ of sodium carbonate solution of unknown concentration was neutralised by $30.0\,cm^3$ of $0.100\,mol\,dm^{-3}$ nitric acid.

$$Na_2CO_{3(aq)} + 2HNO_{3(aq)} \rightarrow$$
$$2NaNO_{3(aq)} + CO_{2(g)} + H_2O_{(l)}$$

c) $25.0\,cm^3$ of $0.250\,mol\,dm^{-3}$ potassium carbonate solution was neutralised by $12.5\,cm^3$ of ethanoic acid of unknown concentration.

$$2CH_3COOH_{(aq)} + K_2CO_{3(aq)} \rightarrow$$
$$2CH_3COOK_{(aq)} + CO_{2(g)} + H_2O_{(l)}$$

8 Lime water is calcium hydroxide solution. In an experiment to find the concentration of calcium hydroxide in lime water, $25.0\,cm^3$ of lime water needed $18.8\,cm^3$ of $0.0400\,mol\,dm^{-3}$ hydrochloric acid to neutralise it. Calculate the concentration of the calcium hydroxide in **a)** $mol\,dm^{-3}$, **b)** $g\,dm^{-3}$. (RAMs: H = 1; O = 16; Ca = 40)

$$Ca(OH)_{2(aq)} + 2HCl_{(aq)} \rightarrow CaCl_{2(aq)} + 2H_2O_{(l)}$$

9 $10.0\,g$ of impure sodium hydrogencarbonate, $NaHCO_3$, was dissolved in pure water and the volume made up to $1000\,cm^3$. $25.0\,cm^3$ of this solution was pipetted into a conical flask. A few drops of methyl orange was added. Sulphuric acid of concentration $0.0500\,mol\,dm^{-3}$ was run in from a burette until the solution became orange. $28.3\,cm^3$ of the acid was needed.

$$2NaHCO_{3(aq)} + H_2SO_{4(aq)} \rightarrow$$
$$Na_2SO_{4(aq)} + 2CO_{2(g)} + 2H_2O_{(l)}$$

a) Calculate the number of moles of sulphuric acid in 28.3 cm^3 of 0.0500 mol dm^{-3} acid.

b) Calculate the number of moles of sodium hydrogencarbonate in 25.0 cm^3 of the solution.

c) Calculate the number of moles of sodium hydrogencarbonate in 1000 cm^3 of the solution.

d) Calculate the mass of pure sodium hydrogencarbonate in 1000 cm^3 of solution. (RAMs: H = 1; C = 12; O = 16; Na = 23)

e) What is the percentage purity of the sodium hydrogencarbonate?

End of Section Questions

1
a) What do you understand by the term *relative atomic mass* of an element? *(2 marks)*

b) Show that the relative atomic mass of chlorine is 35.5, given that an average sample of chlorine contains 75% ^{35}Cl and 25% ^{37}Cl. *(2 marks)*

c) Chlorine gas was bubbled through a solution containing 4.15 g of potassium iodide until no further reaction occurred. Calculate the mass of iodine produced by the reaction:

$$Cl_{2(g)} + 2KI_{(aq)} \rightarrow 2KCl_{(aq)} + I_{2(s)}$$

(Relative atomic masses: K = 39; I = 127)
(4 marks)

d) Calculate the density of chlorine gas at room temperature and pressure in $g\,dm^{-3}$. (Volume of 1 mole of a gas at room temperature and pressure = 24.0 dm^3) *(2 marks)*

Total 10 marks

2 In an experiment to find the empirical formula of some lead oxide, a small porcelain dish was weighed, filled with lead oxide and weighed again. The dish was placed in a tube, and was heated in a stream of hydrogen. The hydrogen reduced the lead oxide to a bead of metallic lead. When the apparatus was cool, the dish with its bead of lead was weighed.

Mass of porcelain dish	= 17.95 g
Mass of porcelain dish + lead oxide	= 24.80 g
Mass of porcelain dish + lead	= 24.16 g

(Relative atomic masses: O = 16; Pb = 207)

a) Calculate the mass of lead in the lead oxide. *(1 mark)*

b) Calculate the mass of oxygen in the lead oxide. *(1 mark)*

c) There are three different oxides of lead: PbO, PbO_2 and Pb_3O_4. Use your results from **a)** and **b)** to find the empirical formula of the oxide used in the experiment. *(3 marks)*

d) Calculate the percentage by mass of lead in the oxide PbO_2. *(2 marks)*

Total 7 marks

3 In an experiment to find the percentage of calcium carbonate in sand from a beach, 1.86 g of sand reacted with an excess of dilute hydrochloric acid to give 0.55 g of carbon dioxide.

$$CaCO_{3(s)} + 2HCl_{(aq)} \rightarrow CaCl_{2(aq)} + CO_{2(g)} + H_2O_{(l)}$$

a) Calculate the number of moles of carbon dioxide present in 0.55 g of CO_2. (Relative atomic masses: C = 12; O = 16) *(2 marks)*

b) How many moles of calcium carbonate must have been present in the sand to produce this amount of carbon dioxide? *(1 mark)*

c) Calculate the mass of calcium carbonate present in the sand. (Relative atomic masses: C = 12; O = 16; Ca = 40) *(2 marks)*

d) Calculate the percentage of calcium carbonate in the sand. *(1 mark)*

Total 6 marks

4
a) Chalcopyrite is a copper-containing mineral with a formula $CuFeS_2$.

i) Calculate the percentage by mass of copper in pure chalcopyrite. *(2 marks)*

(Relative atomic masses: S = 32; Fe = 56; Cu = 64)

ii) Analysis of a copper ore showed that it contained 50% chalcopyrite by mass. Assuming that all the copper can be extracted, what mass of copper could be obtained from 1 tonne (1000 kg) of the copper ore? *(2 marks)*

b) Copper reacts with concentrated nitric acid to give copper(II) nitrate solution and nitrogen dioxide gas.

$$Cu_{(s)} + 4HNO_{3(aq)} \rightarrow Cu(NO_3)_{2(aq)} + 2NO_{2(g)} + 2H_2O_{(l)}$$

i) Calculate the maximum mass of copper(II) nitrate, $Cu(NO_3)_2$, which could be obtained from 8.00 g of copper. (Relative atomic masses: N = 14; O = 16; Cu = 64) *(3 marks)*

ii) Calculate the volume of nitrogen dioxide produced at room temperature and pressure using 8.00 g of copper. (Volume of 1 mole of a gas at room temperature and pressure = 24.0 dm^3) *(2 marks)*

Total 9 marks

5 If pyrite, FeS_2, is heated strongly in air it reacts according to the equation:

$$4FeS_{2(s)} + 11O_{2(g)} \rightarrow 2Fe_2O_{3(s)} + 8SO_{2(g)}$$

Iron can be extracted from the iron(III) oxide produced, and the sulphur dioxide can be converted into sulphuric acid.

a) Calculate the mass of iron(III) oxide which can be obtained from 480 kg of pure pyrite. (Relative atomic masses: O = 16; S = 32; Fe = 56) *(2 marks)*

b) What mass of iron could be obtained by the reduction of the iron(III) oxide formed from 480 kg of pyrite. *(2 marks)*

c) Calculate the volume of sulphur dioxide (measured at room temperature and pressure) produced from 480 kg of pyrite. (Volume of 1 mole of a gas at room temperature and pressure = 24.0 dm^3) *(3 marks)*

d) The next stage of the manufacture of sulphuric acid is to convert the sulphur dioxide into sulphur trioxide.

$$2SO_{2(g)} + O_{2(g)} \rightarrow 2SO_{3(g)}$$

Calculate the volume of oxygen (measured at room temperature and pressure) needed for the complete conversion of the sulphur dioxide produced in part **c)** into sulphur trioxide. *(1 mark)*

Total 8 marks

6 Strontium hydroxide, Sr(OH)$_2$, is only sparingly soluble in water at room temperature. In an experiment to measure its solubility, a student made a saturated solution of strontium hydroxide. She pipetted 25.0 cm^3 of this solution into a conical flask, added a few drops of methyl orange indicator, and then titrated it with 0.100 mol dm^{-3} hydrochloric acid from a burette. She needed to add 32.8 cm^3 of the acid to neutralise the strontium hydroxide.

$$Sr(OH)_{2(aq)} + 2HCl_{(aq)} \rightarrow SrCl_{2(aq)} + 2H_2O_{(l)}$$

a) Calculate the number of moles of HCl in 32.8 cm^3 of 0.100 mol dm^{-3} hydrochloric acid. *(1 mark)*

b) How many moles of strontium hydroxide does that react with? *(1 mark)*

c) Calculate the concentration of the strontium hydroxide in mol dm^{-3}. *(2 marks)*

d) Calculate the concentration of the strontium hydroxide solution in g dm^{-3}. (Relative atomic masses: H = 1; O = 16; Sr = 88) *(2 marks)*

Total 6 marks

7 a) What mass of sodium hydroxide, NaOH, must be dissolved to make 250 cm^3 of solution with a concentration of 0.100 mol dm^{-3}? (Relative atomic masses: H = 1; O = 16; Na = 23) *(2 marks)*

b) 25.0 cm^3 of this 0.100 mol dm^{-3} sodium hydroxide solution was neutralised by 20.0 cm^3 of dilute sulphuric acid. Calculate the concentration of the sulphuric acid in mol dm^{-3}.

$$2NaOH_{(aq)} + H_2SO_{4(aq)} \rightarrow Na_2SO_{4(aq)} + 2H_2O_{(l)}$$

(4 marks)

c) 1.00 dm^3 of this same sulphuric acid was reacted with magnesium.

$$Mg_{(s)} + H_2SO_{4(aq)} \rightarrow MgSO_{4(aq)} + H_{2(g)}$$

i) What is the maximum mass of magnesium which would react with the acid? (Relative atomic mass: Mg = 24) *(2 marks)*

ii) What volume of hydrogen gas would be produced at room temperature and pressure? (Volume of 1 mole of a gas at room temperature and pressure = 24.0 dm^3) *(2 marks)*

Total 10 marks

8 a) During the electrolysis of concentrated copper(II) chloride solution using carbon electrodes, 0.64 g of copper was deposited on the cathode. Calculate the volume of chlorine (measured at room temperature and pressure) produced at the anode.

Cathode equation: $Cu^{2+}_{(aq)} + 2e^- \rightarrow Cu_{(s)}$

Anode equation: $2Cl^-_{(aq)} \rightarrow Cl_{2(g)} + 2e^-$

(Relative atomic mass: Cu = 64. Volume of 1 mole of a gas at room temperature and pressure = 24.0 dm^3) *(2 marks)*

b) Magnesium is manufactured by electrolysing molten magnesium chloride, MgCl$_2$, in electrolytic cells operating at 250 000 amps.

Cathode equation: $Mg^{2+}_{(l)} + 2e^- \rightarrow Mg_{(l)}$

(Relative atomic mass: Mg = 24. 1 faraday = 96 000 coulombs)

i) Calculate the number of moles of electrons required to produce 1.20 tonnes of magnesium. (1 tonne = 1000 kg) *(2 marks)*

ii) Calculate the number of coulombs needed to produce this much magnesium. *(1 mark)*

iii) How long (in hours) would it take to produce 1.20 tonnes of magnesium in a cell operating at 250 000 amps? *(2 marks)*

Total 7 marks

Hazard Warning Symbols

Toxic
These substances can cause death. They may be poisonous when swallowed, breathed in, or absorbed through the skin.

Harmful
These are similar to toxic substances but less dangerous.

Corrosive
These substances attack and destroy material like wood, and living tissue including skin and eyes.

Irritant
These aren't as dangerous as corrosive substances, but still cause reddening or blistering of the skin.

Highly flammable
These substances catch fire easily.

Oxidising
These substances provide oxygen and cause other things to burn more fiercely.

Hazchem Codes

These are used when dangerous substances are transported around the country. The symbols give important information to the emergency services so that they can respond safely and quickly.

code to identify the hazardous substance

code for the type of substance needed to neutralise the risk – in this case, foam

code for the type of breathing apparatus and protective clothing

type of hazard

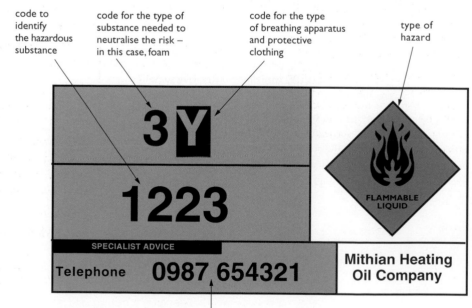

3Y

1223

SPECIALIST ADVICE

Telephone   **0987 654321**

Mithian Heating Oil Company

FLAMMABLE LIQUID

source of further information

Appendix B: Practical Investigations

The practical investigation counts 20% towards your final GCSE grade. It isn't difficult to get full (or very close to full) marks, but it does take time and patience. If you are prepared to spend that time, it means that you need to score fewer marks on the written exam in order to get the grade you want.

The next few pages show you how to gain a high score on an investigation. It is important to listen to your teacher's advice on exactly what you need to do in order to get *full* marks. After all, he or she will be marking your piece of work, and will have up-to-date knowledge of what your examiners want.

The example investigation is:

Investigate the amount of heat produced when magnesium reacts with dilute acids.

You must realise that the version in this book is **incomplete**. You will find it in full on the website supporting the book at www.longman.co.uk/gcsechemistry.

This talks you through finding the information for this particular investigation, but what you need to do will be just the same whatever investigation you are given. Use this, and other GCSE books to get you started. Don't forget the possibility of using A-level textbooks or Data Books to fill in some details.

It might also be worth seeing what you can find on the Awarding Bodies' websites (try all the sites – not just your own!). There may be material designed for teachers that you could make use of. You will find web addresses in the introduction on page vi.

Be very wary of any examples of coursework provided by other students on the internet. They aren't difficult to find, but just because they are on the internet, it doesn't mean that they are any good.

How to Start

See what you can find out about metals reacting with acids by reading Chapter 7 of this book. Page 65 might be particularly useful, because it shows that the reactions between magnesium and both dilute sulphuric acid and dilute hydrochloric acid are represented by the same ionic equation:

$$Mg_{(s)} + 2H^+_{(aq)} \rightarrow Mg^{2+}_{(aq)} + H_{2(g)}$$

Page 64 notes that dilute nitric acid does something different with metals. Use an A level textbook to find out what actually happens.

Reading further through Chapter 7 will lead you to strong and weak acids, and a mention of ethanoic acid which is only slightly ionised in solution. You can find out more about ethanoic acid in Chapter 22. If it is only slightly ionised, how might that affect the amount of heat given off when it reacts with a metal?

Now see what you can find out about heat changes during reactions. If you read Chapter 13, on page 140 you will find some simple apparatus you might use, together with a guide to how you might start a calculation. In the same chapter, on page 136, there is a value for the amount of heat evolved when 1 mole of magnesium reacts with dilute sulphuric acid. That gives you a number which you could use to check the accuracy of your own results.

Don't restrict yourself to this book. Use other GCSE and even A-level books. Nobody expects you to produce an A-level answer, but you should be pushing at the limits of GCSE if you want an A*.

Planning

Your teacher will be marking the planning part of your work by matching it to this checklist. The important terms are explained in the next page or two.

If you can	Mark awarded
outline a simple procedure	2
plan to collect evidence which will be valid plan the use of suitable equipment or sources of evidence	4
use scientific knowledge and understanding to plan and communicate a procedure, to identify key factors to vary, control or take into account, and to make a prediction where appropriate decide a suitable extent and range of evidence to be collected	6
use detailed scientific knowledge and understanding to plan and communicate an appropriate strategy, taking into account the need to produce precise and reliable evidence, and to justify a prediction, when one has been made use relevant information from preliminary work, where appropriate, to inform the plan	8

To score 8 marks, your work must match both the statements in that box, but must also fulfil all the other statements for 6, 4 and 2. In other words, for full marks, you need to do everything in the table. It is important to aim for the highest possible mark. Even if you miss it, you can still score well.

You can score odd-numbered marks if your work falls just short of a level. For example, you might score 7 marks if you satisfied the first statement needed for 8 marks, but didn't do any preliminary experiments when some would have been helpful.

You could use these main headings to help you to get everything in a logical order:

What I am going to do

In this case, it would make sense to react the magnesium with a variety of dilute acids including sulphuric acid, hydrochloric acid, nitric acid (because it reacts differently!) and ethanoic acid (because it is weak). You could aim to find out how much heat was produced per mole of magnesium in each case.

What I already know

To score 8 marks for planning, everything must be based on "detailed scientific knowledge and understanding" – and you have to make it clear how you are going to use that knowledge and understanding. List all the relevant things you have found out from books or other sources, and say why you think they might be useful to you.

The key factors to vary, control or take into account

Don't just list the key factors. Explain why you are choosing to control some things and vary others, and why some things don't matter (if that happens to be the case). Your explanations should again "use detailed scientific knowledge and understanding". For example:

> I am going to use an excess of each acid, so their exact concentrations don't matter. The amount of heat given off will depend only on the amount of magnesium used. I shall use the normal dilute acids found in the lab. I do need to know exactly what volume of acid I am using for each experiment. If I don't know this I can't work out how much heat has been evolved.
>
> I am going to measure the temperature rise of the reaction mixture. Because the temperature has gone up, heat will be lost to the surroundings. I shall have to take care to keep this heat loss as low as possible.

See page 140 to understand why it is important to know the volume of acid.

Warning! None of the examples given are complete. In this case, you also need to consider whether you need to control the amount of magnesium used.

Preliminary work

Preliminary work involves doing rough experiments to find out the best conditions for carrying out your investigation. It is essential if you are going to score 8 marks for planning. Describe your preliminary work carefully, and say exactly how it helped you to decide your final plan.

Again, wherever possible, explain your choices in terms of "detailed scientific knowledge and understanding".

For example, in this case, as well as deciding what sort of apparatus to use, you need to find out how much magnesium and how much acid to use to get a good result. You also need to find out whether it is best to use magnesium ribbon, turnings or powder. For example:

> I started by using a strip of magnesium ribbon which weighed about 0.1 g, and 50 cm^3 of acid (measured roughly with a measuring cylinder). All the magnesium dissolved, and it gave a sensible temperature rise of about 9°C, but it took a very long time using ethanoic acid. That increases the possibility of heat losses to the surroundings. This preliminary experiment showed that my quantities were sensible, but that I needed to speed up the reaction. I tried different forms of magnesium. Magnesium turnings also took quite a long time, but magnesium powder reacted quite quickly with the ethanoic acid, and didn't produce too much spray with the strong acids.
>
> I decided to use about 0.1 to 0.15 g of magnesium powder for my experiments. I weighed this in a weighing bottle, but found that when I tipped it into the acid, not all of it came out. So I would need to reweigh the bottle again to see how much I had actually added.

Again, remember that this isn't a complete version of what you would have to write in your account of your preliminary work.

Producing precise and reliable evidence

"Precise" means that you are measuring things as accurately as possible. Particularly where the quantity you are measuring is small, you should try to measure it using the most accurate equipment you have access to. "Reliable" means that if you repeat the experiments, you will get the same results. For example:

> Because of the small mass of magnesium used, I shall weigh it on a 3 decimal place balance.

I shall repeat each experiment twice to check that my results are reliable. To check whether my results are consistent for each acid used, I shall work out the temperature rise per gram of magnesium after each experiment. If these don't agree, then I shall go on repeating the experiment until there is good agreement.

Safety

List all the safety aspects of your plan in detail, explaining why you need to take each precaution. For example:

Because of the risk of poisonous oxides of nitrogen being given off, the nitric acid experiment should be done in a fume cupboard.

Doing the experiment

Describe what you are going to do in detail, listing all the apparatus you need. Draw diagram(s) to show exactly how the apparatus is used, so that there is no uncertainty about what you mean.

When you have finished describing your method, read it critically and ask yourself whether someone else could carry it out successfully if they did it *exactly* as you have written it. You can assume that they know how to use standard apparatus like pipettes and thermometers, but you could stress any points which are particularly important for accuracy. For example:

Measure 50 cm^3 of one of the acids into the polystyrene cup from a pipette. To get maximum accuracy, you should let the pipette drain naturally, and then touch the tip on the surface of the acid in the cup.

My prediction

Again, your predictions must be justified in terms of "detailed scientific knowledge and understanding". For example:

I predict that the heat given out per mole of magnesium will be the same for the strong acids hydrochloric acid and sulphuric acid, because in both cases the underlying ionic equation is the same.

$$Mg_{(s)} + 2H^+_{(aq)} \rightarrow Mg^{2+}_{(aq)} + H_{2(g)}$$

In both cases, the solid magnesium has turned into ions in solution, and the hydrogen ions have turned into hydrogen molecules. The overall energy change is going to be made up from contributions due to both of these changes. Since these are the same in both cases, the heat evolved ought to be the same.

Don't worry if you think that the nitric acid result is going to be different (and can explain why), but don't know whether it will be larger or smaller. You can leave that open at this point. For example:

With nitric acid, I would expect that the heat evolved would be the same if the reaction produces just hydrogen. If oxides of nitrogen are formed as well as or instead of hydrogen, the heat evolved will be different, because different bonds are being broken and made.

Obtaining Evidence

Your teacher will be using this checklist:

If you can	Mark awarded
collect some evidence using a simple and safe procedure	2
collect appropriate evidence which is adequate for the activity record the evidence	4
collect sufficient systematic and accurate evidence and repeat or check when appropriate record clearly and accurately the evidence collected	6
use a procedure with precision and skill to obtain and record an appropriate range of reliable evidence	8

Draw up a table of results

To get full marks you must record everything which might vary. In this particular case, the exact concentrations of the acids don't really matter, but if someone was going to try to repeat your results, they would have to know approximately what concentrations you were using.

When you draw up your table of results, make sure that everything is in a logical order and that you have labelled everything, *including the correct units*.

Write down the numbers to reflect their accuracy. For example, even if your mass was exactly 0.1 g, you should write it as 0.100 g to show that you were using a 3 decimal place balance.

> The experiments are being done with an excess of acid. That's why it doesn't matter *exactly* how concentrated it is.

Results with hydrochloric acid

Concentration of acid used is approximately 1 mol dm^{-3}.

Volume of acid used = 50.0 cm^3

	Expt 1	Expt 2	Expt 3
Mass of weighing bottle plus Mg (g)	10.807	10.806	10.820
Mass of weighing bottle afterwards (g)	10.684	10.689	10.687
Mass of Mg used (g)	0.123	0.117	0.133
Initial temperature (°C)	17.4	17.5	17.4
Maximum temperature (°C)	27.5	27.4	28.4
Temperature rise (°C)	10.1	9.9	11.0
Accuracy check – temperature rise per gram of Mg (°C/g)	82.1	(84.6)	82.7

Appendix B: Practical Investigations

Analysing your Evidence and Drawing Conclusions

Your teacher will be using this checklist:

If you can	Mark awarded
state simply what is shown by the evidence	2
use simple diagrams, charts or graphs as a basis for explaining the evidence	
identify patterns and trends in the evidence	4
construct and use suitable diagrams, charts, graphs (with lines of best fit where appropriate), or use numerical methods, to process evidence for a conclusion	6
draw a conclusion consistent with the evidence and explain it using scientific knowledge and understanding	
use detailed scientific knowledge and understanding to explain a valid conclusion drawn from processed evidence.	8
explain the extent to which the conclusion supports the prediction, if one has been made	

Notice that to gain 6 or 8 marks you don't necessarily have to draw graphs. You can use "numerical methods" instead. That means doing some reasonably complicated calculations. Although working out a simple average is a "numerical method", it isn't complicated enough to earn you 6 (or 8) marks. The calculations in our example are sophisticated enough to earn full marks. As you will see, however, we shall display the final answers on a simple bar chart to make sure that the requirements for 4 marks are fully covered.

Calculations

We are going to calculate the amount of heat evolved per mole of magnesium.

You would start from the formula given on page 140:

heat given out = mass × specific heat × temperature rise

From there, you can arrive at a general formula for any temperature rise, T, and mass of magnesium, m.

> You would have to show how you arrived at this formula – not just quote it.

$$\text{heat evolved per mole of magnesium} = \frac{50.0 \times 4.18 \times T \times 24.3}{m \times 1000} \text{ kJ}$$

(The numbers in this formula: 50.0 g is the mass of the acid; 4.18 J g^{-1} °C^{-1} is the specific heat; 24.3 is the relative atomic mass of magnesium to 3 significant figures; 1000 converts joules into kilojoules.)

Now you can apply this formula to all your results. For example:

> If you have results that are clearly out of line, it is important that you don't include them in your average. Do, however, say that you are missing them out – and why.

Results from hydrochloric acid

	Mass Mg (g)	Temp rise (°C)	Heat evolved per mole Mg (kJ/mol)	Average value (kJ/mol)
Expt 1	0.123	10.1	417	419
Expt 3	0.133	11.0	420	

The results from Experiment 2 are not being included in the average because the accuracy check showed them to give a rather high value.

Displaying your final results

We haven't used any graphs in this example, because we have used sufficiently complicated "numerical methods". But in the table of marks, to get 4 marks you have to "use simple diagrams, charts or graphs". To get 8 marks, you have to fulfil all the lower marks as well. Display your results as a simple bar chart to cover this.

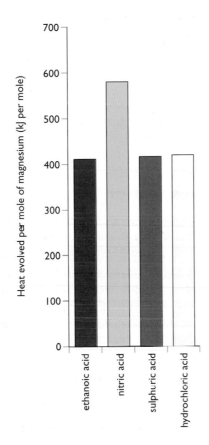

Drawing your conclusions

You have to decide whether your results fit your original prediction. Remind the person marking your work exactly what your prediction was. Don't just refer back to your original scientific explanation. Give it again, together with any modifications or extras that your results show are necessary. For example (talking about the sulphuric acid and hydrochloric acid cases):

The average values in these two cases (419 and 416 kJ per mole) were in very good agreement – certainly to within experimental error. This was exactly in line with my prediction. Both of these are strong acids which means that they are fully ionised in solution. The reaction taking place in each case was simply between the magnesium and the hydrogen ions:

$$Mg_{(s)} + 2H^+_{(aq)} \rightarrow Mg^{2+}_{(aq)} + H_{2(g)}$$

The hydrogen ions were in excess, so that the amount of heat given off was controlled by the amount of magnesium present. The same amount of heat was evolved per mole of magnesium, because the reaction was the same in both cases.

Things aren't so clear cut in the ethanoic acid case.

The results show that the ethanoic acid gave out slightly less heat than the hydrochloric acid and sulphuric acid. The difference could easily be accounted for by experimental error. The ethanoic acid value (411) is only about 2% less than the hydrochloric acid value (419) and that could easily be accounted for by the errors in reading the thermometer or weighing the magnesium.

My results don't give me enough evidence to reliably support my prediction that the heat given out by a weak acid will be less than that for a strong acid. I thought that the heat would be less because...

It doesn't matter if your prediction doesn't work out as you expected, provided you can use your knowledge and understanding to come up with a convincing explanation.

Evaluating your Investigation

If you can	Mark awarded
make a relevant comment about the procedure used or the evidence obtained	2
comment on the quality of the evidence, identifying any anomalies comment on the suitability of the procedure and, where appropriate, suggest changes to improve it	4
consider critically the reliability of the evidence and whether it is sufficient to support the conclusion, accounting for any anomalies describe, in detail, further work to provide additional relevant evidence	6

It is easy to throw marks away in this last stage. You must remember that this part is worth almost as many marks as each of the other three stages, and so you must take just as much care over it. Students frequently get bored at this point, and hope to get full marks with a paragraph of generalised waffle. No chance!

Evaluating the experiment

You need to point out any results which appear wrong, and try to account for them. For example:

I think the errors might be due to the balance I was using. I did the experiment over two days, using a different balance on each day. On the first day, the third decimal place on the balance drifted around quite a lot (perhaps because of draughts).

If you are doing an investigation which produces a definite numerical answer, always try to find out what the accepted answer is. That will give you an immediate idea of how accurate your experiment was. You can then look for reasons to account for your degree of inaccuracy.

> I was confident that my results were good until I compared them with the value for the reaction between magnesium and dilute sulphuric acid I found in my textbook. My figures are about 50 kJ less than the value of −466.9 kJ per mole (the minus sign shows that heat is evolved).
>
> I think this must be caused by large heat losses in the experiment, because of...

Improving the experiment

Make sure that the improvements you suggest are detailed and specific to your investigation. You won't get much credit for general comments like "use more accurate equipment" or "take more care with measurements". There is something to be said for leaving some flaws in your original experiment so that you have something to talk about at this evaluation stage! For example:

> To try to cut down heat losses from the top of the cup, I could use a plastic lid to the cup (like a take-away coffee cup) with a hole punched through it to take the thermometer.
>
> I could improve the temperature reading by using a thermometer calibrated to 0.1°C rather than 0.2°C. I can't do much about the weighing because I am already using the most accurate balance. I should, though, take more care to keep it out of draughts.

Extending my experiment

This is to fulfil the 6 mark statement: "describe, in detail, further work to provide additional relevant evidence". In the case we're looking at, there is nothing like enough evidence to say anything certain about weak acids in general. This would provide an obvious extension:

> I don't have enough results to be sure about the heat given out when weak acids react with magnesium, and so I should repeat the experiment using a selection of other weak acids like citric acid, ascorbic acid (vitamin C) or phosphoric acid. In each case, I would...

You should go on to describe exactly what you would do with these acids. For example, you would have to say what sort of concentrations you would use, whether you would redo the experiment *exactly* as before, whether you would make changes, and what effect those changes might have.

Appendix C: The Periodic Table

Period	Group 1	Group 2															Group 3	Group 4	Group 5	Group 6	Group 7	Group 0
1								1 H Hydrogen														2 He Helium 4
2	3 Li Lithium 7	4 Be Beryllium 9															5 B Boron 11	6 C Carbon 12	7 N Nitrogen 14	8 O Oxygen 16	9 F Fluorine 19	10 Ne Neon 20
3	11 Na Sodium 23	12 Mg Magnesium 24															13 Al Aluminium 27	14 Si Silicon 28	15 P Phosphorus 31	16 S Sulphur 32	17 Cl Chlorine 35.5	18 Ar Argon 40
4	19 K Potassium 39	20 Ca Calcium 40	21 Sc Scandium 45	22 Ti Titanium 48	23 V Vanadium 51	24 Cr Chromium 52	25 Mn Manganese 55	26 Fe Iron 56	27 Co Cobalt 59	28 Ni Nickel 59	29 Cu Copper 64	30 Zn Zinc 65					31 Ga Gallium 70	32 Ge Germanium 73	33 As Arsenic 75	34 Se Selenium 79	35 Br Bromine 80	36 Kr Krypton 84
5	37 Rb Rubidium 85	38 Sr Strontium 88	39 Y Yttrium 89	40 Zr Zirconium 91	41 Nb Niobium 93	42 Mo Molybdenum 96	43 Tc Technetium (99)	44 Ru Ruthenium 101	45 Rh Rhodium 103	46 Pd Palladium 106	47 Ag Silver 108	48 Cd Cadmium 112					49 In Indium 115	50 Sn Tin 119	51 Sb Antimony 122	52 Te Tellurium 128	53 I Iodine 127	54 Xe Xenon 131
6	55 Cs Caesium 133	56 Ba Barium 137	57 La Lanthanum 139	72 Hf Hafnium 178	73 Ta Tantalum 181	74 W Tungsten 184	75 Re Rhenium 186	76 Os Osmium 190	77 Ir Iridium 192	78 Pt Platinum 195	79 Au Gold 197	80 Hg Mercury 201					81 Tl Thallium 204	82 Pb Lead 207	83 Bi Bismuth 209	84 Po Polonium (210)	85 At Astatine (210)	86 Rn Radon (222)
7	87 Fr Francium (223)	88 Ra Radium (226)	89 Ac Actinium (227)																			

58 Ce Cerium 140	59 Pr Praseodymium 141	60 Nd Neodymium 144	61 Pm Promethium (147)	62 Sm Samarium 150	63 Eu Europium 152	64 Gd Gadolinium 157	65 Tb Terbium 159	66 Dy Dysprosium 163	67 Ho Holmium 165	68 Er Erbium 167	69 Tm Thulium 169	70 Yb Ytterbium 173	71 Lu Lutetium 175
90 Th Thorium 232	91 Pa Protoactinium (231)	92 U Uranium 238	93 Np Neptunium (237)	94 Pu Plutonium (242)	95 Am Americium (243)	96 Cm Curium (247)	97 Bk Berkelium (247)	98 Cf Californium (251)	99 Es Einsteinium (254)	100 Fm Fermium (253)	101 Md Mendelevium (256)	102 No Nobelium (254)	103 Lr Lawrencium (257)

Key:

a
X
Name
b

a = atomic number
X = atomic symbol
b = relative atomic mass

(Masses in brackets are the mass numbers of the most stable isotope)

Index

chloroethene 215–16
chlorofluorocarbons (CFCs) 184
chlorophyll 137, 180
chocolates 256
cholesterol 248
citric acid 236
clay 196–7, 200
"closed" conditions 165–6
co-ordinate covalent bonds 71
cobalt chloride paper 88
coding chain lengths 225–7
collecting gases 87–8, 96
collision theory 40
colloids 196–7, 200
combustion 135, 181, 211, 234
compounds
    alkali metals 106–7
    halogens 113–14
    organic 225–7
    structure 29
    transition metals 115
concentration
    rates of reaction 41, 42, 46
    reversible reactions 167–8
    solutions 287–8
conclusion guidelines 302–4
concrete 162
condensation polymerisation 243, 245
condensation reactions 242–3
conduction 22, 25–7, 119
conductivity meters 68–9
conservation of mass law 7
constant composition law 7
contact metamorphism 204
Contact Process 172–3
continuous phases (emulsions) 196
copper
    displacement reactions 56–7
    electrodes 125
    electroplating 155–6
    hydrogen reduction 59
    percentage composition 264
    purification 154–5
copper oxide 54, 66, 80, 268
copper(II) sulphate 56, 60, 80
    anhydrous 88
    dehydration 174
    electrolysis 125, 282
    heating 165
    relative formulae mass 263
coulombs 281
coursework tips 297–305
covalent bonds 11–14, 71
    giant structures 25–7

writing formulae 31
cracking 212–13, 219
crude oil (petroleum) 209–20
cryolite 146
crystallisation 79–80, 269–7
crystals 24, 79–81

Dalton, John 6–7
Daniel cells 129
dative covalent bonds 71
decay 182
decomposition rates 52
dehydration 174, 234–5
deionised water 194
"delocalised" electrons 17
denaturation 256
density 26, 103, 105
descalers 193
detergents 257
diagrams (drawing) 5–6, 136, 137
diamond 25–6
diatomic molecules 12
diesel oil 210
direct combination of salts 83–4
disaccharides 241
disperse phases 196
displacement reactions 54–7,
    111–12
displayed formulae 214, 224
distillation 210–12, 219
Döbereiner, Johann Wolfgang 99
dot and cross diagrams 11–14, 20
drinks industry 231–5, 256–7
drugs 95, 250–4
ductility 22
dynamic equilibria 166–7

E-numbers 249–50
Earth
    air 179–86
    rocks 201–6
    water 187–200
elasticity 22
electrical distortions 18
electricity 119, 129–33, 281
electrochemical cells 129–34
electrodes 120, 123–5, 131, 281–6
electrolysis 119–28
    aluminium 146–8
    anhydrous zinc chloride 120–2
    aqueous solutions 122–6
    carbon electrodes 123–5
    copper(II) sulphate 125, 282
    equations 281–6
    gases 282–3

    heat energy 137
    ion movement 126
    molten compounds 120–2
    sodium chloride 122–6, 159–60
    solutions 122–6, 282–5
electrolytes 119–20, 123
electron half equations 55–6
electron transfer 55–6
electronic arrangements 4–6
electrons 2, 3–6, 15–18
electroplating 155–6
electrovalent bonds see ionic
    bonding
elements (structure) 28
empirical formulae 267–70
employment in the oil
    industry 218
emulsions 196
end points 80
endothermic reactions 135, 136–7,
    168–9
"ene" endings 226–7
energy
    bonds 138–40
    changes in reactions 135–41
    diagrams 136, 137
    heat 136–40, 165, 297
    levels 4
environmental problems 182–4
enzymes 232, 255–8
equations 273–80
    balancing 34–6, 37
    dynamic equilibria 167
    electrolysis 281–6
    ionic 55–6, 66, 69, 82
    molar volumes 276–8
    particles 31–8
    titration 287–93
erosion 201
essential amino acids 245
esters 235, 237, 238
ethanoic acid 75, 236–7
ethanol 231–5
ethene 213–16, 229, 232–3
ethyl ethanoate 237
evaluation guidelines 304–5
evaporation 79
evidence guidelines 299–305
evolution of the atmosphere
    179–82
exothermic reactions 135–6,
    149–50, 168–9
experimental guidelines 300
extraction methods 145–50, 153–4, 161
extrusive rocks 201–2